Environmentally Friendly Polymeric Blends from Renewable Sources

Environmentally Friendly Polymeric Blends from Renewable Sources

Editors

Barbara Gawdzik
Olena Sevastyanova

MDPI • Basel • Beijing • Wuhan • Barcelona • Belgrade • Manchester • Tokyo • Cluj • Tianjin

Editors
Barbara Gawdzik
Department of Polymer Chemistry
Maria Curie-Sklodowska University
Lublin
Poland

Olena Sevastyanova
Department of Fibre and Polymer Technology
KTH-The Royal Institute of Technology
Stockholm
Sweden

Editorial Office
MDPI
St. Alban-Anlage 66
4052 Basel, Switzerland

This is a reprint of articles from the Special Issue published online in the open access journal *Materials* (ISSN 1996-1944) (available at: www.mdpi.com/journal/materials/special_issues/Polymeric_Blends).

For citation purposes, cite each article independently as indicated on the article page online and as indicated below:

LastName, A.A.; LastName, B.B.; LastName, C.C. Article Title. *Journal Name* **Year**, *Volume Number*, Page Range.

ISBN 978-3-0365-2749-9 (Hbk)
ISBN 978-3-0365-2748-2 (PDF)

© 2021 by the authors. Articles in this book are Open Access and distributed under the Creative Commons Attribution (CC BY) license, which allows users to download, copy and build upon published articles, as long as the author and publisher are properly credited, which ensures maximum dissemination and a wider impact of our publications.

The book as a whole is distributed by MDPI under the terms and conditions of the Creative Commons license CC BY-NC-ND.

Contents

About the Editors vii

Barbara Gawdzik and Olena Sevastyanova
Special Issue: "Environmentally Friendly Polymeric Blends from Renewable Sources"
Reprinted from: *Materials* **2021**, *14*, 4858, doi:10.3390/ma14174858 1

Przemysław Paczkowski, Andrzej Puszka, Malgorzata Miazga-Karska, Grażyna Ginalska and Barbara Gawdzik
Synthesis, Characterization and Testing of Antimicrobial Activity of Composites of Unsaturated Polyester Resins with Wood Flour and Silver Nanoparticles
Reprinted from: *Materials* **2021**, *14*, 1122, doi:10.3390/ma14051122 5

Sylwia Członka, Agnė Kairytė, Karolina Miedzińska and Anna Strakowska
Polyurethane Hybrid Composites Reinforced with Lavender Residue Functionalized with Kaolinite and Hydroxyapatite
Reprinted from: *Materials* **2021**, *14*, 415, doi:10.3390/ma14020415 19

Anna Strakowska, Sylwia Członka and Agnė Kairytė
Rigid Polyurethane Foams Reinforced with POSS-Impregnated Sugar Beet Pulp Filler
Reprinted from: *Materials* **2020**, *13*, 5493, doi:10.3390/ma13235493 37

Sylwia Członka, Anna Strakowska and Agnė Kairytė
The Impact of Hemp Shives Impregnated with Selected Plant Oils on Mechanical, Thermal, and Insulating Properties of Polyurethane Composite Foams
Reprinted from: *Materials* **2020**, *13*, 4709, doi:10.3390/ma13214709 53

Tongtong Wang, Hongtao Liu, Cuihua Duan, Rui Xu, Zhiqin Zhang, Diao She and Jiyong Zheng
The Eco-Friendly Biochar and Valuable Bio-Oil from *Caragana korshinskii*: Pyrolysis Preparation, Characterization, and Adsorption Applications
Reprinted from: *Materials* **2020**, *13*, 3391, doi:10.3390/ma13153391 71

Sylwia Członka and Anna Strakowska
Rigid Polyurethane Foams Based on Bio-Polyol and Additionally Reinforced with Silanized and Acetylated Walnut Shells for the Synthesis of Environmentally Friendly Insulating Materials
Reprinted from: *Materials* **2020**, *13*, 3245, doi:10.3390/ma13153245 91

José Carlos Alcántara, Israel González, M. Mercè Pareta and Fabiola Vilaseca
Biocomposites from Rice Straw Nanofibers: Morphology, Thermal and Mechanical Properties
Reprinted from: *Materials* **2020**, *13*, 2138, doi:10.3390/ma13092138 107

Sónia S. Leça Gonçalves, Alisa Rudnitskaya, António J.M. Sales, Luís M. Cadillon Costa and Dmitry V. Evtuguin
Nanocomposite Polymeric Materials Based on Eucalyptus Lignoboost® Kraft Lignin for Liquid Sensing Applications
Reprinted from: *Materials* **2020**, *13*, 1637, doi:10.3390/ma13071637 123

Artur Chabros, Barbara Gawdzik, Beata Podkościelna, Marta Goliszek and Przemysław Paczkowski
Composites of Unsaturated Polyester Resins with Microcrystalline Cellulose and Its Derivatives
Reprinted from: *Materials* **2019**, *13*, 62, doi:10.3390/ma13010062 139

Ferran Serra-Parareda, Quim Tarrés, Marc Delgado-Aguilar, Francesc X. Espinach, Pere Mutjé and Fabiola Vilaseca
Biobased Composites from Biobased-Polyethylene and Barley Thermomechanical Fibers: Micromechanics of Composites
Reprinted from: *Materials* **2019**, *12*, 4182, doi:10.3390/ma12244182 . . . . **153**

Marc Delgado-Aguilar, Quim Tarrés, María de Fátima V. Marques, Francesc X. Espinach, Fernando Julián, Pere Mutjé and Fabiola Vilaseca
Explorative Study on the Use of Curauá Reinforced Polypropylene Composites for the Automotive Industry
Reprinted from: *Materials* **2019**, *12*, 4185, doi:10.3390/ma12244185 . . . . **167**

Jifu Du, Zhen Dong, Zhiyuan Lin, Xin Yang and Long Zhao
Radiation Synthesis of Pentaethylene Hexamine Functionalized Cotton Linter for Effective Removal of Phosphate: Batch and Dynamic Flow Mode Studies
Reprinted from: *Materials* **2019**, *12*, 3393, doi:10.3390/ma12203393 . . . . **181**

Marta Goliszek, Beata Podkościelna, Olena Sevastyanova, Barbara Gawdzik and Artur Chabros
The Influence of Lignin Diversity on the Structural and Thermal Properties of Polymeric Microspheres Derived from Lignin, Styrene, and/or Divinylbenzene
Reprinted from: *Materials* **2019**, *12*, 2847, doi:10.3390/ma12182847 . . . . **197**

About the Editors

Barbara Gawdzik

Barbara Gawdzik obtained her Ph.D. degree in 1986 at Maria Curie-Sklodowska University (UMCS) in Lublin, Poland. She completed her habilitation in 1993, and in 2004 she was appointed as a professor in Polymer Chemistry. Since 2006, Prof. Gawdzik has been the head of the Department of Polymer Chemistry at MCS University.

She is the author and co-author of 170 original papers in international journals, 160 papers, published as conference materials, and 28 patents, and a supervisor of 14 Ph.D. theses. Professor Gawdzik has received numerous awards for research and teaching.

The major scientific interest of Professor Gawdzik is in polymer chemistry, with a special focus on the synthesis of new monomers and polymers, heterogeneous polymerization techniques, the chemical modification of polymers, studies of the porous structure of polymeric materials, preparation of polymer microspheres for chromatography, synthesis and investigations of carbon adsorbents from polymer precursors, polymer composites and biocomposites, environmental protection, and wasteless processes.

Olena Sevastyanova

Olena Sevastyanova obtained her Ph.D. degree in 2006 at KTH Royal Institute of Technology, Stockholm, Sweden. From 2006 to 2007, she completed her postdoctoral fellowship at The University of British Columbia, Department of Forestry, Vancouver, B.C., Canada. In 2007–2008, she led a Lignin laboratory at the biofuel company Lignol Innovations, Ltd., Vancouver, B.C., Canada. Since March 2009, she has been working at the Department of Fibre and Polymer Technology, KTH, Stockholm, Sweden, and since September 2010, at WWSC—a joint research centre of KTH Royal Institute of Technology, Chalmers University of Technology, and Linköping University.

Dr. Sevastyanova is an author of 70 original scientific papers and more than 100 conference presentations, and she supervised and co-supervised nine Ph.D. projects. The main areas of her scientific interests are structure–processing–properties relationships for wood-derived biopolymers, including lignin and hemicelluloses; the development of analytical tools for the characterization of biomaterials; biomass-derived functional materials; and nanomaterials for environmental and medical applications.

Editorial

Special Issue: "Environmentally Friendly Polymeric Blends from Renewable Sources"

Barbara Gawdzik [1,*] and Olena Sevastyanova [2]

1. Department of Polymer Chemistry, Maria Curie-Sklodowska University, 20-614 Lublin, Poland
2. Wallenberg Wood Science Center, Department of Fibre and Polymer Technology, KTH-The Royal Institute of Technology, Teknikringen 56-58, 10044 Stockholm, Sweden; olena@kth.se
* Correspondence: barbara.gawdzik@mail.umcs.pl

1. Introduction

The aim of this Special Issue is to highlight the progress in the manufacturing, characterization, and applications of environmentally friendly polymeric blends from renewable resources.

Materials from renewable resources have attracted increasing attention in past decades as a result of environmental concerns and due to the depletion of petroleum resources. Polymeric materials from renewable sources have a long history. They were already used in ancient times and later accompanied man along with the development of civilization. Currently, they are widespread in many areas of life and used, for example, in packaging, and in the automotive and pharmaceutical industries.

Polymers from renewable resources are generally classified into three groups: (i) natural polymers, such as cellulose, starch and proteins; (ii) synthetic polymers from natural monomers, such as poly(lactic acid), and (iii) polymers from microbial processes, such as poly(hydroxybutyrate). The emergence of new methods and analytical tools provides a new level of understanding of the structure–property relationship of natural polymers and allows the development of materials for new applications.

One of the attractive properties of the natural polymers and synthetic polymers produced from natural monomers is their inherited biodegradability. On the other hand, this is related to their moisture sensitivity, which limits their application. Other important limitations of most polymers from renewable resources are their lower softening temperature and mechanical strength. These and many other properties of polymers can be modified and improved through the blending of two or more compounds, for example two or more polymers, polymers and fibers, polymers and nanoparticles etc.

A blending approach, which may result in both polymer blends or composites, is an effective way to achieve a desirable combination of properties that are often absent in the individual components. Polymer blends and composites are useful as they can be produced from low-cost raw materials, including industrial waste products, without sacrificing their desired properties; they can also be used to prepare high-performance compounds for broader applications due to biodegradability and reusability of the end products.

The final properties can be modified by changing the relative concentration and kind of monomeric units used in the synthesis or by varying the proportion of homopolymers and various additives in a blend composition. Development of effective methods of manufacturing products from blends of renewable polymers and environmental friendly synthetic polymers in a controlled way is the challenge of our time.

Accordingly, in this Special Issue of Materials, which is aimed at recognizing the current state of knowledge and development in the use of environmentally friendly polymeric blends from renewable sources, the following aspects were investigated:

- synthesis of composites based on natural fillers;
- chemical modification of polymers or fillers in order to improve interfacial interactions;

Citation: Gawdzik, B.; Sevastyanova, O. Special Issue: "Environmentally Friendly Polymeric Blends from Renewable Sources". *Materials* **2021**, *14*, 4858. https://doi.org/10.3390/ma14174858

Received: 27 July 2021
Accepted: 1 August 2021
Published: 26 August 2021

Publisher's Note: MDPI stays neutral with regard to jurisdictional claims in published maps and institutional affiliations.

Copyright: © 2021 by the authors. Licensee MDPI, Basel, Switzerland. This article is an open access article distributed under the terms and conditions of the Creative Commons Attribution (CC BY) license (https://creativecommons.org/licenses/by/4.0/).

- potential applications of the biobased materials.

2. Short Description of the Articles Presented in This Special Issue

In this Special Issue, 13 original articles have been published on various topics, including the preparation, characterization, and some examples of uses of polymeric materials from renewable resources. This issue includes five articles on polyurethanes blended with various types of additives [1–5], of which two papers discusses the polyurethane systems partly prepared from naturally-derived components, such as lignin [5] or naturally-derived polyols [4]. Four papers discuss the preparation of synthetic unsaturated polyester resins reinforced with natural fillers [6–9], and two other papers are focused on the modification of polyolefins (PE,PP) to improve their strength properties [10,11]. One work is concerned with the preparation of biochar and bio-oil from biomass, where the potential uses of both types of products are discussed [12]; the other paper considers the modification of cotton linter for use as efficient adsorbent for the phosphate.

Most of above papers pay particular attention to the role of eco-friendly fillers, i.e., microcrystalline cellulose, wood flower, lignin, rice straw nanofibers, lavender residues, sugar beet pulp, barley fibers, hemp shives, walnut shells, and curauá fibers in the improvement of various properties in the final materials.

In the papers of Członka, Strąkowska et al., problems of polyurethane composites with different biofillers were discussed.

In the paper "Polyurethane Hybrid Composites Reinforced with Lavender Residue Functionalized with Kaolinite and Hydroxyapatite" [1], blends were obtained with lavender fillers functionalized with kaolinite and hydroxyapatite. The obtained materials were characterized with enhanced mechanical, thermal, and performance properties.

In the paper "Rigid Polyurethane Foams Reinforced with POSS-Impregnated Sugar Beet Pulp Filler" [2], blends were reinforced with sugar beet pulp impregnated with aminopropyl isobutyl-polyhedral oligomeric silsesquioxanes. The results showed that the greatest improvement in physicomechanical properties was observed at a lower concentration of filler.

In the paper "Impact of Hemp Shives Impregnated with Selected Plant Oils on Mechanical, Thermal, and Insulating Properties of Polyurethane Composite Foams" [3], materials reinforced with hemp shives fillers were synthesized with three different types of fillers: nontreated filler, filler impregnated with sunflower oil, and filler impregnated with tung oil. The incorporation of impregnated biofillers resulted in the improvement of thermal stability and flame retardancy of foams.

In the paper "Rigid Polyurethane Foams Based on Bio-Polyol and Additionally Reinforced with Silanized and Acetylated Walnut Shells for the Synthesis of Environmentally Friendly Insulating" [4], composites produced from walnut shell-derived polyol (20 wt. %) were also reinforced with nontreated, acetylated, and silanized biofiller from walnut shells. The composites with the addition of acetylated and silanized filler exhibited a more uniform structure than foams with the addition of nontreated filler.

In the paper of Gonçalves, Rudnitskaya, Sales, Costa, Evtuguin ("Nanocomposite Polymeric Materials Based on Eucalyptus Lignoboost® Kraft Lignin for Liquid Sensing Applications" [5]) polyurethane–lignin copolymer blended with carbon multilayer nanotubes was used in all-solid-state potentiometric chemical sensors. The interaction between carbon nanotubes and lignin molecules in the polymer enhanced its electrical conductivity.

The papers of Delgado-Aguilar et al. concerned eco-friendly blends derived from polyolefines.

The paper of Delgado-Aguilar, Tarrés, Marques, Espinach, Julián, Mutjé, and Vilaseca ("Explorative Study on the Use of Curauá Reinforced Polypropylene Composites for the Automotive Industry" [10]) studied the properties of composites reinforced with natural fibers (curauá). The results showed that the tensile properties were similar to uncoupled glass fiber-based composites.

In the paper of Serra-Parareda, Tarrés, Delgado-Aguilar, Espinach, Mutjé, and Vilaseca ("Biobased Composites from Biobased-Polyethylene and Barley Thermomechanical Fibers:

Micromechanics of Composites" [11]), barley fibers were used as reinforcement for a polyethylene blend. Grafted polyethylene was used as a coupling agent to increase in interfacial adhesion.

Another two papers were devoted to synthetic resin reinforced with various bio-based fillers.

In the paper of Chabros, Gawdzik, Podkościelna, Goliszek, and Pączkowski ("Composites of Unsaturated Polyester Resins with Microcrystalline Cellulose and Its Derivatives" [7]), modified and unmodified microcrystalline cellulose were used as a filler for unsaturated polyester resin composites. The results showed that especially for modified cellulose, the properties of composites were similar to those with the addition of nanocellulose, but incorporating nanocellulose into the resin was much more difficult.

In the paper of Pączkowski, Puszka, Miazga-Karska, Ginalska, and Gawdzik ("Synthesis, Characterization and Testing of Antimicrobial Activity of Composites of Unsaturated Polyester Resins with Wood Flour and Silver Nanoparticles" [6]) the properties of the wood–resin composites were studied. For composites with wood flour, deterioration of mechanical properties was observed. Antimicrobial activity tests showed that bacterial strains can colonize the surface of the cross-linked unsaturated polyester resin. Composites with the addition of wood flour are a good surface for bacteria colonization and their elimination may occur when nanosilver is added to the composite.

In the paper of Alcántara, González, Pareta, and Vilaseca ("Biocomposites from Rice Straw Nanofibers: Morphology, Thermal and Mechanical Properties" [8]), cellulose nanofibers were extracted from rice straw and reinforced the poly(vinyl alcohol) matrix. The results show that Young's modulus of the polymer increased by three and a half times after adding nanofiller.

In the paper of Wang, Liu, Duan, Xu, Zhang, She, and Zheng ("The Eco-Friendly Biochar and Valuable Bio-Oil from Caragana korshinskii: Pyrolysis Preparation, Characterization, and Adsorption Applications" [12]) biochar and bio-oil were obtained from caragana biomass carbonization in different pyrolysis conditions and the adsorption and pharmaceutical properties of these products were discussed.

The papers of Du at al. and Goliszek at al. concern synthesis of sorbents based on natural components.

In the paper of Du, Dong, Lin, Yang, and Zhao ("Radiation Synthesis of Pentaethylene Hexamine Functionalized Cotton Linter for Effective Removal of Phosphate: Batch and Dynamic Flow Mode Studies" [13]), quaternized cotton linter fibers were prepared by electron beam preirradiation grafting technology. The results showed that the modified cotton linter fibers can be used as sorbents for phosphate removal.

The paper of Goliszek, Podkościelna, Sevastyanova, Gawdzik, and Chabros ("The Influence of Lignin Diversity on the Structural and Thermal Properties of Polymeric Microspheres Derived from Lignin, Styrene, and/or Divinylbenzene" [9]) investigates the impact of lignin influence on the properties of the porous biopolymeric microspheres. It was found that the materials have a high thermal resistance and the incorporation of methacrylated lignin into the microspheres resulted in an increase of specific surface area and porosity.

3. Conclusions

The usage of polymer blends and composites, both prepared by blending, are two strategies used in the "green" requirements of many industries. Green chemistry strategies are mainly accomplished by reducing waste production, reducing raw material usage, reducing nonrenewable energy sources and overall energy demand, reducing risks, hazards, and costs. In this Special Issue, the great potential of agricultural and wood residuals, as they are or chemically modified, for the improvement of broad range of polymeric materials is clearly demonstrated. Cheap and abundant resources such as wood, rice straw, hemp, walnut shells, and other types of biomass and their constituents can improve the strength, thermal and some specific properties of final polymeric materials that have a great potential to be used in automotive, construction, and pharmaceutical and environmental protection industries.

Author Contributions: Conceptualization, B.G. and O.S.; writing—original draft preparation, B.G. and O.S.; writing—review and editing, B.G. and O.S.; supervision, B.G. and O.S. All authors have read and agreed to the published version of the manuscript.

Funding: This research received no external funding.

Institutional Review Board Statement: Not applicable.

Informed Consent Statement: Not applicable.

Data Availability Statement: Not applicable.

Conflicts of Interest: The authors declare no conflict of interest.

References

1. Członka, S.; Kairytė, A.; Miedzińska, K.; Strąkowska, A. Polyurethane Hybrid Composites Reinforced with Lavender Residue Functionalized with Kaolinite and Hydroxyapatite. *Materials* **2021**, *14*, 415. [CrossRef] [PubMed]
2. Strąkowska, A.; Członka, S.; Kairytė, A. Rigid Polyurethane Foams Reinforced with POSS-Impregnated Sugar Beet Pulp Filler. *Materials* **2020**, *13*, 5493. [CrossRef] [PubMed]
3. Członka, S.; Strąkowska, A.; Kairytė, A. The Impact of Hemp Shives Impregnated with Selected Plant Oils on Mechanical, Thermal, and Insulating Properties of Polyurethane Composite Foams. *Materials* **2020**, *13*, 4709. [CrossRef] [PubMed]
4. Członka, S.; Strąkowska, A. Rigid Polyurethane Foams Based on Bio-Polyol and Additionally Reinforced with Silanized and Acetylated Walnut Shells for the Synthesis of Environmentally Friendly Insulating Materials. *Materials* **2020**, *13*, 3245. [CrossRef] [PubMed]
5. Gonçalves, S.S.L.; Rudnitskaya, A.; Sales, A.J.M.; Costa, L.M.C.; Evtuguin, D.V. Nanocomposite Polymeric Materials Based on Eucalyptus Lignoboost® Kraft Lignin for Liquid Sensing Applications. *Materials* **2020**, *13*, 1637. [CrossRef] [PubMed]
6. Pączkowski, P.; Puszka, A.; Miazga-Karska, M.; Ginalska, G.; Gawdzik, B. Synthesis, Characterization and Testing of Antimicrobial Activity of Composites of Unsaturated Polyester Resins with Wood Flour and Silver Nanoparticles. *Materials* **2021**, *14*, 1122. [CrossRef] [PubMed]
7. Chabros, A.; Gawdzik, B.; Podkościelna, B.; Goliszek, M.; Pączkowski, P. Composites of Unsaturated Polyester Resins with Microcrystalline Cellulose and Its Derivatives. *Materials* **2020**, *13*, 62. [CrossRef] [PubMed]
8. Alcántara, J.C.; González, I.; Pareta, M.M.; Vilaseca, F. Biocomposites from Rice Straw Nanofibers: Morphology, Thermal and Mechanical Properties. *Materials* **2020**, *13*, 2138. [CrossRef] [PubMed]
9. Goliszek, M.; Podkościelna, B.; Sevastyanova, O.; Gawdzik, B.; Chabros, A. The Influence of Lignin Diversity on the Structural and Thermal Properties of Polymeric Microspheres Derived from Lignin, Styrene, and/or Divinylbenzene. *Materials* **2019**, *12*, 2847. [CrossRef] [PubMed]
10. Delgado-Aguilar, M.; Tarrés, Q.; Marques, M.F.V.; Espinach, F.X.; Julián, F.; Mutjé, P.; Vilaseca, F. Explorative Study on the Use of Curauá Reinforced Polypropylene Composites for the Automotive Industry. *Materials* **2019**, *12*, 4185. [CrossRef] [PubMed]
11. Serra-Parareda, F.; Tarrés, Q.; Delgado-Aguilar, M.; Espinach, F.X.; Mutjé, P.; Vilaseca, F. Biobased Composites from Biobased-Polyethylene and Barley Thermomechanical Fibers: Micromechanics of Composites. *Materials* **2019**, *12*, 4182. [CrossRef] [PubMed]
12. Wang, T.; Liu, H.; Duan, C.; Xu, R.; Zhang, Z.; She, D.; Zheng, J. The Eco-Friendly Biochar and Valuable Bio-Oil from Caragana korshinskii: Pyrolysis Preparation, Characterization, and Adsorption Applications. *Materials* **2020**, *13*, 3391. [CrossRef] [PubMed]
13. Du, J.; Dong, Z.; Lin, Z.; Yang, X.; Zhao, L. Radiation Synthesis of Pentaethylene Hexamine Functionalized Cotton Linter for Effective Removal of Phosphate: Batch and Dynamic Flow Mode Studies. *Materials* **2019**, *12*, 3393. [CrossRef] [PubMed]

Article

Synthesis, Characterization and Testing of Antimicrobial Activity of Composites of Unsaturated Polyester Resins with Wood Flour and Silver Nanoparticles

Przemysław Pączkowski [1,*], Andrzej Puszka [1], Malgorzata Miazga-Karska [2], Grażyna Ginalska [2] and Barbara Gawdzik [1]

1. Department of Polymer Chemistry, Faculty of Chemistry, Institute of Chemical Sciences, Maria Curie-Sklodowska University in Lublin, Gliniana 33, 20-614 Lublin, Poland; andrzej.puszka@umcs.pl (A.P.); barbara.gawdzik@umcs.pl (B.G.)
2. Chair and Department of Biochemistry and Biotechnology, Faculty of Pharmacy, Medical University of Lublin, Chodzki 1, 20-093 Lublin, Poland; malgorzata.miazga-karska@umlub.pl (M.M.-K.); grazyna.ginalska@umlub.pl (G.G.)
* Correspondence: przemyslaw.paczkowski@umcs.pl

Citation: Pączkowski, P.; Puszka, A.; Miazga-Karska, M.; Ginalska, G.; Gawdzik, B. Synthesis, Characterization and Testing of Antimicrobial Activity of Composites of Unsaturated Polyester Resins with Wood Flour and Silver Nanoparticles. *Materials* 2021, 14, 1122. https://doi.org/10.3390/ma14051122

Received: 3 February 2021
Accepted: 24 February 2021
Published: 27 February 2021

Publisher's Note: MDPI stays neutral with regard to jurisdictional claims in published maps and institutional affiliations.

Copyright: © 2021 by the authors. Licensee MDPI, Basel, Switzerland. This article is an open access article distributed under the terms and conditions of the Creative Commons Attribution (CC BY) license (https://creativecommons.org/licenses/by/4.0/).

Abstract: This paper presents the properties of the wood-resin composites. For improving their antibacterial character, silver nanoparticles were incorporated into their structures. The properties of the obtained materials were analyzed in vitro for their anti-biofilm potency in contact with aerobic Gram-positive *Staphylococcus aureus* and *Staphylococcus epidermidis*; and aerobic Gram-negative *Escherichia coli* and *Pseudomonas aeruginosa*. These pathogens are responsible for various infections, including those associated with healthcare. The effect of silver nanoparticles incorporation on mechanical and thermomechanical properties as well as gloss were investigated for the samples of composites before and after accelerating aging tests. The results show that bacteria can colonize in various wrinkles and cracks on the composites with wood flour but also the surface of the cross-linked unsaturated polyester resin. The addition of nanosilver causes the death of bacteria. It also positively influences mechanical and thermomechanical properties as well as gloss of the resin.

Keywords: antibacterial activity; wood–resin composites; unsaturated polyester resin; recycled PET; wood flour; renewable resources; silver nanoparticles

1. Introduction

All the possibilities of using new products, including composites with the addition of recycled wood, help to reduce the environmental impact and consumption of conventional polymers. Among the various synthetic polymers, unsaturated polyester resins are the most commonly known in preparing composites and exceed 80% of all components [1,2]. The global unsaturated polyester resin market will grow at a steady 5.3% CAGR over the forecast period (2019–2029) [3–5]. Greater concern about health as a result of industrial activities enhances the demand for unsaturated polyester resins. Modern products use a low content of styrene, which reduces the harmful effects of workmen's exposure to poisonous gases, and thus decreases carbon dioxide emissions improving safety standards in many industries [6]. Such materials are widely applied in the production of yachts, kayaks, sailboats, bathtubs, shower cabins, etc., as well as housing goods and medical equipment.

It is known that microorganisms survive on inanimate "touch" surfaces for a long time. This can be particularly troublesome in health care where patient immunity is at a greater risk of infections. Touch surfaces in hospital rooms can serve as a source or a reservoir of bacteria expansion.

A nosocomial infection can be contracted in various ways in hospitals, nursing homes, rehabilitation centers, clinics and even diagnostic laboratories [7–9]. Besides contaminated equipment, bed sheets or air droplets, the infection is also spread by medical personnel themselves. In some cases, the microorganisms may originate from the patient's own skin and become opportunistic after surgery or other procedures that compromise the skin protective barrier. Bacteria can enter the bloodstream as a serious complication of infection during surgery or because of catheters and other foreign bodies entering the arteries or veins.

The common pathogens found in the healthcare settings are *Escherichia coli (E. coli)* and *Pseudomonas aeruginosa (P. aeruginosa)* Gram-negative and rod-shaped bacteria, while *Staphylococcus aureus (S. aureus)* and *Staphylococcus epidermidis (S. epidermidis)* are round Gram-positive bacteria.

One of the most important and common species of Gram-positive bacteria is *Staphylococcus*. These bacteria are normally found on the skin or in the digestive tract and can enter the bloodstream, where *S. aureus* is responsible for the most common healthcare-associated infections.

As a result of infections in the respiratory tract, the genitourinary system, the gastrointestinal tract or the hepatobiliary system, Gram-negative bacteria are able to enter the bloodstream. This type of Gram-negative bacteremia is more commonly associated with the elderly population, where *E. coli* is one of the most common reasons of its formation.

A number of compounds added to materials can reduce the risk of bacteria growing on the surface. The use of silver or copper nanoparticles causes the antimicrobial properties of biocomposites [10–12]. An alternative approach to reducing healthcare-associated pathogens is the use of ultraviolet irradiation [13–16]. Silver nanoparticles (AgNPs) are an important component of nanomaterials in a wide range of industrial and medical applications. There are official EU regulations that allow the use of nanosilver [17]. In another regulation there is the conclusion about the safety of colloidal silver in nano form only at low concentrations [18].

Unfortunately, nanosilver can be risky to human health. According to Ahamed et al. [19], AgNPs produce toxicity that targets various organs, including the lungs, liver, and vascular system. A level of expression of genes involved in cell cycle progression and apoptosis may be induced by AgNP. The induction of ROS (reactive oxygen species), oxidative stress and DNA damage include possible AgNP-induced toxicity mechanisms.

In this paper the antimicrobial properties of wood–resin composites with the addition of silver nanoparticles solution (AgNPs) are presented. Moreover, the influence of the presence of AgNPs on the mechanical and thermomechanical properties as well as gloss of the obtained composites was investigated.

2. Materials and Methods

2.1. Chemicals

The unsaturated polyester resin used for composite preparation is a highly reactive orthophthalic resin of bluish-green color with enhanced chemical resistance properties and high strength parameters (LERG, Pustków, Poland). It was based on recycled poly(ethylene terephthalate) (PET). Luperox DHD-9 (2-butanone peroxide solution) (Sigma Aldrich, St. Louis, MO, USA) and a 4% polymeric cobalt solution (Department of Polymer Chemistry, UMCS Lublin, Poland) were used for its curing.

A virgin wood from spruce and fir commercial softwood JELUXYL WEHO (JELU-WERK, Rosenberg-Ludwigsmühle, Rosenberg, Germany) was used as raw material for modification of the unsaturated polyester resin. The wood particles were characterized by the following technical parameters: pH = 5.5, light yellow color and Alpine air sieve fraction: 75 μm (~35%), 100 μm (~20%) and 180 μm (traces). The manufacturer also declares that the moisture content is about 10%.

Aqueous solutions of silver nanoparticles (AgNPs) NL-100 aqua (NANOLAB, Katowice, Poland) with antibacterial properties were used. Images of all components used in synthesis are presented at Figure 1.

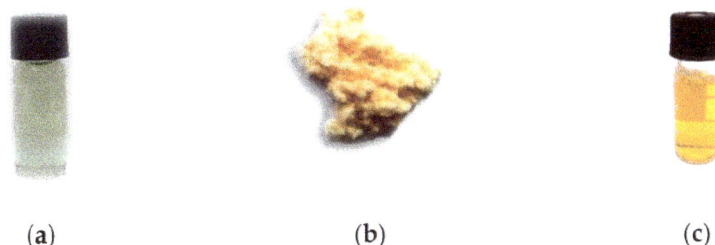

(a) (b) (c)

Figure 1. Images of components used in synthesis: (**a**) unsaturated polyester resin; (**b**) softwood flour; (**c**) silver nanoparticles concentrate.

2.2. Preparation of Composites

A detailed description of the synthetic method and the curing procedure was described previously [20]. The wood flour (WF) or the extract of silver nanoparticles in the aqueous solution (AgNPs) was added to pre-accelerated resin with 1.1 wt% of Luperox and 0.25 wt% of 4% polymeric cobalt solution, while mixing. All ingredient amounts were calculated for the pure resin. The prepared mixtures were poured into Petri dishes (50 mm diameter). Curing was conducted in these dishes at room temperature for 24 h and then at 80 °C for 10 h for additional post-curing. The same procedure was applied for the starting resin and their composites with WF or AgNPs. The compositions of the prepared composites are presented in Table 1.

Table 1. The compositions of the prepared composites.

Sample	UPR (wt%)	WF (wt%)	AgNPs aq. sol. (wt%)
1	100.0	-	-
2	99.0	1.0	-
3	98.0	2.0	-
4	95.0	5.0	-
5	99.8	-	0.2
6	99.5	-	0.5
7	99.0	-	1.0
8	94.0	5.0	1.0

2.3. Research Methods

2.3.1. Preparation of Composite Samples

To study properties of the prepared composites, the samples of suitable sizes were prepared. The CNC milling machine MFG 8037P (Ergwind, Poland) was used for the preparation of samples in the shape of a cuboid (10 mm × 10 mm × 2.5 mm) for antibacterial tests, whereas for mechanical and thermomechanical studies molds of cuboid shapes with the dimensions 80 mm × 10 mm × 4 mm and 65 mm × 10 mm × 4 mm were applied.

2.3.2. Accelerated Aging Test

Accelerated aging tests were performed using a Xenon Arc Lamp Atlas Xenotest Alpha + simulator (Chicago, IL, USA). The source of irradiation in the test chamber was a xenon lamp, emitting radiation similar to natural sunlight. Accelerating aging was carried out for 1000 h according to the standard EN ISO 4892-2:2013 [21]. The parameters imitated typical outdoor weather conditions to which materials can be exposed: irradiance 60 W m^{-2}, daylight filter system, chamber temperature 38 °C, black standard temperature 65 °C, and relative humidity of 50%, spraying switch-on periods (18 min of spraying and 102 min of rest).

2.3.3. Mechanical and Thermomechanical Properties

Mechanical and thermomechanical properties were determined using a mechanical testing machine, a dynamic mechanical analyzer and a hardness tester.

The mechanical properties of the composites were determined based on the Zwick-Roell Z010 mechanical testing machine from ZWICK GmbH Co (Ulm, Germany). Determination of mechanical properties was based on the three-point bending test, where the samples of 80 mm × 10 mm × 4 mm diameter were used with a span of 64 mm between the supports. The bending speed was 5 mm min^{-1}. The test procedure was in accordance with the standard EN ISO 178:2019 [22]. Finally, the arithmetic averaging of five measurements was taken both before and after the accelerating aging test samples.

Using a GYZJ 934-1 Barcol hardness tester from the Barber-Colman Company (Loves Park, IL, USA) the values of composite hardness were determined. The measurements were made according to the standard ASTM D2583 [23]. The final result was the average of ten measurements for the samples before and after the accelerated aging test.

All data were subjected to analysis of variance using Origin 8 (OriginLab, Northampton, MA, USA) applications. One-way analysis of variance (one-way ANOVA) was used to detect significant differences among the tested mechanical parameters (flexural strength and hardness) depending on the wood flour or silver nanoparticles content.

To determine thermomechanical properties of the prepared composites, the Q800 Dynamic Mechanical Analyzer (DMA) from TA Instruments (New Castle, NY, USA) equipped with a dual-cantilever device was used. Suitably prepared samples with the dimensions of 65 mm × 10 mm × 4 mm were tested. A temperature scanning from 0 to 200 °C was made with a constant heating rate of 3 °C min^{-1} at a sinusoidal strain with an amplitude of 10 µm and frequency 1 Hz. The test procedure was in accordance with the standard EN ISO 6721-1:2019 [24]. From the obtained data the glass-transition temperature, mechanical loss factor, values of storage modulus and loss modulus were determined. According to the standard [24] and due to the fact that the analysis was carried out over a wide range of temperatures, only one specimen before and after the accelerated aging test was measured.

2.3.4. Determination of Gloss

Measurements of samples' gloss were made using a Zehntner ZGM 1110 triple-angle gloss meter from Zehntner GmbH Testing Instruments (Sissach, Switzerland). This device operates simultaneously in one of three geometric units in which the angles 20°, 60° and 85° correspond from a high-gloss to matte surface (standard gloss: 20° (86.8 GU), 60° (93.4 GU) and 85° (99.7 GU)). These determinations were made according to the standard ASTM D2457 [25]. The final result was the mean value of ten measurements for the samples before and after the accelerated aging.

2.3.5. Bacterial Strains

All tested samples were analyzed for their anti-biofilm potency in contact with aerobic Gram-positive *Staphylococcus aureus* ATCC 25923 and *Staphylococcus epidermidis* ATCC 12228; and aerobic Gram-negative *Escherichia coli* ATCC 25992 and *Pseudomonas aeruginosa* ATCC 27853. In the microbiological assay we used Mueller–Hinton broth (MH-broth) (BioMaxima, Lublin, Poland). After 24 h bacterial growth (at 37 °C) on agar, an inoculum in 5 mL of 0.9% was prepared, obtaining a density 0.5 McFarland (1.5×10^8 CFU mL^{-1} (CFU: colony forming unit)).

2.3.6. Anti-Biofilm Activity of Tested Materials

Seeding of materials with bacterial strains for biofilm formation determination was made to visualize viability of the biofilm structure [26]. The materials samples (unmodified one as a control (1); modified materials (2, 3, 4, 5, 6, 7, 8)) were washed in ethanol (2000 µL) and M-H broth (2000 µL) and moved to the bottoms of fresh 12-well polystyrene plates (NEST Biotechnology, Wuxi, Jiangsu, China). Then, these materials were covered with 2000 µL of M-H broth. Finally, 6.4 µL of tested bacteria inoculum (1.5×10^8 CFU mL^{-1})

was added. Wells with broth alone were also included in the experiment as a control for the sterility of the experiment. Thus, the obtained plates with material, broth and inoculum were incubated twice as long (48 h, 37 °C) to allow bacterial plankton cells to eventually attach to the material surface and form colonies and biofilm. The tests were performed using three replicates.

Then, the tested material was taken out into a fresh 12-well dish. It was then washed gently with 2 mL of 0.9% NaCl to eliminate loosely adhered planktonic cells of bacteria. The materials obtained in this way were subjected to a biofilm test using the confocal laser scanning microscopy technique (CLSM).

2.3.7. Confocal Laser Scanning Microscopy—Biofilm Visualization

Using the procedure of double fluorescence staining of dead and living bacterial cells, the presence, architecture and structure of the viability of the biofilm can be demonstrated. In this assay, the Viability/Cytotoxicity Assay kit for Bacteria LIVE/DEAD Cells (Biotium, Fremont, CA, USA) was used [26,27]. After 48 h incubation with the tested strain, the biomaterials were moved to new plates and covered with 500 µL of 0.9% NaCl and live/dead dye (prepared by mixing 1 µL of DMAO with 1 µL of EthD-III in 8 µL of 0.9% NaCl). Five microliters live/dead dye solution obtained in such a way and 500 µL of PBS were added to each well containing the tested biomaterial. The stained biofilm was visualized after 15 min of dark incubation of the material with the dye in 0.9% NaCl. The whole material area was inspected to verify the presence of biofilms, then the most representative location (3180 µm × 3180 µm) was scanned for the experiment. The images were acquired using a confocal laser scanning microscope Olympus IX81 (Olympus, Tokyo, Japan) equipped with Olympus Fluoview FV1000 scanning head by Olympus Fluoview ver. 4.2c software (Olympus, Tokyo, Japan). Pictures were visualized with Imaris ver. 7.2.3 software (Oxford Instruments, Abingdon, UK).

3. Results and Discussion

Generally, unsaturated polyester resins as cross-linked polymers are treated to resist biodegradation [28,29]. In the case of unsaturated polyester resins, there is no case when biological agents such as bacteria, fungi and their enzymes destroy the polymer; thus, its original form disappears. In order to increase the susceptibility of resins to microorganisms, many authors propose the use of biodegradable fillers [30–32]. That is why in our research resin composites with wood flour and composites with the addition of colloidal silver were used. The unsaturated polyester resin without any additives was used as a reference material.

In Tables 2–4 the results of mechanical and thermomechanical studies of these composites before and after the accelerating aging test are presented. It was assumed that the addition of wood flour into the resin would facilitate the existence of microorganisms on the surface. The data for the unmodified resin (sample 1) showed that after aging, the flexural strength decreased, whereas the hardness and mechanical loss factor significantly increased. This is related to additional hardening of the resin, which is also confirmed by the value of full width at half-maximum (FWHM). For the composites with wood flour (samples 2–4), a decrease in flexural strength was visible before and after the accelerating aging studies. The larger the wood content, the smaller their flexural strength is. A similar tendency can be observed for the values of mechanical loss factor. In turn, hardness increased after the wood was incorporated into the resin. It was particularly visible for the samples before aging. After aging, the composites lost their hardness. Even with the slightest addition of wood, they did not achieve hardness similar to that of the pure resin. Full widths at half-maximum for this series of composites became narrower [20]. Samples 5–7 were obtained by adding the colloidal solution of nanosilver into the resin. These samples were prepared to create materials completely immune to the action of bacteria. For these samples a significant increase in flexural strength before aging can be observed. After aging, its values decreased with increasing silver content. Similarly, the hardness of the obtained composites was much higher than that for pure resin. Their hardness increased evidently after aging. The mechanical loss modulus and full widths at half-maximum decreased insignificantly with the increasing content of silver. After aging, their values were lower. Sample 8 contained both wood and silver. Its flexural strength and mechanical loss factor were much lower compared to those of the pure resin, but similar to those of the wood composites. Due to the presence of silver it had greater hardness. It is also characterized by increased heterogeneity (FWHM).

Table 2. Mechanical and thermomechanical data for the studied composite samples before and after the accelerated aging test.

Sample	Flexural Strength, σ_f (MPa)		Barcol Hardness, HBa (°B)		Mechanical Loss Factor, Tan δ_{max}		Full Width at Half Maximum, FWHM (°C)	
	before	after	before	after	before	after	before	after
1	108.19 ± 3.77	69.26 ± 4.35	36.0 ± 1.0	56.4 ± 0.4	0.4579	0.4839	46.1	39.7
2	81.96 ± 2.71	50.03 ± 3.33	39.5 ± 0.5	55.7 ± 0.7	0.4419	0.4510	45.9	39.6
3	76.83 ± 2.65	40.95 ± 4.27	40.3 ± 0.5	54.1 ± 0.6	0.4368	0.4397	43.4	40.0
4	74.61 ± 1.48	39.43 ± 4.39	45.2 ± 0.7	51.3 ± 0.5	0.4227	0.4131	42.4	38.5
5	119.05 ± 2.65	72.31 ± 4.82	48.8 ± 0.8	57.0 ± 1.0	0.4543	0.4575	40.6	41.9
6	121.81 ± 1.97	70.71 ± 4.45	49.2 ± 0.2	57.5 ± 0.5	0.4513	0.4427	42.0	41.2
7	128.74 ± 1.27	70.03 ± 3.69	53.3 ± 0.3	58.8 ± 0.8	0.4479	0.4408	42.7	41.1
8	76.69 ± 1.90	51.95 ± 3.51	55.0 ± 1.0	60.6 ± 0.6	0.3383	0.3328	47.6	46.2

Table 3. Glass transition temperature and storage modulus in the glassy and rubbery regions (from the storage modulus curve) for the studied composites before and after the accelerated aging test.

Sample	Glass-Transition Temperature, T_g (°C)				Storage Modulus, E'			
	From tan δ		From Loss Modulus Curve, E"		Glassy, E' (20 °C) (GPa)		Rubbery, E' (180 °C) (MPa)	
	before	after	before	after	before	after	before	after
1	127	132	100	107	2.91	2.81	24.06	25.24
2	126	131	101	108	3.17	2.78	31.45	23.78
3	126	131	103	108	2.94	2.81	34.68	28.50
4	126	130	105	110	2.86	2.90	49.08	40.04
5	128	131	106	108	2.79	2.73	27.13	22.05
6	129	131	108	110	2.78	2.83	27.53	23.33
7	130	132	109	111	2.75	2.86	27.75	25.75
8	126	129	104	107	2.77	2.83	46.11	41.20

Table 4. Gloss measurement data of the studied composites before and after the accelerated aging test.

Sample	Gloss (GU)					
	20°	60°	85°	20°	60°	85°
	before Aging			after Aging		
1	101.3	112.4	100.4	30.6	50.2	81.1
2	96.8	110.4	98.5	63.3	92.2	96.0
3	96.4	101.8	99.2	57.5	84.0	90.3
4	76.8	95.3	96.1	55.3	81.0	87.6
5	113.3	121.8	99.9	85.9	97.3	98.1
6	128.3	133.9	102.9	83.9	96.8	98.1
7	107.1	121.1	102.0	83.4	91.6	94.6
8	94.7	100.4	100.3	73.5	90.2	95.5

It seems that the lack of chemical binding between the inorganic silver nanoparticles and resin itself is the main cause of decreasing the mechanical properties. However, samples 5–8 can be treated as nanocomposites. The properties of such materials not only depend on those of their components but also crucially on their interfacial and morphological characteristics [33].

Analysis of variance showed that the values of flexural strength and Barcol hardness for samples with different percentages of the wood flour before and after accelerated aging were statistically significant, which was confirmed using one-way analysis of variance (ANOVA) at the significance level $p < 0.05$. The same observations occurred for composites with different percentages of silver nanoparticles.

In Table 3 the comparison of glass transition and storage modulus values for the composites before and after the accelerating aging test is presented. From these data one can see that, regardless of the method of glass transition temperature determination, for all the studied samples they became larger after the aging. The addition of both wood flour and silver caused an increase in T_g. Increasing the wood content of the sample increased the transition temperature from 101 to 105 °C before aging and from 108 to 110 °C after aging. On the other hand, with the increase in the amount of silver nanoparticles in the material, an increase in T_g was observed from 106 to 109 °C before aging and from 108 to 111 °C after aging (Figures 2 and 3). In this Table also the values of storage modulus in the glassy and rubbery regions, characterizing viscoelastic behavior of the sample, are presented. In the glassy region its values decreased slightly with the increase in filler content before aging. After aging, an insignificant increase in its value can be observed, especially for the samples with silver. In the rubbery region all samples had a greater storage modulus compared to that of the original resin before aging. It is particularly noticeable for the samples with the increasing amounts of wood flour. After aging, its values decreased for all samples except the unmodified resin.

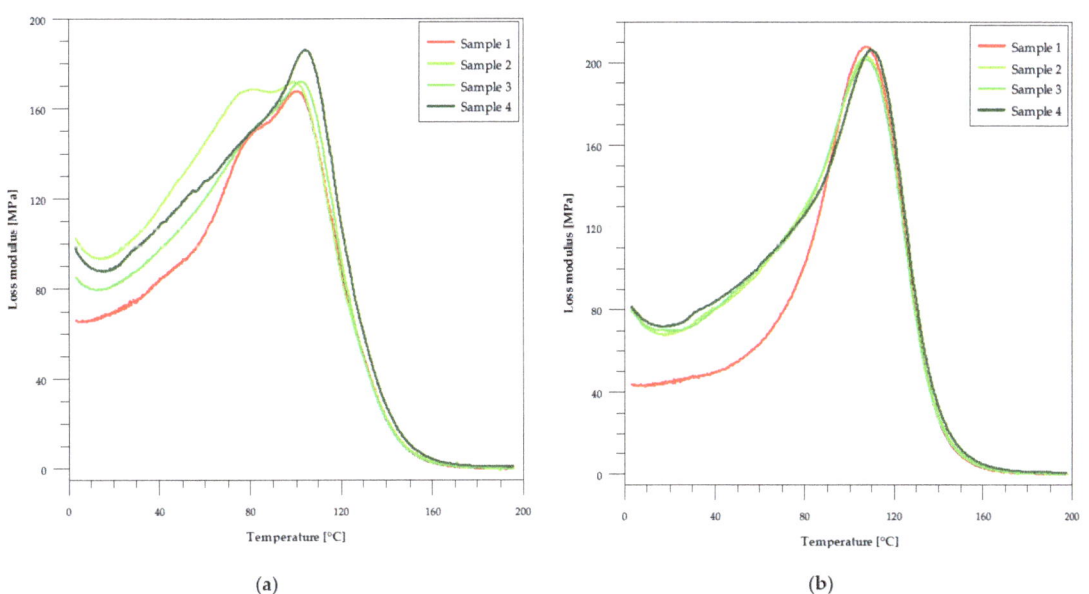

Figure 2. The temperature-dependent graph of loss modulus (E″) for samples with wood flour. (**a**) Before the accelerated aging test and (**b**) after the accelerated aging test.

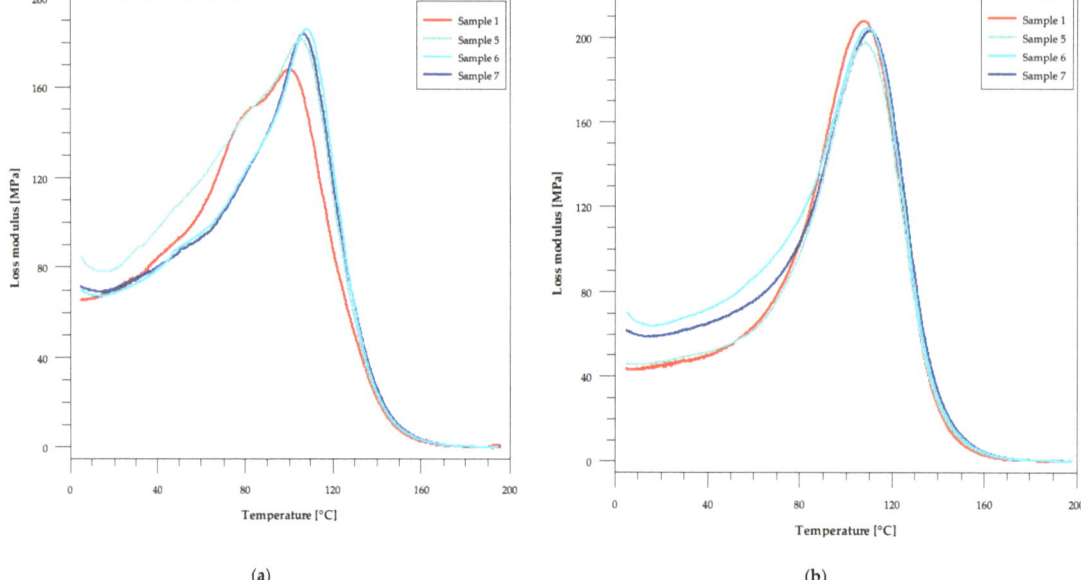

Figure 3. The temperature-dependent graph of loss modulus (E″) for samples with silver nanoparticles. (**a**) Before the accelerated aging test and (**b**) after the accelerated aging test.

As can be seen from Figures 2 and 3, the peaks in the loss modulus curves before ag-ing are asymmetric, and two glass transition temperatures can be determined from their shape. This indicates some inhomogeneity in these samples. Generally, the aging process of these composites resulted in improving homogeneity, which can be seen in both the FWHM values and the shape of the loss modulus curves.

The incorporation of silver nanoparticles has a beneficial effect on the mechanical and thermomechanical properties of the unsaturated polyester resin. In the case of composites, both before and after accelerated aging, the addition of AgNP caused an increase in the hardness of the glass transition temperatures compared to the original resin.

Table 4 presents the results of gloss measurements. According to Zhao et al. [34], gloss is defined as the specular reflection ability of the material surface under a particular standard source and at a certain angle of incidence. This is an important parameter characterizing the surface optical properties of different materials. Gloss determination enables the evaluation of changes taking place on the surface of resins after aging [20]. Gloss is reduced when cracks and pitting appear due to the surface exposition to accelerated weathering conditions. This primary destruction can become a source of sample degradation. From the data in Table 4 one can see that before aging, all the obtained composites can be treated as high-gloss materials. Their values with 60° geometry, which is typically used, are much greater than 70 GU. For very high gloss materials such as unsaturated polyester resins, an angle of 20° is more recommended. For measurements at 20° the case is similar. In fact, a deterioration in gloss as the wood content increases and improvement in gloss with an increase in the silver content can be observed. After aging, the gloss values at 20° for the samples with wood decreased significantly. However, the samples with nanosilver still had a gloss above 70 GU. This indicates that the addition of AgNPs to the resin (samples 5–7) allows to obtain a high-gloss material after accelerated weathering. Similar properties indicate the wood–resin composite with silver nanoparticles (sample 8). Generally, it was found that AgNPs improved the gloss of the composites.

The effect of resin modifications on the power of biofilm formation was also tested. For this purpose, Gram-positive (Figure 4) and Gram-negative (Figure 5) strains were used. They were incubated with the materials long enough to allow the transformation of planktonic forms into an organized biofilm on the surface of the tested materials. Then, after washing away loose planktonic forms the samples were stained and imaged in a confocal microscope.

We noted that all tested strains (Gram-positive *Staphylococcus aureus* ATCC 25923, *Staphylococcus epidermidis* ATCC 12228; and aerobic Gram-negative *Escherichia coli* ATCC 25992, *Pseudomonas aeruginosa* ATCC 27853) could form biofilm architecture on the control unmodified material (sample 1). Green live bacterial cells merged into the forms and the groups that populated a large part of the control material. This is a completely unexpected result.

Whereas, on samples 2 and 3 surfaces a thin layer of biofilm was visible, regardless of the bacterial strain. Such live green colonies were uneven and formed wrinkles of the material. Material 4 showed on its surface a few single colonies of living cells with a thin layer of red dead bacterial cells. In sample 4 these bacteria also accumulate at the wrinkles.

Importantly, the data presented in Figures 4 and 5 indicate clearly that the composites modified with silver compounds (**5–8**) were completely resistant to biofilm formation under the tested conditions. Thus, materials 5–8 were resistant to biofilm formation across a broad spectrum of both Gram-negative and Gram-positive strains, regardless of the additional chemical components.

Certainly, silver ions inhibit the viability of bacteria by bacterial membrane damage [35]. Many authors show use of silver nanoparticles as a beneficial way to prevention of infection [36–39]. Our observations are confirmed by the authors pointing to the anti-biofilm nature of the materials modified with silver-nanoparticles (AgNP). Significant reduction in biofilm density and increase in dead bacterial cells (especially *E. coli* and *P. aeruginosa*) on AgNP-modified materials, associated with destabilization of the bacterial membrane, was observed by Singh et al. [40]. Similar anti-biofilm potentials of AgNP structures were noticed by Hussain et al. and Pal et al. [41,42].

Sanyasi et al. reported that their AgNP inhibited biofilm formation, which notably caused bacterial membrane damage [35]. The adhesion of living cells, both *Eukaryotic* or *Prokaryotic*, to abiotic surfaces may be influenced by the chemical structure of these materials, size, shape, capping layer, functional groups and smoothness/wrinkles. The features of the composition and surface type of the material affect initial interactions and, consequently, adhesion to the cell membrane.

Many authors point to the possibility of using colloidal or ions or silver nanoparticles in vivo, where other antibiotic therapies are not effective. However, studies showing the optimal (safe for the patient and antibacterial-effective) doses of such silver compounds are still needed to obtain a therapeutic effect against Gram-negative and Gram-positive bacteria [43].

Therefore, it seems to be advantageous to obtain materials with the addition of silver compounds, which would enable their applicability not only in the household but also in pharmacy, analytical and medical as well as food industries.

Biocides, such as our silver-modified materials, have a broad spectrum of activity; they can inhibit the viability of not only microbes but also other organisms, including eukaryotic cells [43]. Thus, in the next step of our research we will study cytotoxicity against normal eukaryotic fibroblasts lines, to possess information about the safety of using our materials in close contact with human skin.

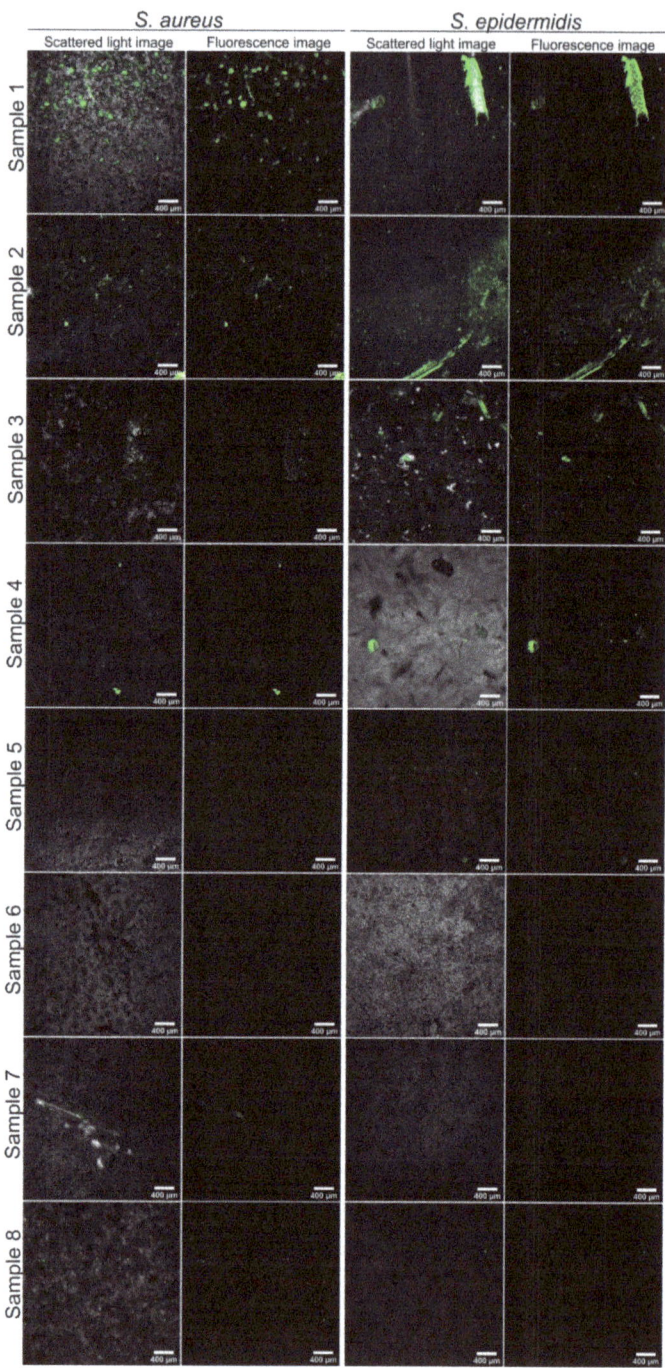

Figure 4. Confocal laser scanning microscope CLSM images showing Gram-positive strains biofilm formation (after 48 h of incubation) on tested materials: 1, control unmodified; 2–8, modified materials; magnification 40×; scale bar = 400 μm.

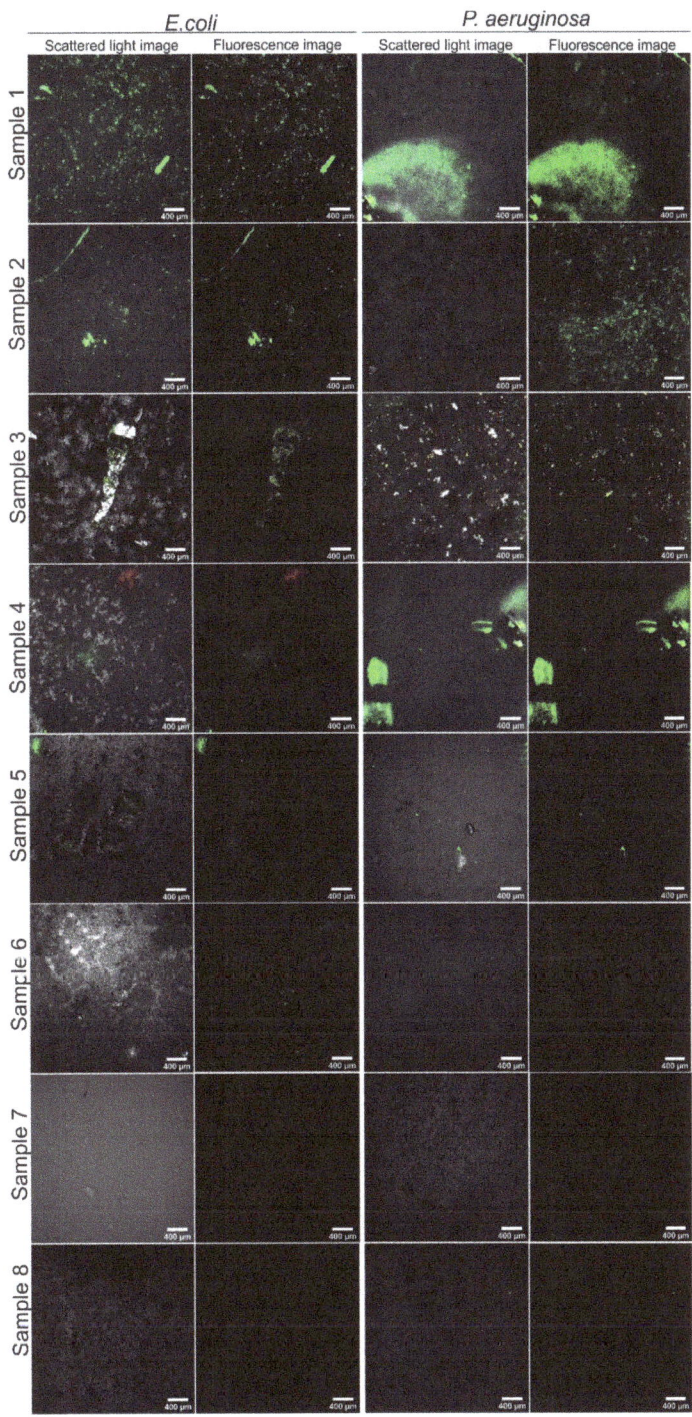

Figure 5. Confocal laser scanning microscope CLSM images showing Gram-negative strains biofilm formation (after 48 h of incubation) on tested materials: 1, control unmodified; 2–8, modified materials; magnification 40×; scale bar = 400 μm.

4. Conclusions

As it was assumed, Gram-negative and Gram-positive bacteria colonize mainly in various wrinkles as well as cracks on the composites with wood flour. In turn, the resins modified with silver nanoparticles (AgNP) were completely resistant to biofilm formation. Importantly, the obtained results indicate that bacterial strains can colonize the surface of the cross-linked unsaturated polyester resin. This can change the current views on the disposal of products made of unsaturated polyester resins.

Favorable effects of silver nanoparticles incorporation on mechanical and thermomechanical properties as well as gloss of unsaturated polyester resin were observed. For the composites before and after accelerated aging, the addition of AgNPs caused an increase in flexural strength and hardness compared to those of the original resin. A significant impact on the glass transition temperatures was observed.

In the case of the composites with wood flour, deterioration of mechanical properties was observed. As the wood flour content increased, they became brittle, and their gloss diminished.

As was expected, silver nanoparticles modification promoted antimicrobial activity of the wood–resin composites in contact with pathogens such as S. aureus, S. epidermidis, E. coli and P. aeruginosa. The materials characterized by such properties are promising for medical field applications.

Importantly, the addition of nanosilver caused the death of bacteria not only on the surface of the unsaturated polyester resin but also on the surface of the composite with biofiller. This may be an indication for manufacturers of devices for medical purposes. In order to reduce the cost and consumption of the resin, the addition of biofillers can be put into practice, provided that the product will contain nanosilver.

Author Contributions: Conceptualization, P.P. and B.G.; methodology, B.G. and M.M.-K. validation, P.P., A.P. and M.M.-K.; formal analysis, P.P., A.P., M.M.-K., B.G. and G.G.; investigation, P.P., A.P., M.M.-K., B.G. and G.G.; resources, B.G. and G.G.; writing—original draft preparation, P.P.; writing—review and editing, P.P., B.G. and M.M.-K.; visualization, P.P., A.P. and M.M.-K.; supervision, B.G. and G.G. All authors have read and agreed to the published version of the manuscript.

Funding: Financial assistance was partially (microbiology research) provided by Ministry of Science and Higher Education in Poland within DS2 project of Medical University of Lublin.

Institutional Review Board Statement: Not applicable.

Informed Consent Statement: Not applicable.

Data Availability Statement: Data sharing is not applicable to this article.

Acknowledgments: The authors would like to express their appreciation to LERG S.A. (Pustków, POLAND) for supplying the unsaturated polyester resin based on recycled PET, JELU-WERK (Rosenberg, GERMANY) for supplying the softwood flour, and NANOLAB (Katowice, POLAND) for supplying the silver nanoparticle concentrate in the form of an aqueous solution. The authors would like to thank Michal Wojcik from the Chair and Department of Biochemistry and Biotechnology at the Medical University of Lublin (Poland) for Confocal Laser Scanning Microscope analyses.

Conflicts of Interest: The authors declare no conflict of interest.

References

1. Penczek, P.; Czub, P.; Pielichowski, J. Unsaturated polyester resins: Chemistry and technology. *Adv. Polym. Sci.* **2005**, *184*. [CrossRef]
2. Johnson, K.G.; Yang, L.S. Preparation, in Properties and Applications of Unsaturated Polyesters. In *Modern Polyesters: Chemistry and Technology of Polyesters and Copolyesters*; Scheirs, J., Long, T.E., Eds.; John Wiley & Sons Ltd.: Chichester, UK, 2003; pp. 699–714.
3. Unsaturated Polyester Resin Market Forecast, Trend Analysis & Competition Tracking—Global Market Insights 2020 to 2030. Available online: https://www.factmr.com/report/4731/unsaturated-polyester-resin-market (accessed on 12 January 2021).
4. Cherian, B.; Thachil, E.T. Synthesis of Unsaturated Polyester Resin—Effect of Sequence of Addition of Reactants. *Polym. Plast. Technol. Eng.* **2007**, *44*, 931–938. [CrossRef]

5. Fink, J.K. Unsaturated Polyester Resins. In *Reactive Polymers Fundamentals and Applications*, 2nd ed.; Fink, J.K., Ed.; William Andrew Inc.: San Francisco, CA, USA, 2005; pp. 1–67.
6. Chabros, A.; Gawdzik, B.; Podkościelna, B.; Goliszek, M.; Pączkowski, P. Composites of unsaturated poliester resins with cellulose and its derivatives. *Materials* **2020**, *13*, 62. [CrossRef] [PubMed]
7. Khan, H.A.; Baig, F.K.; Mehboob, R. Nosocomial infections: Epidemiology, prevention, control and surveillance. *Asian Pac. J. Trop. Biomed.* **2017**, *7*, 478–482. [CrossRef]
8. Khan, H.; Ahmad, A.; Mehboob, R. Nosocomial infections and their control strategies. *Asian Pac. J. Trop. Biomed.* **2015**, *5*, 509–514. [CrossRef]
9. Joshi, M.; Kaur, S.; Kaur, H.P.; Mishra, T. Nosocomial infection: Source and prevention. *Int. J. Pharm. Sci. Res.* **2019**, *10*, 1613–1624.
10. Kalwar, K.; Shan, D. Antimicrobial effect of silver nanoparticles (AgNPs) and their mechanism—A mini review. *Micro Nano Lett.* **2018**, *13*, 277–280. [CrossRef]
11. Deshmukh, S.P.; Patil, S.M.; Mullani, S.B.; Delekar, S.D. Silver nanoparticles as an effective disinfectant: A review. *Mater. Sci. Eng. C* **2019**, *97*, 954–965. [CrossRef]
12. Kim, J.S.; Kuk, E.; Yu, K.N.; Kim, J.-H.; Park, S.J.; Lee, H.J.; Kim, S.H.; Park, Y.K.; Park, Y.H.; Hwang, C.-Y.; et al. Antimicrobial effects of silver nanoparticles. *Nanomedicine* **2007**, *3*, 95–101. [CrossRef]
13. Dai, T.; Vrahas, M.S.; Murray, C.K.; Hamblin, M.R. Ultraviolet C irradiation: An alternative antimicrobial approach to localized infections? *Expert Rev. Anti Infect. Ther.* **2012**, *10*, 185–195. [CrossRef]
14. Memarzadeh, F.; Olmsted, R.N.; Bartley, J.M. Applications of ultraviolet germicidal irradiation disinfection in health care facilities: Effective adjunct, but not stand-alone technology. *Am. J. Infect. Control.* **2010**, *38*, S13–S24. [CrossRef] [PubMed]
15. Yang, J.H.; Wu, U.I.; Tai, H.M.; Sheng, W.H. Effectiveness of an ultraviolet-C disinfection system for reduction of healthcare-associated pathogens. *J. Microbiol. Immunol. Infect.* **2019**, *52*, 487–493. [CrossRef] [PubMed]
16. Weber, D.J.; Rutala, W.A.; Anderson, D.J.; Chen, L.F.; Sickbert-Bennett, E.E.; Boyce, J.M. Effectiveness of ultraviolet devices and hydrogen peroxide systems for terminal room decontamination: Focus on clinical trials. *Am. J. Infect. Control* **2016**, *44*, e77–e84. [CrossRef]
17. SCENIHR (Scientific Committee on Emerging and Newly Identified Health Risks). *Opinion on Nanosilver: Safety, Health and Environmental Effects and Role in Antimicrobial Resistance*; European Union: Brussels, Belgium, 2013; ISBN 978-92-79-30132-2. [CrossRef]
18. SCCS (Scientific Committee on Consumer Safety). *Opinion on Colloidal Silver*; European Union: Brussels, Belgium, 2018; ISBN 978-92-76-00236-9. [CrossRef]
19. Ahamed, M.; AlSalhi, M.S.; Siddiqui, M.K.J. Silver nanoparticle applications and human health. *Clin. Chim. Acta* **2010**, *411*, 1841–1848. [CrossRef] [PubMed]
20. Pączkowski, P.; Puszka, A.; Gawdzik, B. Green Composites Based on Unsaturated Polyester Resin from Recycled Poly (Ethylene Terephthalate) with Wood Flour as Filler—Synthesis, Characterization and Aging Effect. *Polymers* **2020**, *12*, 2966. [CrossRef] [PubMed]
21. EN ISO. *4892-2:2013 Plastics—Methods of Exposure to Laboratory Light Sources—Part 2: Xenon-arc Lamps*; International Organization of Standardization: Geneva, Switzerland, 2013.
22. EN ISO. *178:2019 Plastics—Determination of Flexural Properties*; International Organization of Standardization: Geneva, Switzerland, 2019.
23. ASTM. *D2583, Standard Test Method for Indentation Hardness of Rigid Plastics by Means of a Barcol Impressor*; ASTM International: West Conshohocken, PA, USA, 2013.
24. EN ISO. *6721-1:2019 Plastics—Determination of Dynamic Mechanical Properties—Part 1: General Principles*; International Organization of Standardization: Geneva, Switzerland, 2019.
25. ASTM. *D2457, Standard Test Method for Specular Gloss of Plastic Films and Solid Plastics*; ASTM International: West Conshohocken, PA, USA, 2013.
26. Miazga-Karska, M.; Michalak, K.; Ginalska, G. Anti-Acne Action of Peptides Isolated from Burdock Root—Preliminary Studies and Pilot Testing. *Molecules* **2020**, *25*, 2027. [CrossRef]
27. Kommerein, N.; Doll, K.; Stumpp, N.S.; Stiesch, M. Development and characterization of an oral multispecies biofilm implant flow chamber model. *PLoS ONE* **2018**, *13*, e0196967. [CrossRef]
28. Chandra, R.; Rustgi, R. Biodegradable polymers. *Prog. Polym. Sci.* **1998**, *23*, 1273–1335. [CrossRef]
29. Siracusa, V. Microbial Degradation of Synthetic Biopolymers Waste. *Polymers* **2019**, *11*, 1066. [CrossRef]
30. Shenov, M.A.; D'Melo, D.J. Evaluation of mechanical properties of unsaturated polyester-guar gum/hydroxypropyl guar gum composites. *EXPRESS Polym. Lett.* **2007**, *1*, 622–628. [CrossRef]
31. Mohammed, M.H.; Dauda, B. Unsaturated Polyester Resin Reinforced with Chemically Modified Natural Fibre. *IOSR J. Polymer. Text. Eng.* **2014**, *1*, 31–38.
32. Rahman, M.R.; Hamdan, S.; Hasan, M.; Baini, R.; Salleh, A.A. Physical, mechanical, and thermal of wood flour reinforced maleic anhydride grafted unsaturated polyester (UP) biocomposites. *BioResources* **2015**, *10*, 4557–4568. [CrossRef]
33. Sen, M. Nanocomposite materials. In *Nanotechnology and the Environment*; Sen, M., Ed.; IntechOpen: London, UK, 2020.
34. Zhao, Q.; Jia, Z.; Li, X.; Ye, Z. Surface degradation of unsaturated poliester resin in Xe artificial 309 weathering environment. *Mater. Design.* **2010**, *31*, 4457–4460. [CrossRef]

35. Sanyasi, S.; Majhi, R.K.; Kumar, S.; Mishra, M.; Ghosh, A.; Suar, M.; Satyam, P.V.; Mohapatra, H.; Goswami, C.; Goswami, L. Polysaccharide-capped silver Nanoparticles inhibit biofilm formation and eliminate multi-drug-resistant bacteria by disrupting bacterial cytoskeleton with reduced cytotoxicity towards mammalian cells. *Sci. Rep.* **2016**, *6*, 24929. [CrossRef]
36. Kim, S.H.; Lee, H.S.; Ryu, D.S.; Choi, S.J.; Lee, D.S. Antibacterial activity of silver-nanoparticles against Staphylococcus aureus and *Escherichia coli*. *Korean J. Microbiol. Biotechnol.* **2011**, *39*, 77–85.
37. Franci, G.; Falanga, A.; Galdiero, S.; Palomba, L.; Rai, M.; Morelli, G.; Galdiero, M. Silver nanoparticles as potential antibacterial agents. *Molecules* **2015**, *20*, 8856–8874. [CrossRef]
38. Palza, H. Antimicrobial polymers with metal nanoparticles. *Int. J. Mol. Sci.* **2015**, *16*, 2099–2116. [CrossRef]
39. Furno, F.; Morley, K.S.; Wong, B.; Sharp, B.L.; Arnold, P.L.; Howdle, S.M.; Bayston, R.; Brown, P.D.; Winship, P.D.; Reid, H.J. Silver nanoparticles and polymeric medical devices: A new approach to prevention of infection? *J. Antimicrob. Chemother.* **2004**, *54*, 1019–1024. [CrossRef]
40. Singh, P.; Pandit, S.; Garnæs, J.; Tunjic, S.; Mokkapati, V.R.; Sultan, A.; Thygesen, A.; Mackevica, A.; Mateiu, R.V.; Daugaard, A.E.; et al. Green synthesis of gold and silver nanoparticles from Cannabis sativa (industrial hemp) and their capacity for biofilm inhibition. *Int. J. Nanomed.* **2018**, *13*, 3571–3591. [CrossRef]
41. Hussain, A.; Alajmi, M.F.; Khan, M.A.; Pervez, S.A.; Ahmed, F.; Amir, S.; Husain, F.M.; Khan, M.S.; Shaik, G.M.; Hassan, I.; et al. Biosynthesized Silver Nanoparticle (AgNP) From Pandanus odorifer Leaf Extract Exhibits Anti-metastasis and Anti-biofilm Potentials. *Front. Microbiol.* **2019**, *10*, 8. [CrossRef]
42. Pal, S.; Tak, Y.K.; Song, J.M. Does the anti-bacterial activity of silver nanoparticles depend on the shape of the nanoparticle? A study of the Gram-negative bacterium *Escherichia coli*. *Appl. Environ. Microbiol.* **2007**, *73*, 1712–1720. [CrossRef] [PubMed]
43. Pazos-Ortiz, E.; Roque-Ruiz, J.H.; Hinojos-Márquez, E.A.; López-Esparza, J.; Donohué-Cornejo, A.; Cuevas-González, J.C.; Espinosa-Cristóbal, L.F.; Reyes-López, S.Y. Dose-dependent Antimicrobial Activity of Silver Nanoparticles on Polycaprolactone Fibers against Gram-positive and Gram-negative Bacteria. *J. Nanomater.* **2017**, *6*. [CrossRef]

Article

Polyurethane Hybrid Composites Reinforced with Lavender Residue Functionalized with Kaolinite and Hydroxyapatite

Sylwia Członka [1,*], Agnė Kairytė [2], Karolina Miedzińska [1] and Anna Strąkowska [1]

1. Institute of Polymer & Dye Technology, Lodz University of Technology, 90-924 Lodz, Poland; karolina.miedzinska@dokt.p.lodz.pl (K.M.); anna.strakowska@p.lodz.pl (A.S.)
2. Laboratory of Thermal Insulating Materials and Acoustics, Institute of Building Materials, Faculty of Civil Engineering, Vilnius Gediminas Technical University, Linkmenu st. 28, LT-08217 Vilnius, Lithuania; agne.kairyte@vgtu.lt
* Correspondence: sylwia.czlonka@dokt.p.lodz.pl

Citation: Członka, S.; Kairytė, A.; Miedzińska, K.; Strąkowska, A. Polyurethane Hybrid Composites Reinforced with Lavender Residue Functionalized with Kaolinite and Hydroxyapatite. *Materials* **2021**, *14*, 415. https://doi.org/10.3390/ma14020415

Received: 29 December 2020
Accepted: 14 January 2021
Published: 15 January 2021

Publisher's Note: MDPI stays neutral with regard to jurisdictional claims in published maps and institutional affiliations.

Copyright: © 2021 by the authors. Licensee MDPI, Basel, Switzerland. This article is an open access article distributed under the terms and conditions of the Creative Commons Attribution (CC BY) license (https://creativecommons.org/licenses/by/4.0/).

Abstract: Polyurethane (PUR) composites were modified with 2 wt.% of lavender fillers functionalized with kaolinite (K) and hydroxyapatite (HA). The impact of lavender fillers on selected properties of PUR composites, such as rheological properties (dynamic viscosity, foaming behavior), mechanical properties (compressive strength, flexural strength, impact strength), insulation properties (thermal conductivity), thermal characteristic (temperature of thermal decomposition stages), flame retardancy (e.g., ignition time, limiting oxygen index, heat peak release) and performance properties (water uptake, contact angle) was investigated. Among all modified types of PUR composites, the greatest improvement was observed for PUR composites filled with lavender fillers functionalized with kaolinite and hydroxyapatite. For example, on the addition of functionalized lavender fillers, the compressive strength was enhanced by ~16–18%, flexural strength by ~9–12%, and impact strength by ~7%. Due to the functionalization of lavender filler with thermally stable flame retardant compounds, such modified PUR composites were characterized by higher temperatures of thermal decomposition. Most importantly, PUR composites filled with flame retardant compounds exhibited improved flame resistance characteristics—in both cases, the value of peak heat release was reduced by ~50%, while the value of total smoke release was reduced by ~30%.

Keywords: polyurethane composites; lavender; kaolinite; hydroxyapatite; high-ball milling process; thermal conductivity

1. Introduction

Polyurethanes (PUR) are dynamically developing groups of polymers [1–3]. Polyurethanes can be used in a variety of applications thanks to the ability to control their mechanical, physical, and chemical properties. Due to this, polyurethane materials are used in many industries, for instance automotive, building and construction, furniture, or industrial insulation [4]. Polyurethanes are obtained by polyaddition reaction between polyols and polyisocyanates, during which the urethane bond, which is the backbone of the resulting polymer, is formed. The first polyurethane was synthesized by Wurtz in 1849, then Otto Bayer in 1937 obtained PUR in the reaction between polyol and polyisocyanate known to this day [5]. The reaction can be carried out in the presence of catalysts, chain extenders, and other additives such as blowing agents or flame retardants [6]. Due to the large selection of substrates used in the reaction, their various mutual ratios, and additives used in the production of PUR materials, it is possible to design the products for a specific application. This allows PUR materials to exist in such forms as foams (rigid and flexible), adhesives, coating, films, sealants, and fibers [7,8]. Among different types of PUR materials, the most commonly used are PUR foams (PUFs), which dominated the market, reaching about 67% of the world's production of PUR [4].

Rigid polyurethane foams are strongly cross-linked materials. They are characterized by the values of the thermal conductivity coefficient in the range from 0.018 to 0.025 $Wm^{-1} K^{-1}$, which are lower than the values of this parameter, which are achieved by other thermal insulation materials. Due to their thermal insulation and mechanical properties rigid PUR foams are used as low-cost products that find varied applications across many areas such as building insulation, packaging, furniture, transportation, and automotive [4,9,10].

Increasingly stringent environmental requirements for polymeric materials have led to the development of research to improve their mechanical and physical properties by adding ingredients of natural origin [11]. Due to the low price and ecological aspects, special attention is paid to materials of wood and plant origin [12]. One of the methods of modification of rigid PUR foams is the use of bio-fillers which affect their environmentally friendly character. In recent years, there has been a growing interest in the application as natural fillers of raw materials being waste biomass from food processing, including walnut [13] or hazelnut [14] shells and cinnamon [15], coffee, or cocoa [16] extracts. The applied bio-fillers can improve the compressive strength, dimensional stability in critical conditions [17], and thermal insulation capabilities [18] of modified products. It can also change the apparent density and cell structure of modified polyurethane foams and reduce their water absorption [19]. An interesting bio-filler seems to be the distilled lavender residue, which remains after the production of essential oils. Lavender residue is generally considered as waste materials and has been used for composting or burnt to generate energy. An alternative to the traditional uses of solid lavender waste could be its application as a bio-filler in the process of obtaining rigid polyurethane foams. This will allow for the management of the generated waste in a different way [20]. The chemical composition of lavender includes lignin, hemicellulose, and cellulose, significantly affect the properties of the obtained polymer products [21]. As described in the literature, extract of distilled lavender contains interesting volatile molecules including terpenoids and terpenes (e.g., geraniol, linalyl acetate, linalool, and borneol), as well as flavonoids (e.g., luteolin and isoquercitrin), nitrogen compounds (e.g., amino acids, chlorophyll derivatives, and alkaloids) and non-volatile phenolic compounds (e.g., caffeic acid and rosmarinic acid) [22]. It was reported that the extract obtained from distilled lavender shows antioxidant and antimicrobial activity [23]. This suggests that lavender distillation residues can be used as additives to obtain products with better mechanical properties, in parallel, with antioxidant and antibacterial properties [24–26].

The main disadvantage that limits the use of PUR foams in many engineering applications is their easy ignition and high flame spreadability [27]. It is known that the degradation of the urethane bond of PUR foams starts at 200 °C [28]. However, it is possible to reduce their combustibility by applying various anti-pyrenes, also known as flame retardants [29]. There are known many types of anti-pyrenes, such as melamine compounds [30], bromine compounds [31], phosphorus compounds [32], inorganic salts [33], or expandable graphite [31,34]. Furthermore, inorganic metal hydroxides and oxides are also mentioned, including compounds of aluminum, magnesium, and silicon, which also play an important role as flame retardants [35–38]. In the past, halogen compounds were often used as anti-pyrenes, but the growing environmental requirements resulted in the limitation of the application of these substances, due to the release of harmful and toxic gases during the combustion process [10,39]. Therefore, in recent years, the attention of researchers has been focused on the analysis and use of halogen-free flame retardants that do not harm the environment.

One of the flame retardant modifiers of obtained products used in this work is kaolin clay (K) described with the chemical composition of $Al_2O_3 \cdot 2SiO_2 \cdot 2H_2O$. It has the form of hexagonal platelets and is a two-layer hydrated alumina silicate consisting of chemically bonded layers of hydrous alumina and silica [40]. Kaolin can be used in many industries due to its low thermal conductivity, chemically inert in a wide range of pH, and exquisite covering power when it is applied as a pigment [41]. Kaolinite is mainly used in the ceramic and paper industry; however, there are studies describing its use in polymer materials [42–44]. Kaolin clay can be used also as a flame retardant. Ullah et al. [45] have shown that it is a material that creates ceramic like a protective barrier that reduces the heat transfer to the modified products.

Another flame retardant modifier used in this work is hydroxyapatite ($Ca_5(OH)(PO_4)_3$) (HA) which is a bio-based polycrystalline calcium phosphate with a hexagonal structure. Its chemical composition includes 39 wt.% Ca, 18.5 wt.% P and 3.38 wt.% OH [46]. The antiflammable properties of hydroxyapatite have been investigated and proven in various polymer matrices. Researchers investigated the mechanical and thermal performances of HA-modified polymers such as polyvinyl alcohol [47], cellulose [48], and polylactide [49]. For example, Akindoyo et al. [50] reported that the incorporation of 10 wt.% HA into PLA increased the mass residue from 0.35% to 6.17% at 750 °C. Composites of poly(butylene succinate-co-lactate) (PBSL), poly(lactic acid) (PLA) and hydroxyapatite (HA) were investigated by Behera et al. [51]. The results showed that HA was uniformly dispersed in polymeric matrix, while the size of PBSL domain decreased after the addition of HA. In another study, Nabipour et al. [46] have formed a coating containing hydroxyapatite to enhance the fire safety feature of flexible PUR foams and have demonstrated their increased fire resistance.

In the present study, the influence of used modifiers on flammability and rigid polyurethane foams properties was assessed. The fillers used were lavender, kaolin modified lavender, and hydroxyapatite modified lavender.

Although many studies have investigated the influence of cellulosic fillers on PUR foams, there is no studies examined the polyurethane composites filled with lavender filler. Keeping in view, the main disadvantage of PUR foams, which is their high flammability, the current study will focus on the physical functionalization of lavender filler with selected natural, flame retardant compounds—kaolinite (K) and hydroxyapatite (HA) using a high-energy ball milling process. Considering the beneficial properties of lavender and natural flame retardants (kaolinite and hydroxyapatite), it is predicted that the PUR composites developed in this study will exhibit the outstanding mechanical and thermal characteristics, extending their application in the building and construction industry. Because of this, the impact of such developed lavender residue on the mechanical, thermal, insulation, and performance properties of PUR composites will be clearly defined.

2. Materials and Methods
2.1. Materials

- Polyether polyol with a brand name of Stapanpol PS-2352 was purchased from Stepan Company (Northfield, IL, USA),
- Polymeric diphenylmethane diisocyanate with a brand name of Purocyn B was purchased from Purinova Company (Bydgoszcz, Poland).
- Potassium octoate with a brand name of Kosmos 75 and potassium acetate with a brand name of Kosmos 33 were purchased from Evonik Industry (Essen, Germany),
- Silicone-based surfactant with a brand name of Tegostab B8513 was purchased from Evonik Industry (Essen, Germany),
- Pentane, cyclopentane, sodium hydroxide (pellets, anhydrous), kaolinite (aluminum silicate, powder), hydroxyapatite (nanopowder, <200 nm) were purchased from Sigma-Aldrich Corporation (Saint Louis, MO, USA)
- Lavender residue was obtained from a local company (Lodz, Poland).

2.2. Methods and Instruments

Cell size distribution and morphology of analyzed foams were examined on the basis of the cellular structure pictures of foams that were taken using JEOL JSM-5500 LV scanning electron microscopy (JEOL LTD, Akishima, Japan). The apparent density of foams was determined in accordance with the standard ASTM D1622 (equivalent to ISO 845). The compressive strength ($\sigma_{10\%}$) of analyzed foams was determined in accordance with the standard ASTM D1621 (equivalent to ISO 844) using Zwick Z100 Testing Machine (Zwick/Roell Group, Ulm, Germany). Three-point bending test of analyzed foams was examined in accordance with the standard ASTM D7264 (equivalent to ISO 178) using Zwick Z100 Testing Machine (Zwick/Roell Group, Ulm, Germany). The impact examination was carried out in accordance with the standard ASTM D4812. Surface hydrophobicity was determined by contact angle measurements using the sessile drop method using a manual contact angle goniometer with an optical system OS-45D (Oscar, Taiwan). Water absorption of analyzed foams was analyzed in accordance with the standard ASTM D2842 (equivalent to ISO 2896). The thermal stability of analyzed foams was analyzed using a Mettler Toledo thermogravimetric analyzer TGA/DSC1 (Columbus, OH, USA). Antibacterial properties of PUR composites against selected bacteria and fungi (*Escherichia coli, Staphylococcus aureus, Bacillus subtilis, Candida albicans,* and *Aspergillus niger*) were determined according to the National Committee for Clinical Laboratory Standards. Bacteria were cultured on Tryptic Soy Agar (TSA) medium at 30 °C for 48 h, and fungi on Malt Extract Agar (MEA) medium at 25 °C for 5 days. After incubation, the zones of growth inhibition under and around the film strips were determined. The fire behavior of analyzed foams was performed in accordance with the standard ISO 5660 using the cone calorimeter apparatus in S.Z.T.K. 'TAPS'—Maciej Kowalski Company (Saugus, Poland).

2.3. Filler Functionalization and Production Process of PUR Composites

Before adding to the polyol system, the lavender fillers were alkali-treated, according to the procedure described in [52]. After that, the lavender powder was functionalized with kaolinite (K) and hydroxyapatite (HA). In order to lavender functionalization, a selected amounts of lavender powder and kaolinite/hydroxyapatite (1:1 w/w) were weighed up and mixed intensively, using a high-energy ball milling process (1 h, 3000 rpm). Such developed lavender fillers (Figure 1) were used as a reinforcing fillers in the synthesis of PUR composites.

Figure 1. Optical image of (**a**) non-functionalized lavender filler, (**b**) lavender filler functionalized with kaolinite, and (**c**) lavender fillers functionalized with hydroxyapatite.

In this regard, the calculated amounts of lavender fillers, polyol, catalysts, surfactant, and blowing agent were placed in a beaker and mixed vigorously (60 s, 2000 rpm). Subsequently, an isocyanate compound was poured into a beaker with vigorous stirring (30 s, 2000 rpm). According to the supplier information, the isocyanate was mixed in the ratio of 100:160 (ratio of polyol to isocyanate) to provide a complete reaction between hydroxyl and isocyanate groups. PUR composites were cured at room temperature for 48 h. The schematic procedure of the synthesis of PUR composites is presented in Figure 2. All formulations of prepared PUR composites are listed in Table 1.

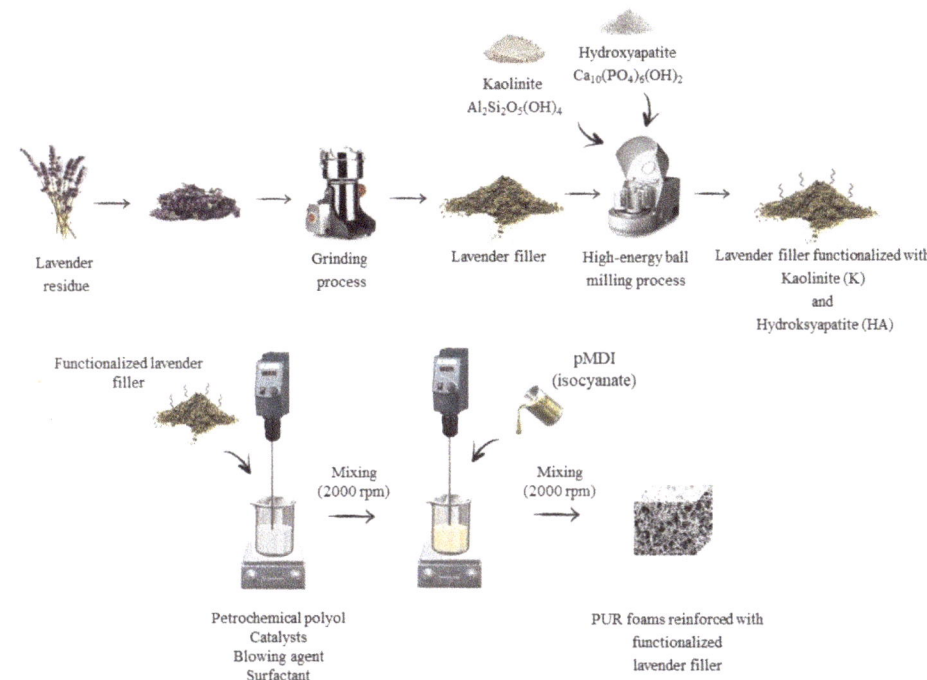

Figure 2. Schematic procedure of the synthesis of polyurethane (PUR) composites filled with non-functionalized and functionalized lavender fillers.

Table 1. Composition of PUR composite foams.

Component	PUR_REF	PUR_L	PUR_L_K	PUR_L_HA
	Parts by Weight (wt.%)			
STEPANPOL PS-2352	100	100	100	100
PUROCYN B	160	160	160	160
Kosmos 75	6	6	6	6
Kosmos 33	0.8	0.8	0.8	0.8
Tegostab B8513	2.5	2.5	2.5	2.5
Water	0.5	0.5	0.5	0.5
Pentane/cyclopentane	11	11	11	11
Lavender non-functionalized	0	2	0	0
Lavender functionalized with Kaolinite (K)	0	0	2	0
Lavender functionalized with Hydroxyapatite (HA)	0	0	0	2

3. Results and Discussion

3.1. Filler Characterization

The functionalization of lavender using a high-energy ball milling process affects the external morphology and size of filler particles. As presented in Figure 3, the external morphology of non-functionalized lavender filler is quite rough, while the size of lavender particles oscillates between 950 nm and 3 μm with an average value at ~1.5 μm. Due to the functionalization of the filler with kaolinite and hydroxyapatite compounds, the overall structure of lavender particles becomes more uniform and smooth, while the average size of filler decreases to 712 and 615 nm, respectively. This may be connected with the fact, that during the continuing ball milling process, the filler particles break into smaller particles, which contribute to the formation of powder fillers with narrow size distribution and a uniform structure. Furthermore, the size of lavender particles affects the viscosity of PUR systems (Table 2). Due to the incorporation of bigger particles of non-functionalized lavender, the viscosity increases from 860 mPa·s (for PUR_REF) to 1215 mPa·s. After the incorporation of lavender filler functionalized with kaolinite and hydroxyapatite, the value of viscosity increases to 1015 and 1050 mPa·s, respectively.

Figure 3. SEM images of lavender fillers: (**a**,**b**) Non-functionalized lavender, (**c**,**d**) lavender functionalized with kaolinite, (**e**,**f**) lavender functionalized with hydroxyapatite.

Table 2. Rheological and structural properties of PUR composites.

	PUR_REF	PUR_L	PUR_L_K	PUR_L_HA
Dynamic viscosity at 10 rpm (mPa·s)	860 ± 7	1215 ± 9	1015 ± 8	1050 ± 6
Cream time (s)	40 ± 4	56 ± 2	49 ± 2	48 ± 3
Expansion time (s)	268 ± 3	335 ± 8	315 ± 6	310 ± 4
Tack-free time (s)	345 ± 5	330 ± 9	310 ± 7	315 ± 8
Cell size (μm)	485 ± 6	470 ± 5	450 ± 6	455 ± 7
Closed-cell content (%)	87.2 ± 0.7	85.4 ± 0.5	87.9 ± 0.6	88.2 ± 0.4
Apparent density (kg m^{-3})	36.8 ± 0.8	37.4 ± 0.7	38.9 ± 0.5	38.6 ± 0.5

3.2. PUR Composites Characterization

As presented in Figure 4, the incorporation of non-functionalized and functionalized lavender fillers affects the processing times in both cases. When compared with PUR_REF, on the addition of non-functionalized, kaolinite-functionalized, and hydroxyapatite-functionalized lavender fillers the start time increases by ~40, ~22, and ~20% while the expansion time increases by ~25, ~17, and ~15%. According to the results presented in previous works, the addition of organic/inorganic fillers may affect the proper stoichiometry of the reaction between isocyanate and hydroxyl groups of polyurethane systems [53]. Due to the presence of filler particles, some highly reactive isocyanate groups react with the active groups introduced by the filler (e.g., hydroxyl groups), limiting the possibility of the reaction between the isocyanate and water and reducing the formation of carbon dioxide (CO_2). Moreover, the expansion of the cells is additionally reduced by higher viscosity of the modified systems, which explains the extended expansion time of the PUR composites modified with lavender fillers. The analog tendency has been reported in previous studies, which concern the formation of PUR composites modified with another type of organic and inorganic fillers [54].

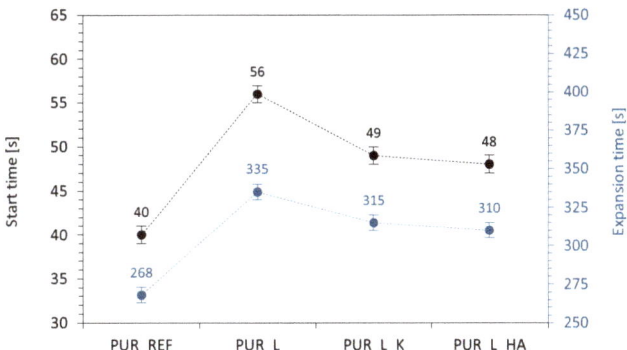

Figure 4. The results of start time and expansion time measured for PUR systems.

The impact of lavender filler addition on the cellular morphology of PUR composites was evaluated by SEM. As presented in Figure 5a,b, PUR_REF shows the typical, polyhedral structure with a high content of closed-cells. When lavender fillers are added to the PUR systems, an average size of closed-cells tends to be smaller, which is connected with a nucleating effect of the added fillers. The average size of cells decreases from 485 µm (for PUR_REF) to 470, 450, and 455 µm for PUR_L, PUR_L_K, and PUR_L_HA, respectively. According to the SEM results, the incorporation of non-functionalized lavender filler results in the formation of PUR composites with a higher number of open-cells—the content of closed-cells decreases from 87.2 to 85.4% (Figure 5c,d). This may be connected with poor compatibility between the surface of filler particles and the PUR matrix, which results in rupturing and collapsing of the PUR structure and opening of the cells, which in turn, weakens the final structure of PUR composites. The addition of lavender fillers functionalized with kaolinite and hydroxyapatite compounds results in the production of PUR composites with a little more regular structure when compared with PUR composites filled with non-functionalized lavender filler (Figure 5e–h). When compared with PUR_REF, the content of closed-cells increases from 87.2% to 87.9 and 88.2% for PUR_L_K and PUR_L_HA, respectively. This result indicates that the functionalization of lavender filler using a high-energy ball milling process promotes the formation of PUR composites with a higher-crosslinking degree, which prevents the deterioration of the structure during the foaming process. This is additionally enhanced by the presence of solid particles of functionalized lavender fillers which are successfully build into the PUR structure. As discussed previously, the application of a high-energy ball milling process leads to the fracturing of the filler particles, which results in the formation of powder filler with a reduced size of the particles. Due to this, the particles easily build into the PUR matrix, forming new edges for blowing agent encapsulation and enhancing the stability of the overall cellular morphology of PUR composites. Such effect is disturbed in the case of PUR composites filled with non-functionalized lavender fillers, due to the larger size of filler particles. Previous studies have shown, that the application of the filler particles with larger diameters, results in rupturing of the cells, due to incomplete incorporation of the filler particles into the PUR matrix [55]. A similar explanation may be found in our study as well.

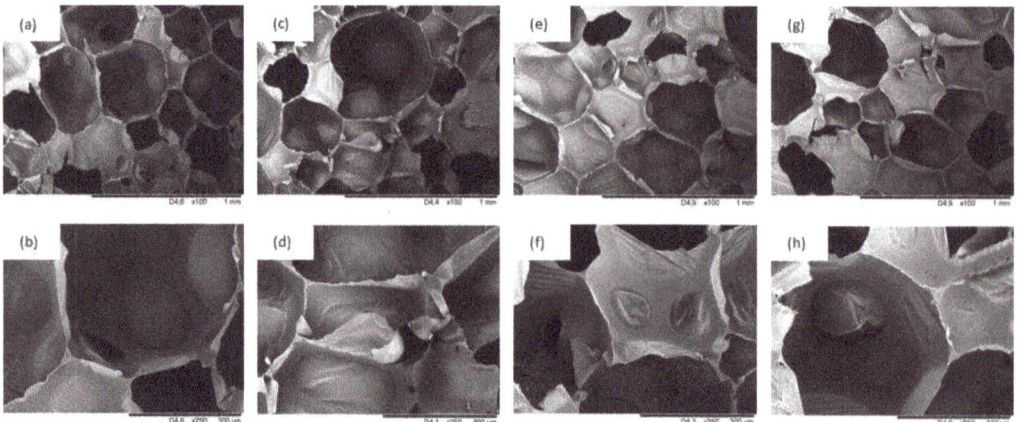

Figure 5. Cellular morphology of (**a,b**) PUR_REF, (**c,d**) PUR_L, (**e,f**) PUR_L_K, (**g,h**) PUR_L_HA.

As presented in Figure 6, on the addition of lavender fillers, the value of apparent density increases from 36.8 kg m^{-3} (for PUR_REF) to 37.4, 38.9, and 38.6 kg m^{-3}, for PUR_L, PUR_L_K, and PUR_L_HA, respectively. This may be explained by the fact, that the incorporation of lavender fillers, which are characterized by greater density than the PUR matrix. On the other hand, as shown in Table 2, the incorporation of lavender fillers increased the viscosity of PUR systems, limiting the expansion of the PUR systems and resulting in the formation of smaller cells. Therefore, the density of PUR composites is enhanced.

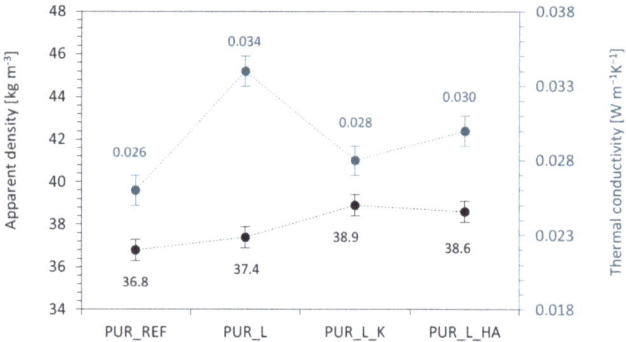

Figure 6. The results of apparent density and thermal conductivity measured for PUR composites.

According to the results given in Figure 6, on the incorporation of non-functionalized lavender filler the value of thermal conductivity (λ) increases from 0.026 Wm^{-1} K^{-1} (for PUR_REF) to 0.034 Wm^{-1} K^{-1}, while after the incorporation of functionalized lavender fillers, the value of λ increases insignificantly to 0.028 and 0.030 Wm^{-1} K^{-1}. Such an insignificant increase of λ may be attributed to the incorporation of particles of lavender fillers, which increases the heat transfer through the solid particles. Moreover, the most noticeable increase in λ is observed for PUR composites filled with non-functionalized lavender filler. As discussed previously, comparing to PUR_REF, those samples, are characterized by a higher number of open cells. Due to the diffusion through the PUR structure, the gas inside the PUR cells changes from CO_2 (0.014 Wm^{-1} K^{-1}) to atmospheric air (0.025 Wm^{-1} K^{-1}), increasing the overall value of λ. A similar trend was observed in previous studies as well [56,57]. For example, Paciorek-Sadowska et al. [58] reported that the addition of rapeseed cake filler in the amount of 30–60 wt.% did not affect the thermal insulation properties of PUR composites—on the addition of 60 wt.% of rapeseed cake filler, the value of λ increased insignificantly from 0.0341 to 0.0348 Wm^{-1} K^{-1}.

Thermal stability of lavender fillers and PUR composites was examined using thermogravimetric (TGA) and derivative thermogravimetry (DTG) analysis. The obtained results are presented in Figure 7 and Table 3.

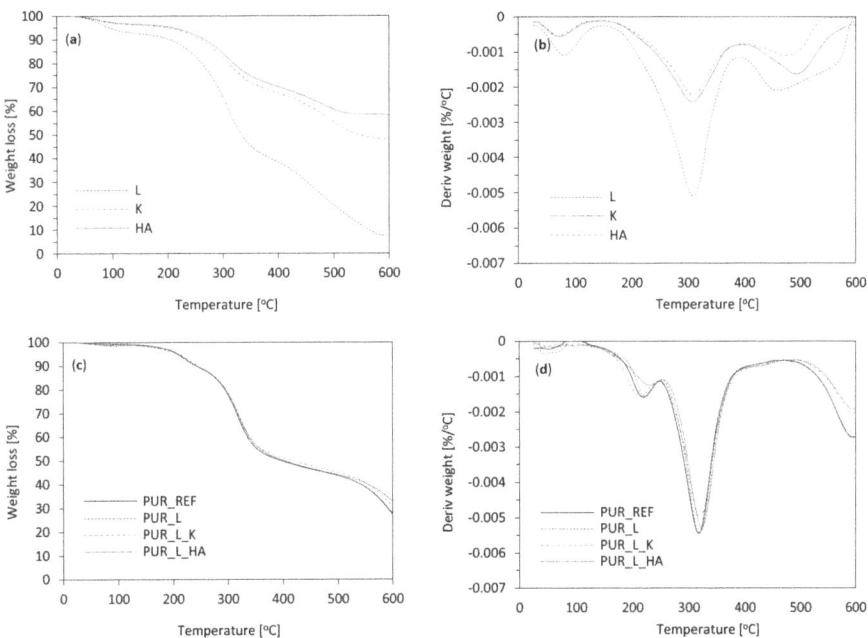

Figure 7. Thermogravimetric (TGA) and derivative thermogravimetry (DTG) results obtained for (a,b) lavender fillers, and (c,d) PUR composites.

Table 3. The results of thermal stability of PUR composites.

Sample	T$_{max}$ (°C)			Residue (at 600 °C) (wt.%)
	1st Stage	2nd Stage	3rd Stage	
PUR_REF	220	322	580	28.0
PUR_L	222	325	585	30.0
PUR_L_K	220	334	590	33.1
PUR_L_HA	235	335	592	35.2

In the case of non-functionalized lavender filler, three stages of thermal decompositions are observed. The first stage of mass loss occurs at relatively low temperature (~100 °C) and refers to the evaporation of the moisture absorbed by the filler and volatile compounds (low molecular weight esters and fatty acids) which are inherent to the filler [59]. The second stage representing 30% of mass loss occurs between 300 and 500 °C. The maximum rate at ~313 °C refers to the thermal degradation of cellulose and hemicellulose [60,61]. The last step of thermal decomposition, representing 42% of mass loss occurs between 400 and 500 °C with the maximum rate at ~460 °C and refers to the thermal decomposition of lignin [61]. When comparing with non-functionalized lavender, the lavender fillers functionalized with kaolinite and hydroxyapatite also revealed three main stages of thermal decomposition; however, the maximum rate of hemicellulose and lignin decomposition is displaced to higher temperatures. This behavior may be connected with the presence of thermal protective layers created by kaolinite and hydroxyapatite, improving the thermal stability of functionalized fillers.

TGA and DTG results of PUR composites are presented in Figure 7. All composites showed three stages of mass loss—T_{max1}, T_{max2}, and T_{max3}. The first stage of mass loss (T_{max1}) occurs at relatively low temperatures, between 200 and 250 °C and refers to the thermal decomposition of low molecular weight compounds, which are inherent in lavender fillers [62]. The incorporation of non-functionalized and functionalized lavender fillers results in higher values of T_{max1}, indicating the partial crosslinking between lavender fillers and isocyanate groups. T_{max2} occurs between 300 and 350 °C and corresponds to the thermal degradation of hard segments of polyurethane structure and thermal decomposition of lavender fillers [63,64]. Due to the addition of lavender fillers, the maximum temperature of thermal decomposition is displaced to higher temperatures. The greatest improvement is observed for functionalized lavender fillers—the value for T_{max2} increases from 322 °C to 334 and 335 °C, for PUR_L_K and PUR_L_HA. Previous studies have shown, that the addition of filler particles may act as a barrier for heat transfer, effectively inhibiting the further degradation of composites [62]. T_{max3} occurs between 500 and 600 °C and refers to the thermal degradation of lignocellulosic compounds—cellulose, hemicellulose, and lignin [65,66]. When compared with PUR_REF, on the addition of non-functionalized and functionalized lavender fillers, the value of T_{max3} slightly increases, due to the presence of cellulosic fillers and incomplete miscibility of PUR segments (soft and hard segments) [67]. Furthermore, the improved thermal stability of PUR composites was confirmed by the amount of char residue, measured at 600 °C. When compared with PUR_REF, the value increases from 28% to 30, 33, and 35%, for PUR_L, PUR_L_K, and PUR_L_HA, respectively. It may be concluded that filler particles may act as cross-linker points between PUR chains, reducing the heat transfer through the composite structure. The higher cross-linked structure of PUR composites effectively reduces the amount of volatile compounds, which are releasing during the thermal degradation process. Such an explanation may be found in previous studies, as well [62].

The impact of lavender fillers addition on mechanical characteristics was evaluated by measuring the compressive strength (labeled as $\sigma_{10\%}$), flexural strength (labeled as σ_f) and impact strength (labeled as σ_I). As presented in Figure 8a, the addition of non-functionalized and functionalized lavender fillers affects the value of $\sigma_{10\%}$. When compared with PUR_REF, $\sigma_{10\%}$ (measured parallel to the direction of foam expansion) increases by ~7, ~15, and ~17%, for PUR_L, PUR_L_K, and PUR_L_HA, respectively. An analog trend is observed in the case of $\sigma_{10\%}$ measured perpendicular to the direction of the foam expansion—the value of $\sigma_{10\%}$ increases by ~8, ~18, and ~16% for PUR_L, PUR_L_K and PUR_L_HA, respectively. To avoid the impact of apparent density on mechanical properties of PUR composites, the specific compressive strength was measured as well. Moreover, on the addition of lavender fillers, the specific strength of PUR composites slightly increases—the specific strength (measured parallel) calculated for PUR_REF is 6.5 MPa/kg/m^3, while due to the incorporation of lavender fillers, the value increases to 6.8, 7.1 and 7.3% for PUR_L, PUR_L_K, and PUR_L_HA. Such improvement may be connected with the morphology features of PUR foams. According to SEM images (see Figure 5), PUR composites reinforced with lavender fillers exhibit a more regular morphology. Thanks to this, the external force is encountered by a greater number of cells which improves the mechanical resistance of PUR composites. The mechanical properties of PUR composites are additionally supported by their more cross-linked structure and the presence of filler particles, which successfully support the load-bearing process [57,68,69]. Most importantly, the obtained results confirm the reinforcing effect of lavender fillers and they are in line with international requirements for constructive materials [70].

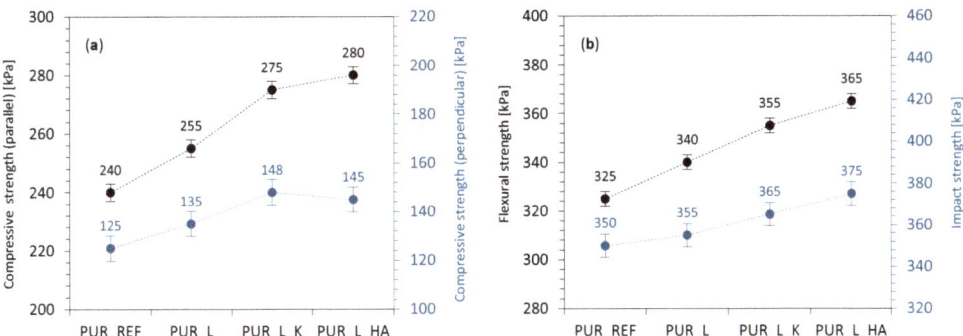

Figure 8. The mechanical performances of PUR foams—(**a**) compressive strength, (**b**) flexural and impact strength.

A reinforcing effect of lavender fillers was also confirmed by the results of σ_f and σ_I. As presented in Figure 8b, when compared with PUR_REF, the addition of non-functionalized lavender filler increases the value of σ_f by ~5%, while the addition of lavender filler functionalized with kaolinite and hydroxyapatite increases the value of σ_f by ~9 and ~12%, respectively. A similar trend is observed for σ_I. The greatest improvement is observed for PUR_L_K and PUR_L_HA—the value of σ_I increases by ~4 and ~7%, respectively. As discussed previously, due to the greater number of smaller cells of PUR composite structure, the crack propagation, which is formed under the action of an external load is reduced. Moreover, due to the higher cross-linking degree, the more rigid structure, of PUR composites may absorb more energy, increasing the mechanical resistance of PUR composites [71,72].

In agreement with the literature, lavender residue contains compounds with antibacterial and antioxidative properties, such as phenolic compounds and flavonoids [22]. Antibacterial properties have already been described for kaolinite and hydroxyapatite as well [73,74]. Because of this, the antibacterial properties of PUR composites filled with lavender fillers have been evaluated. The results of the bacterial activity of PUR composites against *Escherichia coli*, *Staphylococcus aureus*, *Bacillus subtilis*, *Candida albicans*, and *Aspergillus niger* are presented in Table 4. The obtained results confirmed the antibacterial activity of PUR composites against bacteria, but no activity against fungi was observed. Low antibacterial activity against fungi, may be connected with low concertation of lavender fillers in the PUR composites. Antibacterial activity of composites filled with lavender has been reported in previous works, however, most studies concern the application of lavender extract. For example, the antimicrobial activity of gelatin-based films containing lavender oils derived from lavender leaves and flowers was investigated by Martucci et al. [26]. Such developed composites exhibited by antibacterial activity against selected bacteria, e.g., *Staphylococcus aureus*, *Salmonella typhimurium*, *Escherichia coli*, and *Bacillus subtilis*. Similar results were confirmed in the case of composites filled with lavender oil embedded in sol-gel hybrid matrices [75].

Table 4. Antibacterial properties of PUR composites against selected bacteria and fungi.

Sample	Bacteria			Fungi	
	E. coli	*S. aureus*	*B. subtilis*	*C. albicans*	*A. niger*
PUR_REF	−	−	−	−	−
PUR_L	+	+	+	-	-
PUR_L_K	+	+	+	-	-
PUR_L_HA	+	+	+	-	-

Based on the results reported in previous works, the water uptake of porous materials depends not only on the morphology features of porous materials (mostly type of cells—closed or open) but, also depends on the hydrophilic character of the system components, including incorporated fillers [76–79]. As presented in Figure 9, the water uptake of PUR composites increases by ~3, ~16, and ~21%, for PUR_L, PUR_L_K, and PUR_L_HA, respectively. Based on the SEM results, the PUR composites possess a well-developed structure with a dominant number of closed-cells, which are not able to accommodate the water. Therefore, it seems that an increased water uptake capacity results from the hydrophilic character of incorporated fillers—lavender, as well as lavender functionalized with kaolinite and hydroxyapatite, which also possess a hydrophilic character [80–82]. The more hydrophilic character of PUR composites is confirmed by the results of contact angle (Figures 10 and 11)—on the incorporation of lavender fillers, the contact angle decreases from 123° (for PUR_REF) to 119, 115, and 110° for PUR_L. PUR_L_K, and PUR_L_HA, respectively.

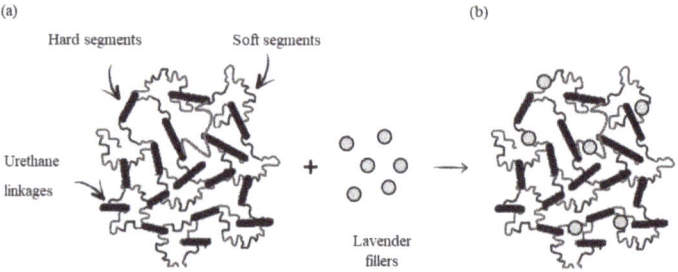

Figure 9. The impact of lavender filler addition on the cross-linking of PUR composites—(a) PUR_REF, (b) PUR composites with the addition lavender fillers.

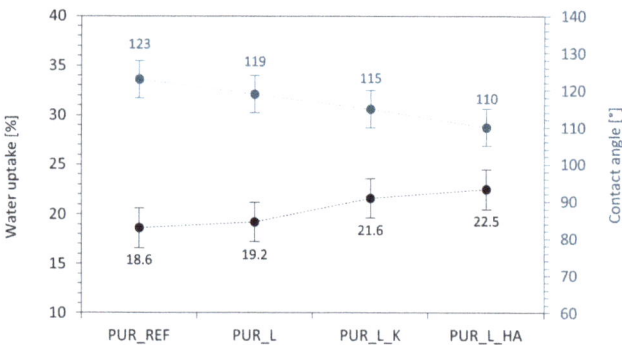

Figure 10. Selected properties of PUR composites—water uptake and contact angle results.

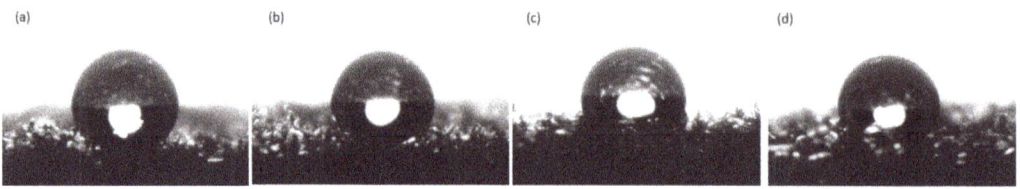

Figure 11. The images of contact angles measured for (a) PUR_REF, (b) PUR_L, (c) PUR_L_K, and (d) PUR_L_HA.

The flame retardant abilities of PUR composites were performed using a cone calorimeter. The results of ignition time (IT), peak heat release rate (pHRR), total smoke release (TSR), total heat release (THR), average yield of CO (COY) and CO_2 (CO_2Y), and limiting oxygen index (LOI) are presented in Table 5.

Table 5. Flame retardant properties of PUR composites.

Sample	IT (s)	pHRR (kW m^{-2})	TSR (m^2 m^{-2})	THR (MJ m^{-2})	COY (kg kg^{-1})	CO_2Y (kg kg^{-1})	COY/CO_2Y (−)	LOI (%)
PUR_REF	4	263	1500	21.5	0.210	0.240	0.875	20.2
PUR_L	4	203	1400	21.1	0.190	0.225	0.844	20.5
PUR_L_K	6	144	1060	20.5	0.140	0.190	0.736	22.2
PUR_L_HA	6	130	1055	19.8	0.142	0.180	0.788	22.7

Comparing the modified foams to the reference one PUR_REF, it can be noticed that the modifications do not significantly affect the ignition time (IT). As shown in Figure 12a, all series of analyzed PUR composites expose one peak of HRR, which corresponds to the release of low molecular weight compounds, like amines, olefins or isocyanate. Compared to the PUR_REF, which pHRR value is 263 kW m^{-2}, modified foams achieve much lower values of this parameter and these the values are respectively 203 kW m^{-2} for PUR_L, 144 kW m^{-2} for PUR_L_K and 130 kW m^{-2} for PUR_L_HA. This may be connected with the formation of a protective char layer, which becomes a physical barrier hindering the flow of heat [83] on the surface of PUR composites. Among all modified series of PUR composites, the lowest value of the pHRR parameter is observed for the PUR_L_HA, which is over 50% lower than for the PUR_REF. As shown in Figure 12b, the incorporation of each filler results in a lower value of total smoke release (TSR). When compared with the PUR_REF, the value of TSR decreases by ~7% for PUR_L, ~29% for PUR_L_K, and ~30% for PUR_L_HA. This suggests that incorporation of lavender fillers protects the PUR structure from further combustion and prevents the heat transfer [84] through the PUR matrix. Furthermore, the incorporation of lavender fillers can decrease the value of total heat release (THR). Comparing to the PUR_REF, for which the value of THR is 21.5 MJ m^{-2}, the addition of lavender fillers decreases the value of this parameter to 21.1 MJ m^{-2}, 20.5 MJ m^{-2}, and 19.8 MJ m^{-2} for PUR_L, PUR_L_K, and PUR_L_HA respectively. As presented in Table 5, the incorporation of lavender fillers decreases the carbon monoxide (CO) to carbon dioxide (CO_2) ratio, which is related to the foam toxicity. Generally, a higher value of this ratio reveals the incomplete combustion of PUR composites and an increased amount of toxic gases. The application of lavender fillers increases the value of the (CO/CO_2) ratio, which indicates that the release of toxic gases during the combustion of PUR composites is increased.

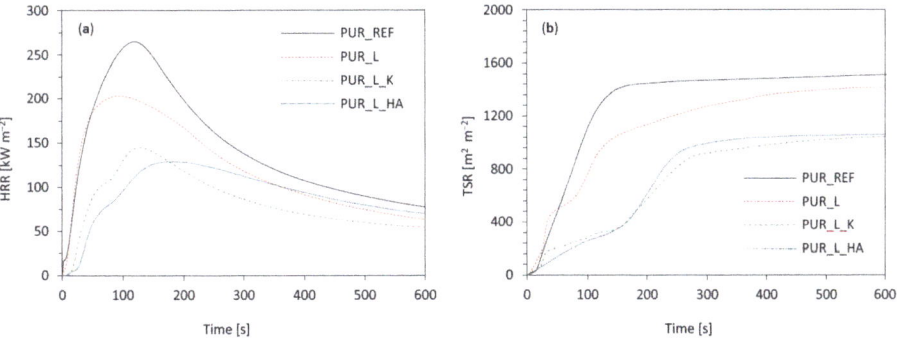

Figure 12. The results of (a) peak heat release rate (pHRR) and (b) total smoke release (TSR) measured for PUR composites.

As shown in Table 5 the incorporation of lavender fillers effectively increases the value of limiting oxygen index (LOI). The most significant improvement is observed for PUR_L_HA and PUR_L_K—the value of LOI increases from 20.2% (for PUR_REF) to 22.7% for PUR_L_HA and to 22.2% for PUR_L_K. A less noticeable improvement is observed for PUR_L—the value of LOI increases to 20.5%.

SEM images of char residue of PUR composites after the combustion process is presented in Figure 13. As shown in Figure 13a, after the combustion process, the char residue of PUR_REF seems to be loose and possess a few fragments which were formed during the decomposition process, due to the releasing of the flammable gases. On the other hand, PUR composites filled with lavender fillers, present more compact char residue (Figure 13b–d), which may act as a physical barrier, effectively limiting the heat transfer through the PUR structure. Therefore, the combustion process of PUR composites is successfully inhibiting.

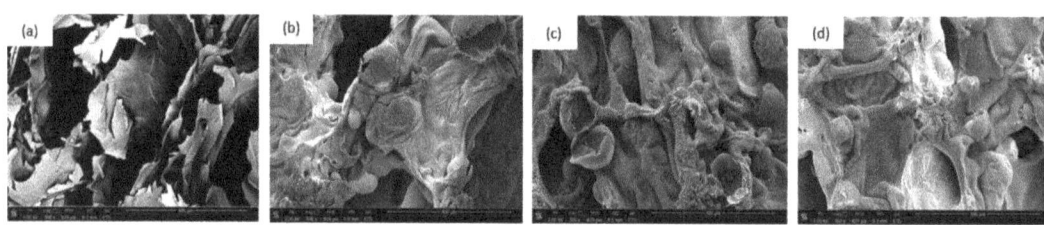

Figure 13. SEM images of char residue of (**a**) PUR_REF, (**b**) PUR_L, (**c**) PUR_L_K, and (**d**) PUR_L_HA (obtained after the cone calorimeter test).

4. Conclusions

Polyurethane (PUR) composites were successfully reinforced with 2 wt.% of lavender fillers functionalized with kaolinite (K) and hydroxyapatite (HA). The impact of lavender fillers on selected properties of PUR composites, such as rheological properties (dynamic viscosity, foaming behavior), mechanical properties (compressive strength, flexural strength, impact strength), insulation properties (thermal conductivity), thermal characteristic (temperature of thermal decomposition stages), flame retardancy (e.g., ignition time, limiting oxygen index) and performance properties (water uptake, contact angle) was investigated. Among all modified types of PUR composites, the best properties exhibited PUR composites filled with lavender fillers functionalized with kaolinite and hydroxyapatite. For example, on the addition of functionalized lavender fillers, the compressive strength was enhanced by ~16–18%, flexural strength by ~9–12%, and impact strength by ~7%. Due to the functionalization of lavender filler with thermally stable flame retardant compounds, such modified PUR composites were characterized by higher temperatures of thermal decomposition. Most importantly, PUR composites filled with flame retardant compounds exhibited improved flame resistance characteristics—when compared with the reference foam, in both cases, the value of peak heat release was reduced by ~50%, while the value of total smoke release was reduced by ~30%. The results reported in the following study confirmed that the application of the high-ball milling process may be an easy and successful attempt in the functionalization of cellulosic filler. The application of such developed fillers in the production of PUR materials is an effective way in the synthesis of PUR composites with enhanced mechanical, thermal, and performance properties.

Author Contributions: Methodology, S.C., A.K., K.M. and A.S.; Investigation, S.C., A.K., K.M. and A.S.; Data Curation, S.C.; Writing—Original Draft, S.C.; Writing—Review and Editing, S.C.; Visualization, S.C. All authors have read and agreed to the published version of the manuscript.

Funding: This research received no external funding.

Institutional Review Board Statement: Not applicable.

Informed Consent Statement: Not applicable.

Conflicts of Interest: The authors reported no conflicts of interest related to this study.

References

1. Kurańska, M.; Beneš, H.; Sałasińska, K.; Prociak, A.; Malewska, E.; Polaczek, K. Development and Characterization of "Green Open-Cell Polyurethane Foams" with Reduced Flammability. *Materials* **2020**, *13*, 5459. [CrossRef]
2. Kurańska, M.; Barczewski, R.; Barczewski, M.; Prociak, A.; Polaczek, K. Thermal Insulation and Sound Absorption Properties of Open-Cell Polyurethane Foams Modified with Bio-Polyol Based on Used Cooking Oil. *Materials* **2020**, *13*, 5673. [CrossRef]
3. Kurańska, M.; Malewska, E.; Polaczek, K.; Prociak, A.; Kubacka, J. A Pathway toward a New Era of Open-Cell Polyurethane Foams—Influence of Bio-Polyols Derived from Used Cooking Oil on Foams Properties. *Materials* **2020**, *13*, 5161. [CrossRef] [PubMed]
4. Gama, N.V.; Ferreira, A.; Barros-Timmons, A. Polyurethane foams: Past, present, and future. *Materials* **2018**, *11*, 1841. [CrossRef]
5. Arévalo-Alquichire, S.; Valero, M. Castor Oil Polyurethanes as Biomaterials. In *Elastomers*; Çankaya, N., Ed.; InTech: Rijeka, Croatia, 2017.
6. Engels, H.W.; Pirkl, H.G.; Albers, R.; Albach, R.W.; Krause, J.; Hoffmann, A.; Casselmann, H.; Dormish, J. Polyurethanes: Versatile materials and sustainable problem solvers for today's challenges. *Angew. Chemie Int. Ed.* **2013**, *52*, 9422–9441. [CrossRef]
7. Joshi, M.; Adak, B.; Butola, B.S. Polyurethane nanocomposite based gas barrier films, membranes and coatings: A review on synthesis, characterization and potential applications. *Prog. Mater. Sci.* **2018**, *97*, 230–282. [CrossRef]
8. Ionescu, M. *Chemistry and Technology of Polyols for Polyurethanes*; Rapra Technology: Shropshire, UK, 2005; ISBN 9781847350350.
9. Wang, S.X.; Zhao, H.B.; Rao, W.H.; Huang, S.C.; Wang, T.; Liao, W.; Wang, Y.Z. Inherently flame-retardant rigid polyurethane foams with excellent thermal insulation and mechanical properties. *Polymer* **2018**, *153*, 616–625. [CrossRef]
10. Zhang, G.; Lin, X.; Zhang, Q.; Jiang, K.; Chen, W.; Han, D. Anti-flammability, mechanical and thermal properties of bio-based rigid polyurethane foams with the addition of flame retardants. *RSC Adv.* **2020**, *10*, 32156–32161. [CrossRef]
11. Cichosz, S.; Masek, A. Superiority of Cellulose Non-Solvent Chemical Modification over Solvent-Involving Treatment: Application in Polymer Composite (part II). *Materials* **2020**, *13*, 2901. [CrossRef] [PubMed]
12. Herrán, R.; Amalvy, J.I.; Chiacchiarelli, L.M. Highly functional lactic acid ring-opened soybean polyols applied to rigid polyurethane foams. *J. Appl. Polym. Sci.* **2019**, *136*, 1–13. [CrossRef]
13. Członka, S.; Strąkowska, A.; Kairytė, A. Effect of walnut shells and silanized walnut shells on the mechanical and thermal properties of rigid polyurethane foams. *Polym. Test.* **2020**, *87*, 106534. [CrossRef]
14. Barbu, M.C.; Sepperer, T.; Tudor, E.M.; Petutschnigg, A. Walnut and hazelnut shells: Untapped industrial resources and their suitability in lignocellulosic composites. *Appl. Sci.* **2020**, *10*, 6340. [CrossRef]
15. Liszkowska, J.; Moraczewski, K.; Borowicz, M.; Paciorek-Sadowska, J.; Czupryński, B.; Isbrandt, M. The effect of accelerated aging conditions on the properties of rigid polyurethane-polyisocyanurate foams modified by cinnamon extract. *Appl. Sci.* **2019**, *9*, 2663. [CrossRef]
16. Liszkowska, J.; Borowicz, M.; Paciorek-Sadowska, J.; Isbrandt, M.; Czupryński, B.; Moraczewski, K. Assessment of photodegradation and biodegradation of RPU/PIR foams modified by natural compounds of plant origin. *Polymers* **2020**, *12*, 33. [CrossRef] [PubMed]
17. Zhou, X.; Sethi, J.; Geng, S.; Berglund, L.; Frisk, N.; Aitomäki, Y.; Sain, M.M.; Oksman, K. Dispersion and reinforcing effect of carrot nanofibers on biopolyurethane foams. *Mater. Des.* **2016**, *110*, 526–531. [CrossRef]
18. Paberza, A.; Cabulis, U.; Arshanitsa, A. Wheat straw lignin as filler for rigid polyurethane foams on the basis of tall oil amide. *Polimery/Polymers* **2014**, *59*, 477–481. [CrossRef]
19. Zieleniewska, M.; Leszczyński, M.K.; Szczepkowski, L.; Bryśkiewicz, A.; Krzyżowska, M.; Bień, K.; Ryszkowska, J. Development and applicational evaluation of the rigid polyurethane foam composites with egg shell waste. *Polym. Degrad. Stab.* **2016**, *132*, 78–86. [CrossRef]
20. Ratiarisoa, R.V.; Magniont, C.; Ginestet, S.; Oms, C.; Escadeillas, G. Assessment of distilled lavender stalks as bioaggregate for building materials: Hygrothermal properties, mechanical performance and chemical interactions with mineral pozzolanic binder. *Constr. Build. Mater.* **2016**, *124*, 801–815. [CrossRef]
21. Lesage-Meessen, L.; Bou, M.; Ginies, C.; Chevret, D.; Navarro, D.; Drula, E.; Bonnin, E.; Del Río, J.C.; Odinot, E.; Bisotto, A.; et al. Lavender- and lavandin-distilled straws: An untapped feedstock with great potential for the production of high-added value compounds and fungal enzymes. *Biotechnol. Biofuels* **2018**, *11*, 1–13. [CrossRef]
22. Lesage-Meessen, L.; Bou, M.; Sigoillot, J.C.; Faulds, C.B.; Lomascolo, A. Essential oils and distilled straws of lavender and lavandin: A review of current use and potential application in white biotechnology. *Appl. Microbiol. Biotechnol.* **2015**, *99*, 3375–3385. [CrossRef]
23. Park, C.H.; Park, Y.E.; Yeo, H.J.; Chun, S.W.; Baskar, T.B.; Lim, S.S.; Park, S.U. Chemical compositions of the volatile oils and antibacterial screening of solvent extract from downy lavender. *Foods* **2019**, *8*, 132. [CrossRef] [PubMed]
24. Smigielski, K.; Prusinowska, R.; Stobiecka, A.; Kunicka-Styczyńska, A.; Gruska, R. Biological Properties and Chemical Composition of Essential Oils from Flowers and Aerial Parts of Lavender (Lavandula angustifolia). *J. Essent. Oil-Bearing Plants* **2018**, *21*, 1303–1314. [CrossRef]

25. Rota, C.; Carramiñana, J.J.; Burillo, J.; Herrera, A. In vitro antimicrobial activity of essential oils from aromatic plants against selected foodborne pathogens. *J. Food Prot.* **2004**, *67*, 1252–1256. [CrossRef] [PubMed]
26. Martucci, J.F.; Gende, L.B.; Neira, L.M.; Ruseckaite, R.A. Oregano and lavender essential oils as antioxidant and antimicrobial additives of biogenic gelatin films. *Ind. Crops Prod.* **2015**, *71*, 205–213. [CrossRef]
27. Kaur, R.; Kumar, M. Addition of anti-flaming agents in castor oil based rigid polyurethane foams: Studies on mechanical and flammable behaviour. *Mater. Res. Express* **2020**, *7*, 015333. [CrossRef]
28. Jiao, L.; Xiao, H.; Wang, Q.; Sun, J. Thermal degradation characteristics of rigid polyurethane foam and the volatile products analysis with TG-FTIR-MS. *Polym. Degrad. Stab.* **2013**, *98*, 2687–2696. [CrossRef]
29. Wrześniewska-Tosik, K.; Zajchowski, S.; Bryśkiewicz, A.; Ryszkowska, J. Feathers as a flame-retardant in elastic polyurethane foam. *Fibres Text. East. Eur.* **2014**, *103*, 119–128.
30. Thirumal, M.; Khastgir, D.; Nando, G.B.; Naik, Y.P.; Singha, N.K. Halogen-free flame retardant PUF: Effect of melamine compounds on mechanical, thermal and flame retardant properties. *Polym. Degrad. Stab.* **2010**, *95*, 1138–1145. [CrossRef]
31. Ye, L.; Meng, X.-Y.; Liu, X.-M.; Tang, J.-H.; Li, Z.-M. Flame-Retardant and Mechanical Properties of High-Density Rigid Polyurethane Foams Filled with Decabrominated Dipheny Ethane and Expandable Graphite. *J. Appl. Polym. Sci.* **2009**, *111*, 2373–2380. [CrossRef]
32. Xu, W.; Wang, G.; Zheng, X. Research on highly flame-retardant rigid PU foams by combination of nanostructured additives and phosphorus flame retardants. *Polym. Degrad. Stab.* **2015**, *111*, 142–150. [CrossRef]
33. Lindholm, J.; Brink, A.; Wilen, C.-E.; Hupa, M. Cone Calorimeter Study of Inorganic Salts as Flame Retardants in Polyurethane Adhesive with Limestone Filler. *J. Appl. Polym. Sci.* **2012**, *123*, 1793–1800. [CrossRef]
34. Shi, L.; Li, Z.-M.; Xie, B.-H.; Wang, J.-H.; Tian, C.-R.; Yang, M.-B. Flame retardancy of different-sized expandable graphite particles for high-density rigid polyurethane foams. *Polym. Int.* **2006**, *55*, 862–871. [CrossRef]
35. Fanglong, Z.; Qun, X.; Qianqian, F.; Rangtong, L.; Kejing, L. Influence of nano-silica on flame resistance behavior of intumescent flame retardant cellulosic textiles: Remarkable synergistic effect? *Surf. Coatings Technol.* **2016**, *294*, 90–94. [CrossRef]
36. Zhao, P.; Guo, C.; Li, L. Exploring the effect of melamine pyrophosphate and aluminum hypophosphite on flame retardant wood flour/polypropylene composites. *Constr. Build. Mater.* **2018**, *170*, 193–199. [CrossRef]
37. Chen, L.; Wang, Y.Z. A review on flame retardant technology in China. Part I: Development of flame retardants. *Polym. Adv. Technol.* **2010**, *21*, 1–26. [CrossRef]
38. Ai, L.; Chen, S.; Zeng, J.; Yang, L.; Liu, P. Synergistic Flame Retardant Effect of an Intumescent Flame Retardant Containing Boron and Magnesium Hydroxide. *ACS Omega* **2019**, *4*, 3314–3321. [CrossRef] [PubMed]
39. Zhou, F.; Zhang, T.; Zou, B.; Hu, W.; Wang, B.; Zhan, J.; Ma, C.; Hu, Y. Synthesis of a novel liquid phosphorus-containing flame retardant for flexible polyurethane foam: Combustion behaviors and thermal properties. *Polym. Degrad. Stab.* **2020**, *171*, 109029. [CrossRef]
40. Shehata, A.B.; Hassan, M.A.; Darwish, N.A. Kaolin modified with new resin-iron chelate as flame retardant system for polypropylene. *J. Appl. Polym. Sci.* **2004**, *92*, 3119–3125. [CrossRef]
41. Murray, H.H. Traditional and new applications for kaolin, smectite, and palygorskite: A general overview. *Appl. Clay Sci.* **2000**, *17*, 207–221. [CrossRef]
42. Batistella, M.; Otazaghine, B.; Sonnier, R.; Petter, C.; Lopez-Cuesta, J.M. Fire retardancy of polypropylene/kaolinite composites. *Polym. Degrad. Stab.* **2016**, *129*, 260–267. [CrossRef]
43. Ansari, D.M.; Price, G.J. Correlation of mechanical properties of clay filled polyamide mouldings with chromatographically measured surface energies. *Polymer* **2004**, *45*, 3663–3670. [CrossRef]
44. Vahabi, H.; Batistella, M.A.; Otazaghine, B.; Longuet, C.; Ferry, L.; Sonnier, R.; Lopez-Cuesta, J.M. Influence of a treated kaolinite on the thermal degradation and flame retardancy of poly(methyl methacrylate). *Appl. Clay Sci.* **2012**, *70*, 58–66. [CrossRef]
45. Ullah, S.; Ahmad, F.; Shariff, A.M.; Bustam, M.A. Synergistic effects of kaolin clay on intumescent fire retardant coating composition for fire protection of structural steel substrate. *Polym. Degrad. Stab.* **2014**, *110*, 91–103. [CrossRef]
46. Nabipour, H.; Wang, X.; Song, L.; Hu, Y. A fully bio-based coating made from alginate, chitosan and hydroxyapatite for protecting flexible polyurethane foam from fire. *Carbohydr. Polym.* **2020**, *246*, 116641. [CrossRef]
47. Guo, W.; Liu, J.; Zhang, P.; Song, L.; Wang, X.; Hu, Y. Multi-functional hydroxyapatite/polyvinyl alcohol composite aerogels with self-cleaning, superior fire resistance and low thermal conductivity. *Compos. Sci. Technol.* **2018**, *158*, 128–136. [CrossRef]
48. Guo, W.; Wang, X.; Zhang, P.; Liu, J.; Song, L.; Hu, Y. Nano-fibrillated cellulose-hydroxyapatite based composite foams with excellent fire resistance. *Carbohydr. Polym.* **2018**, *195*, 71–78. [CrossRef]
49. Khalili, P.; Liu, X.; Zhao, Z.; Blinzler, B. Fully biodegradable composites: Thermal, flammability, moisture absorption and mechanical properties of Natural fibre-reinforced composites with nano-hydroxyapatite. *Materials* **2019**, *12*, 1–13. [CrossRef]
50. Akindoyo, J.O.; Beg, M.D.H.; Ghazali, S.; Heim, H.P.; Feldmann, M. Effects of surface modification on dispersion, mechanical, thermal and dynamic mechanical properties of injection molded PLA-hydroxyapatite composites. *Compos. Part A Appl. Sci. Manuf.* **2017**, *103*, 96–105.
51. Behera, K.; Sivanjineyulu, V.; Chang, Y.H.; Chiu, F.C. Thermal properties, phase morphology and stability of biodegradable PLA/PBSL/HAp composites. *Polym. Degrad. Stab.* **2018**, *154*, 248–260.
52. Członka, S.; Strakowska, A.; Pospiech, P.; Strzelec, K. Effects of chemically treated eucalyptus fibers on mechanical, thermal and insulating properties of polyurethane composite foams. *Materials* **2020**, *13*, 1781. [CrossRef]

53. Cao, X.; James Lee, L.; Widya, T.; Macosko, C. Polyurethane/clay nanocomposites foams: Processing, structure and properties. *Polymer* **2005**, *46*, 775–783. [CrossRef]
54. Lee, L.J.; Zeng, C.; Cao, X.; Han, X.; Shen, J.; Xu, G. Polymer nanocomposite foams. *Compos. Sci. Technol.* **2005**, *65*, 2344–2363. [CrossRef]
55. Sung, G.; Kim, J.H. Influence of filler surface characteristics on morphological, physical, acoustic properties of polyurethane composite foams filled with inorganic fillers. *Compos. Sci. Technol.* **2017**, *146*, 147–154. [CrossRef]
56. Barczewski, M.; Kurańska, M.; Sałasińska, K.; Michałowski, S.; Prociak, A.; Uram, K.; Lewandowski, K. Rigid polyurethane foams modified with thermoset polyester-glass fiber composite waste. *Polym. Test.* **2020**, *81*, 106190. [CrossRef]
57. Członka, S.; Bertino, M.F.; Strzelec, K. Rigid polyurethane foams reinforced with industrial potato protein. *Polym. Test.* **2018**, *68*, 135–145. [CrossRef]
58. Paciorek-Sadowska, J.; Borowicz, M.; Isbrandt, M.; Czupryński, B.; Apiecionek, Ł. The use of waste from the production of rapeseed oil for obtaining of new polyurethane composites. *Polymers* **2019**, *11*, 1431. [CrossRef]
59. Cichosz, S.; Masek, A. Thermal Behavior of Green Cellulose-Filled Thermoplastic Elastomer Polymer Blends. *Molecules* **2020**, *25*, 1279. [CrossRef]
60. Cichosz; Masek Cellulose Fibers Hydrophobization via a Hybrid Chemical Modification. *Polymers* **2019**, *11*, 1174. [CrossRef]
61. Cichosz, S.; Masek, A.; Rylski, A. Cellulose Modification for Improved Compatibility with the Polymer Matrix: Mechanical Characterization of the Composite Material. *Materials* **2020**, *13*, 5519. [CrossRef]
62. Tian, H.; Wu, J.; Xiang, A. Polyether polyol-based rigid polyurethane foams reinforced with soy protein fillers. *J. Vinyl Addit. Technol.* **2018**, *24*, E105–E111. [CrossRef]
63. Mizera, K.; Ryszkowska, J.; Kurańska, M.; Prociak, A. The effect of rapeseed oil-based polyols on the thermal and mechanical properties of ureaurethane elastomers. *Polym. Bull.* **2020**, *77*, 823–846. [CrossRef]
64. Kurańska, M.; Polaczek, K.; Auguścik-Królikowska, M.; Prociak, A.; Ryszkowska, J. Open-cell rigid polyurethane bio-foams based on modified used cooking oil. *Polymer* **2020**, *190*, 1–7. [CrossRef]
65. Luo, X.; Xiao, Y.; Wu, Q.; Zeng, J. Development of high-performance biodegradable rigid polyurethane foams using all bioresource-based polyols: Lignin and soy oil-derived polyols. *Int. J. Biol. Macromol.* **2018**, *115*, 786–791. [CrossRef] [PubMed]
66. Mahmood, N.; Yuan, Z.; Schmidt, J.; Xu, C. Preparation of bio-based rigid polyurethane foam using hydrolytically depolymerized Kraft lignin via direct replacement or oxypropylation. *Eur. Polym. J.* **2015**, *68*, 1–9. [CrossRef]
67. Gómez-Fernández, S.; Ugarte, L.; Calvo-Correas, T.; Peña-Rodríguez, C.; Corcuera, M.A.; Eceiza, A. Properties of flexible polyurethane foams containing isocyanate functionalized kraft lignin. *Ind. Crops Prod.* **2017**, *100*, 51–64. [CrossRef]
68. Mosiewicki, M.A.; Dell'Arciprete, G.A.; Aranguren, M.I.; Marcovich, N.E. Polyurethane foams obtained from castor oil-based polyol and filled with wood flour. *J. Compos. Mater.* **2009**, *43*, 3057–3072. [CrossRef]
69. Finlay, K.A.; Gawryla, M.D.; Schiraldi, D.A. Effects of fiber reinforcement on clay aerogel composites. *Materials* **2015**, *8*, 5440–5451. [CrossRef]
70. Federation of European Rigid Polyurethane Foam Associations. *Thermal Insulation Materials Made of Rigid Polyurethane Foam (PUR/PIR) Properties-Manufacture*; European Rigid Polyurethane Foam Associations: Brussels, Belgium, 2006. Available online: http://highperformanceinsulation.eu/wpcontent/uploads/2016/08/Thermal_insulation_materials_made_of_rigid_polyurethane_foam.pdf (accessed on 14 January 2020).
71. Ciecierska, E.; Jurczyk-Kowalska, M.; Bazarnik, P.; Gloc, M.; Kulesza, M.; Kowalski, M.; Krauze, S.; Lewandowska, M. Flammability, mechanical properties and structure of rigid polyurethane foams with different types of carbon reinforcing materials. *Compos. Struct.* **2016**, *140*, 67–76. [CrossRef]
72. Gu, R.; Konar, S.; Sain, M. Preparation and characterization of sustainable polyurethane foams from soybean oils. *J. Am. Oil Chem. Soc.* **2012**, *89*, 2103–2111. [CrossRef]
73. Holešová, S.; Hundáková, M.; Pazdziora, E. Antibacterial Kaolinite Based Nanocomposites. *Procedia Mater. Sci.* **2016**, *12*, 124–129. [CrossRef]
74. Seyedmajidi, S.; Rajabnia, R.; Seyedmajidi, M. Evaluation of antibacterial properties of hydroxyapatite/bioactive glass and fluorapatite/bioactive glass nanocomposite foams as a cellular scaffold of bone tissue. *J. Lab. Physicians* **2018**, *10*, 265–270. [CrossRef] [PubMed]
75. Vega, O.; Araya, J.J.; Chavarría, M.; Castellón, E. Antibacterial biocomposite materials based on essential oils embedded in sol–gel hybrid silica matrices. *J. Sol-Gel Sci. Technol.* **2016**, *79*, 584–595. [CrossRef]
76. Aranberri, I.; Montes, S.; Wesołowska, E.; Rekondo, A.; Września-Tosik, K.; Grande, H.-J. Improved Thermal Insulating Properties of Renewable Polyol Based Polyurethane Foams Reinforced with Chicken Feathers. *Polymers* **2019**, *11*, 2002. [CrossRef] [PubMed]
77. Sair, S.; Oushabi, A.; Kammouni, A.; Tanane, O.; Abboud, Y.; El Bouari, A. Mechanical and thermal conductivity properties of hemp fiber reinforced polyurethane composites. *Case Stud. Constr. Mater.* **2018**, *8*, 203–212. [CrossRef]
78. de Avila Delucis, R.; Magalhães, W.L.E.; Petzhold, C.L.; Amico, S.C. Forest-based resources as fillers in biobased polyurethane foams. *J. Appl. Polym. Sci.* **2018**, *135*, 1–7. [CrossRef]
79. Joanna, P.S.; Bogusław, C.; Joanna, L. Application of waste products from agricultural-food industry for production of rigid polyurethane-polyisocyanurate foams. *J. Porous Mater.* **2011**, *18*, 631–638. [CrossRef]

80. Saada, A.; Siffert, B.; Papirer, E. Comparison of the hydrophilicity/hydrophobicity of illites and kaolinites. *J. Colloid Interface Sci.* **1995**, *174*, 185–190. [CrossRef]
81. Bouiahya, K.; Oulguidoum, A.; Laghzizil, A.; Shalabi, M.; Nunzi, J.M. Hydrophobic chemical surface functionalization of hydroxyapatite nanoparticles for naphthalene removal. *Colloids Surfaces A Physicochem. Eng. Asp.* **2020**, *595*, 124706. [CrossRef]
82. Wolski, K.; Cichosz, S.; Masek, A. Surface hydrophobisation of lignocellulosic waste for the preparation of biothermoelastoplastic composites. *Eur. Polym. J.* **2019**, *118*, 481–491. [CrossRef]
83. Chattopadhyay, D.K.; Webster, D.C. Thermal stability and flame retardancy of polyurethanes. *Prog. Polym. Sci.* **2009**, *34*, 1068–1133. [CrossRef]
84. Cheng, J.; Wang, H.; Wang, X.; Li, S.; Zhou, Y.; Zhang, F.; Wang, Y.; Qu, W.; Wang, D.; Pang, X. Effects of flame-retardant ramie fiber on enhancing performance of the rigid polyurethane foams. *Polym. Adv. Technol.* **2019**, *30*, 3091–3098. [CrossRef]

Article

Rigid Polyurethane Foams Reinforced with POSS-Impregnated Sugar Beet Pulp Filler

Anna Strąkowska [1], Sylwia Członka [1,*] and Agnė Kairytė [2]

1. Institute of Polymer & Dye Technology, Lodz University of Technology, 90-924 Lodz, Poland; anna.strakowska@p.lodz.pl
2. Laboratory of Thermal Insulating Materials and Acoustics, Institute of Building Materials, Faculty of Civil Engineering, Vilnius Gediminas Technical University, Linkmenu st. 28, LT-08217 Vilnius, Lithuania; agne.kairyte@vgtu.lt
* Correspondence: sylwia.czlonka@dokt.p.lodz.pl

Received: 11 November 2020; Accepted: 30 November 2020; Published: 2 December 2020

Abstract: Rigid polyurethane (PUR) foams were reinforced with sugar beet pulp (BP) impregnated with Aminopropylisobutyl-polyhedral oligomeric silsesquioxanes (APIB-POSS). BP filler was incorporated into PUR at different percentages—1, 2, and 5 wt.%. The impact of BP filler on morphology features, mechanical performances, and thermal stability of PUR was examined. The results revealed that the greatest improvement in physico-mechanical properties was observed at lower concentrations (1 and 2 wt.%) of BP filler. For example, when compared with neat PUR foams, the addition of 2 wt.% of BP resulted in the formation of PUR composite foams with increased compressive strength (~12%), greater flexural strength (~12%), and better impact strength (~6%). The results of thermogravimetric analysis (TGA) revealed that, due to the good thermal stability of POSS-impregnated BP filler, the reinforced PUR composite foams were characterized by better thermal stability—for example, by increasing the content of BP filler up to 5 wt.%, the mass residue measured at 600 °C increased from 29.0 to 31.9%. Moreover, the addition of each amount of filler resulted in the improvement of fire resistance of PUR composite foams, which was determined by measuring the value of heat peak release (pHRR), total heat release (THR), total smoke release (TSR), limiting oxygen index (LOI), and the amount of carbon monoxide (CO) and carbon dioxide (CO_2) released during the combustion. The greatest improvement was observed for PUR composite foams with 2 wt.% of BP filler. The results presented in the current study indicate that the addition of a proper amount of POSS-impregnated BP filler may be an effective approach to the synthesis of PUR composites with improved physico-mechanical properties. Due to the outstanding properties of PUR composite foams reinforced with POSS-impregnated BP, such developed materials may be successfully used as thermal insulation materials in the building and construction industry.

Keywords: rigid polyurethane foams; sugar beet pulp; bio-filler; thermal conductivity; mechanical properties

1. Introduction

Rigid polyurethane (PUR) foams are porous materials, which are commonly used in many industries, including building engineering, automotive, construction, and furniture [1,2]. Currently, environmental awareness is increasing interest in sustainable development goals (SDGs) including sustainability and environmental protection. Because of this, the production of PUR composites with bio-based, organic fillers has attracted increased attention in the academic environment. Due to the low price, unlimited availability, and good physico-mechanical properties, such as low density and high stiffness, the cellulosic materials seem to be the most promising fillers for polymeric composites [3–7].

Many previous studies have shown, that the incorporation of cellulosic fillers may successfully improve the physico-mechanical characteristics of polymeric composites [8–11], including PUR foams [12,13]. For example, Silva et al. [14] produced rigid PUR foams reinforced with cellulose fiber residue from bleached eucalyptus pulp. The incorporation of cellulose fiber up to 16 wt.% has changed the properties of the composites, including the foam morphology and reducing cell size. PUR foams reinforced with 5, 10, and 15 wt.% of wood ash and fly ash fillers were prepared by Hejna et al. [15]. The introduction of both types of fillers caused some deterioration of the mechanical properties of PUR foams, such as compressive strength and flexural strength, but the obtained results were still considered satisfactory for most industrial applications. Moreover, the addition of fly/wood ash fillers improved the thermal stability of PUR foams reducing their thermal degradation. Interesting results were presented by Olszewski et al. [16] in the case of PUR composites modified with glass and sisal fibers–a significant improvement in mechanical properties of PUR composites was observed. Similar results were reported by Kurańska et al. [17]–the addition of 3–40 wt.% of basalt waste fillers resulted in the synthesis of PUR foams with improved mechanical characteristics. PUR foams reinforced with bamboo fibers (BF) with a particle size of 250–500 µm were produced by Li et al. [18]. When compared with neat PUR foams, the addition of BF increased the mechanical and flexural strength by ~47% and ~16%, respectively. Moreover, the results have shown that alkali treatment of BF results in better interaction between the BF surface and PUR matrix, leading to greater mechanical performances and better thermal stability of PUR foams. The impact of selected plant fillers, such as cinnamon extract, cocoa extract, and green coffee extract on the physico-mechanical properties of PUR foams was examined by Liszowska et al. [19]. It has been found that due to the incorporation of bio-based fillers, all series of PUR foams were more susceptible to biodegradation than the neat PUR foams. PUR foams reinforced with algal cellulose were produced by Jonjaroen et al. [20]. Due to the incorporation of algal cellulose, PUR foams were characterized by increased apparent density, greater stiffness and lower loss modulus when compared with neat PUR foams.

Sugar beet is one of the most commonly grown plants which is used in the sugar industry [21]. According to the Food and Agriculture Organization (FAO), sugar production from sugar beet is the second largest production in the world. The greatest amount of sugar beet is produced in Europe, Asia, and North America [21]. According to FAO, the production of sugar beet in Europe was 2772 million tonnes in 2016. The main problem with sugar production from sugar beets is the formation of a huge amount of waste-sugar beet pulp (BP), which consisting of polysaccharides (~70–80%), cellulose (~20%), hemicelluloses (~30%), and pectin (~25%) [22]. BP is mostly used as feed for animals, however, multiple different applications of BP have already been proposed–BP was used as an energy raw material, as a new source for bio-polyols for polyurethane synthesis as well as for paper manufacture [23]. For example, polyurethane foams were prepared from microwave liquefied sugar beet pulp (LSBP) and polymethylene polyphenyl isocyanate (PAPI) by Zheng et al. [24]. The effect of [NCO]/[OH] ratio on the mechanical, thermal, and microstructural properties of the LSBP–PU foams were studied. It has been shown, that as the [NCO]/[OH] ratio increased from 0.6 to 1.2, the LSBP-PU foams with the less regular structure deteriorated mechanical performances were produced. In another study, rigid polyurethane foams of significant renewable content (up to 50%) were produced using biomass biopolyols obtained via crude-glycerol mediated solvothermal liquefaction of sugar beet pulp and commercial diphenylmethane diisocyanate [25]. The produced foams exhibited higher apparent densities 43–160 kg m^{-3} and compressive strengths 34–254 kPa compared to tested commercial analogues. Biopolyol foams exhibited higher thermal stability and the non-flame retarded foams showed lower potential for fire spread due to lower pyrolysis gas combustion heat release rates and total released amounts of heat. Besides this, BP has not been used as a reinforcing filler for rigid polyurethane foams. Taking into account, the amount of BP, their availability, and production efficiency it seems logical and well-argued to use BP as a reinforcing filler in the PUR industry.

Previous studies have reported, that in the case of PUR foams reinforced with natural fillers the main concern is connected with the low thermal stability of bio-based organic fillers, which consequently, deteriorates the thermal stability of final products [26]. The incompatibility between the hydrophilic surface of cellulose fillers and the hydrophobic structure of PUR results in interfacial separation, leading to deteriorated performances of PUR foams [27]. Because of this, the modification of the filler surface seems to be a crucial step before the incorporation of organic fillers into the PUR matrix. Up to now, various attempts have been reported to improve the compatibility of the fillers with the polymeric matrix [28–32]. Several authors have reported different surface modifications for organic fillers, which involve chemical modifications, such as acetylation [33], alkalization [34], benzoylation [35], grafting [35], and silane treatment [36]. The results have shown, that the surface modification of organic fillers results in the production of polymeric composites with improved thermal and mechanical characteristics.

Many previous works have studied the impact of natural fillers on the mechanical and thermal characteristics of polymeric composites, however, no studies have been devoted to the examination of the polyurethane composite foams reinforced with beet pulp (BP) filler physically impregnated with thermally stable inorganic compounds, such as Polyhedral Oligomeric Silsesquioxanes (POSSs). POSS compounds are organic-inorganic compounds. The chemical structure of POSS includes a silica core and oxygen atoms at the edges of the molecule [37,38]. Our previous studies have shown that the incorporation of POSS molecules into PUR foams may improve the thermal stability and degradation behavior of PUR foams, as well as their mechanical performances [39–42]. Keeping in view the advantageous properties of beet pulp filler and POSS compounds, it seems logical to use POSS-impregnated BP filler as a reinforcing, hybrid filler for new bio-based polyurethane composite foams. The preparation of novel materials from POSS-impregnated BP filler may successfully improve the mechanical and thermal properties of the polyurethane materials. Therefore, in this study, rigid PUR foams were reinforced with 1, 2, and 5 wt.% of BP filler impregnated with Aminopropylisobutyl-POSS (APIB-POSS). The impact of POSS-impregnated BP filler on the morphology and physico-mechanical performances of PUR composite foams was examined.

2. Materials and Methods

2.1. Materials

Polyether polyol (commercial name: Stapanpol PS-2352, Northfield, MN, USA), polymeric diphenylmethane diisocyanate (commercial name: Purocyn B, Purinova Sp. z o.o., Bydgoszcz, Poland), potassium octoate (commercial name: Kosmos 75, Evonik Industries AG., Essen, Germany), and potassium acetate (commercial name: Kosmos 33 (Evonik Industries AG., Essen, Germany) were used as catalysts. Tegostab B8513 (Evonik Industries AG., Essen, Germany) was used as a silicone-based surfactant. A mixture of pentane and cyclopentane (50:50 v/v%) was used as a blowing agent. Aminopropylisobutyl-POSS (APIB-POSS) was purchased from Hybrid Plastics Inc (Hattiesburg, MS, USA). Sodium hydroxide (pellets, anhydrous), was purchased from Sigma-Aldrich Corporation (Saint Louis, MO, USA). Sugar beet pulp was obtained from a local company (Lodz, Poland).

2.2. Filler Preparation and Synthesis of PUR Composite Foams

PUR composite foams were synthesized according to the procedure described previously [43]. The selected amount of beet pulp (BP) filler was added to the polyol system. Before the addition, the BP filler was alkali-treated with 5 wt.% solution of sodium hydroxide solution using the method presented in [26]. After the alkali treatment, such prepared BP filler was mixed with Aminopropylisobutyl-POSS (APIB-POSS) (1:1 w/w) using a planetary ball mill for 1 h (3000 rpm). Then, the calculated amount of other ingredients, such as a surfactant, catalyst, and blowing agent was added to the polyol system and the mixture was intensively mixed by mechanical stirrer at 1000 rpm for 60 s. After the complete dispersion of BP filler, a calculated amount of isocyanate was added to the mixture and mixed at 2000 rpm for 30 s. Such prepared composite foams were left to expand freely at room temperature. To provide complete curing, PUR composite foams were conditioned at room temperature for 48 h.

The formulations of PUR composite foams containing BP filler are presented in Table 1. The schematic procedure of PUR composite foams preparation is presented in Figure 1. Different concentrations of BP fillers were used to determine the impact of filler content on the further properties of PUR composite foams. The reference PUR foams were prepared without the addition of BP filler.

Table 1. Formula of neat polyurethane (PUR) foams and PUR composite foams.

Component	PUR_0	PUR_BP_1	PUR_BP_2	PUR_BP_5
	Parts by Weight (wt.%)			
STEPANPOL PS-2352	100	100	100	100
PUROCYN B	160	160	160	160
Kosmos 75	6	6	6	6
Kosmos 33	0.8	0.8	0.8	0.8
Tegostab B8513	2.5	2.5	2.5	2.5
Water	0.5	0.5	0.5	0.5
Pentane/cyclopentane	11	11	11	11
Beet pulp impregnated with polyhedral oligomeric silsesquioxanes (POSS)	0	1	2	5

Figure 1. Schematic procedure of the synthesis of PUR composite foams.

2.3. Sample Characterization

The dynamic viscosity was determined according to ISO 2555 [44] using Viscometer DVII+ (Viscometer DVII+, Brookfield, Berlin, Germany). The morphology of PUR composite foams was determined using a scanning electron microscope (SEM) (JEOL JSM 5500 LV, JEOL Ltd., Peabody, MA, USA). The cell sizes of PUR composite foams was determined by ImageJ software (Java 1.8.0, Media Cybernetics Inc., Rockville, MD, USA). The apparent density of PUR composite foams was determined in accordance with ISO 845 [45]. The number of closed-cells was evaluated according to ISO 4590 [46]. Thermal conductivity (λ) of PUR composite foams was measured using LaserComp 50 (TA Instruments Inc., New Castle, DE, USA). The mechanical properties (compressive strength, flexural strength) of PUR composite foams were performed using Zwick Z100 Testing Machine (Zwick/Roell Group, Ulm, Germany) according to ISO 844 [47] and ISO 178 [48]. The impact strength was determined using the Charpy Impact Strength Test Machine according to ISO 180 [49]. The thermogravimetric analysis (TGA) test was performed in the function of temperature (0–600 °C) using the STA 449 F1 Jupiter Analyzer (Netzsch Group, Selb, Germany). Surface hydrophobicity of PUR composite foams was measured using contact angle goniometer OEC-15EC (DataPhysics Instruments GmbH, Filderstadt, Germany). A cone calorimeter test was performed using the cone calorimeter apparatus according to ISO 5660 [50].

3. Results and Discussion

3.1. Filler Characterization

An external topography of neat BP and BP modified with POSS is presented in Figure 2. Comparing SEM images, it is clear that the physical modification of BP filler with POSS compounds affected the overall structure of the filler. After the modification, the external surface of BP filler became more rough and non-uniform. The crystals of POSS particles are visible on the surface of the BP filler. The size of neat BP filler is in the range of 400–800 nm, however, the filler particles tended to agglomerate and some bigger clusters with an average diameter of ~3 µm are presented. After the modification, the size of filler particles increased to ~2 µm due to the impregnation of BP filler's surface with POSS molecules. When compared with neat BP filler, the size distribution of modified BP filler is more uniform, which indicates that the physical modification with POSS prevented the filler particles from agglomerating. Moreover, the addition of modified BP filler affected the viscosity of PUR systems (Table 2)—when compared with neat PUR system, the viscosity increases from 800 to 1550 mPa·s, for PUR system with 1, 2, and 5 wt.% of BP filler, respectively.

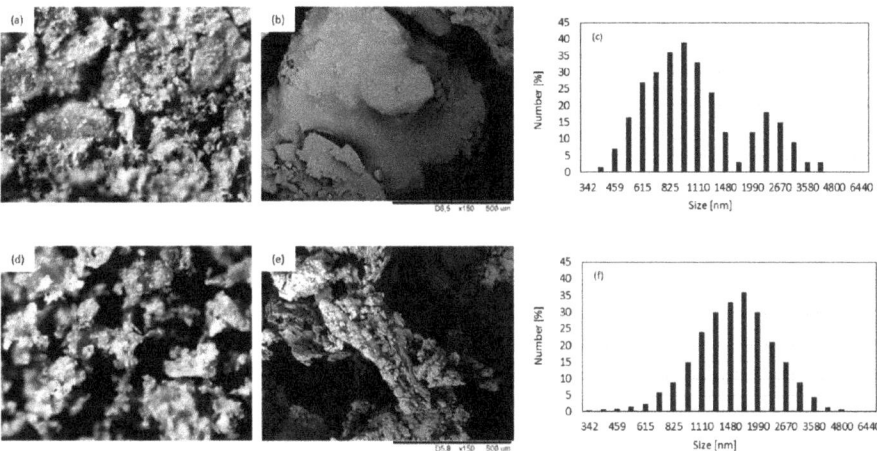

Figure 2. External surface and particle size of neat BP filler (**a–c**) and BP filler impregnated with POSS (**d–f**).

Table 2. Dynamic viscosity and processing times of PUR systems with BP filler.

Sample	Dynamic Viscosity η (mPa·s)	Processing Times (s)		
		Cream Time	Free-Rise Time	Tack-Free Time
PUR_0	800 ± 9	39 ± 3	282 ± 9	360 ± 10
PUR_BP_1	980 ± 10	45 ± 2	308 ± 8	345 ± 9
PUR_BP_2	1250 ± 10	49 ± 2	322 ± 6	350 ± 7
PUR_BP_5	1550 ± 11	55 ± 1	365 ± 7	355 ± 7

3.2. Foaming Behavior of PUR Systems

The viscosity of PUR systems has a great impact on the foaming behavior of porous materials. The foaming behavior of PUR foams was examined by measuring start, free-rise, and tack-free times. According to the results given in Table 2, the processing times tended to increase with the addition of BP filler. When compared with neat PUR systems, the addition of 1, 2, and 5 wt.% of BP filler increased the start time by ~15, ~25, and ~41%, while the free-rise time increased by ~9, ~14, and ~29%, respectively. The main reason for extended times may be attributed to the increased viscosity of PUR systems with BP filler, which limits the proper expansion of the cells and consequently extends the processing times of PUR systems. This effect is more noticeable in the case of PUR systems with 5 wt.% of BP filler. In general, the higher the BP filler content, the greater the viscosity of the PUR systems and the longer their processing times. On the other hand, the addition of BP filler may decrease the reactivity of PUR systems, due to the incorporation of the additional groups of BP filler, which are able to react with highly reactive isocyanate groups. Due to this, a greater amount of isocyanate is consumed, which affects the proper stoichiometry of the PUR synthesis and reduces the amount of carbon dioxide (CO_2) produced. A reduced amount of blowing agent results in limited expansion of the cells, extending the processing times of PUR synthesis. Similar results have also been found in previous works [51,52].

3.3. Cellular Structure of PUR Composite Foams

Physical and mechanical properties of porous materials are affected by their cellular structure; therefore, the examination of foams' structure is vital. In order to determine the microscopic morphology of PUR composite foams with BP filler, SEM analysis was examined.

According to the results presented in Table 3, the average cell diameter of neat PUR foams is 492 μm. The cell diameter increases from 450 to 530 μm with the increase in BP content from 1 to 5 wt.%. As presented in Figure 3, the cell size distribution of PUR composite foams containing 1 and 2 wt.% is uniform and it is distributed between 200 and 700 μm. With the addition of 5 wt.% of BP filler, the cell size distribution becomes wider and some bigger cells are visible in the PUR structure. It can be concluded that lower content of BP filler can be mixed with polyurethane system homogenously, leading to the formation of a PUR structure with uniform cell size distribution. Moreover, the incorporation of BP fillers provides additional nucleation sites which result in the formation of a higher number of smaller cells when compared with neat PUR foams. The addition of BP fillers increases the viscosity of PUR systems, reducing the increase in cell size. Increasing the content of BP filler up to 5 wt.% deteriorates the structure of PUR composite foams. This may be connected with the agglomeration of BP filler after a certain content due to the high viscosity of the PUR system. The agglomeration of filler particles results in inhomogeneous distribution and the formation of uneven foam structures with bigger cells.

Table 3. Selected properties of PUR composite foams with BP filler.

Sample	Average Cell Diameter (μm)	Content of Closed-Cells (%)	Apparent Density (kg m^{-3})	Thermal Conductivity (W m^{-1} K^{-1})	Contact Angle (°)	Water Uptake (%)
PUR_0	492 ± 6	90.2 ± 0.4	37.1 ± 0.7	0.026 ± 0.001	123 ± 1	20.4 ± 0.6
PUR_BP_1	450 ± 5	89.4 ± 0.3	39.2 ± 0.6	0.027 ± 0.001	120 ± 1	18.6 ± 0.5
PUR_BP_2	445 ± 5	88.1 ± 0.4	41.4 ± 0.5	0.030 ± 0.001	128 ± 1	18.9 ± 0.4
PUR_BP_5	530 ± 6	82.3 ± 0.3	44.6 ± 0.6	0.035 ± 0.001	130 ± 1	22.4 ± 0.5

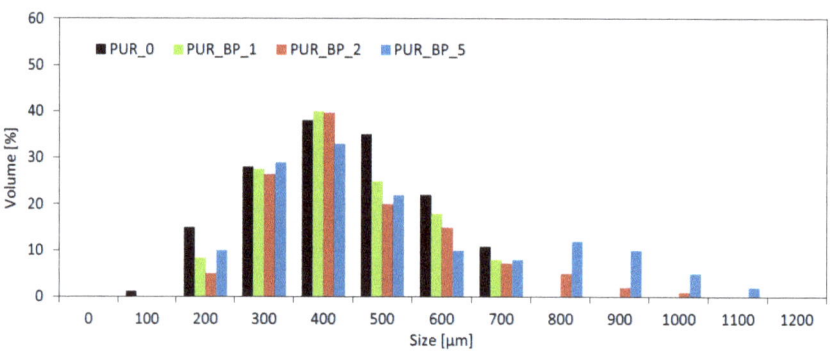

Figure 3. The cell size distribution of PUR composite foams with BP filler.

Figure 4 presents the overall structure of PUR composite foams with various BP filler content. Generally speaking, by increasing the content of BP filler, the overall structure of PUR composite foams becomes less uniform with a higher number of cracked cells. This may be ascribed to the following reasons: BP filler particles act as nucleation sites for the gas phase, affecting the rheology around the growing air bubbles and changing the nucleation character from homogenous to heterogeneous [53]. This results in the creation of a higher number of finer cells, which tend to collapse, resulting in a less homogeneous structure of PUR composite foams. Moreover, the poor interphase compatibility between filler surface and PUR matrix promotes earlier collapsing of the cells, leading to the formation of a more defective structure of PUR composite foams [54]. This effect is most prominent in the case of PUR composite foams with higher loading of BP filler, due to the significant agglomeration of the filler, which promotes the rupturing of the cells and creating a weakened PUR structure.

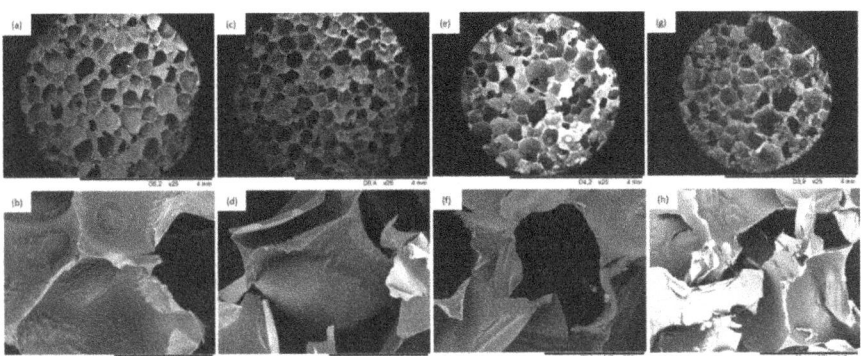

Figure 4. Cellular morphologies at different magnifications of (**a,b**) PUR_0, (**c,d**) PUR_BP_1, (**e,f**) PUR_BP_2, (**g,h**) PUR_BP_5.

3.4. Apparent Density of PUR Composite Foams

The results of the apparent density of PUR composite foams with different BP content are presented in Figure 5a. The density of PUR composite foams increased from 37 kg m^{-3} to 40, 41, and 44 kg m^{-3} after the addition of 1, 2, and 5 wt.% of BP filler, respectively, due to the following reasons. Firstly, the density of modified PUR composite foams is enhanced due to the incorporation of BP filler which possesses a higher density than neat PUR foams. Therefore, the density of the modified PUR composite foams increased with the increasing content of BP filler. Furthermore, the addition of BP filler increases the viscosity of the PUR systems and reduces the expansion of the cells. Thus, the density of PUR composite foams increased with the increase in the BP filler content.

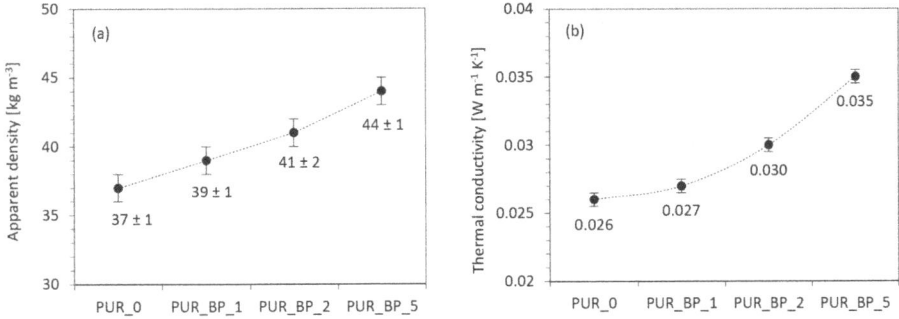

Figure 5. The results of (**a**) apparent density and (**b**) thermal conductivity of PUR composite foams.

3.5. Thermal Conductivity of PUR Composite Foams

The cellular structure and apparent density of PUR composite foams affect their insulating properties. The value of thermal conductivity (λ) of neat PUR_0 is 0.025 Wm^{-1} K^{-1}. The addition of 1 and 2 wt.% of BP filler had no effect on λ, however, the addition of 5 wt.% of BP filler increased the value of λ to 0.035 Wm^{-1} K^{-1} (Figure 5b). In general, thermal conductivity is determined as a combination of λ_{gas}, λ_{solid}, and $\lambda_{convection}$. Due to the incorporation of solid particles of BP filler, which are built in the PUR matrix, the value of λ_{solid} increases. This effect is most prominent in the case of PUR composite foams with 5 wt.% of BP filler. As discussed previously, BP filler particles tend to agglomerate, resulting in the formation of a greater number of open cells. This, in turn, increases the value of λ_{gas}, worsening the insulating properties of PUR composite foams. Similar behavior was observed in previous works, concerning the PUR composite foams modified with the addition of selected organic and/or inorganic fillers. For example, an increased value of λ was reported for PUR foams with 3–40 wt.% of basalt waste filler by Kurańska et. al. [17]. It has been shown that besides the well-developed, closed-cell structure of modified PUR foams, the insulating properties are deteriorated, however, following industrial standards, they are still considered at an acceptable level.

3.6. Contact Angle and Water Uptake of PUR Composite Foams

Water uptake depends on morphological features of PUR foams (e.g., the content of open/closed-cells) as well as on the hydrophobic nature of the filler [55]. According to the results presented in Table 3, the addition of BP filler affects the water uptake of PUR composite foams. When compared with neat PUR_0, the addition of 1 and 2 wt.% of BP filler decreased the water uptake by ~9 and ~7%, while for PUR composite foams with 5 wt.% of BP filler, the value increased by ~10%. As presented in SEM images (Figure 4), the addition of 5 wt.% results in the opening of cells which are able to store a greater amount of water. Reduced water uptake of PUR composite foams with 1 and 2 wt.% of BP filler may be connected with a greater number of closed-cells that are not able to accommodate water. Furthermore, this effect may be also enhanced by the hydrophobic character of incorporated filler, which was also confirmed by the results of the contact angle (Table 3). However, the impact of the cellular structure and the content of open/closed-cells seems to be a more dominant factor, which determines water uptake ability.

3.7. Mechanical Performances of PUR Composite Foams

Mechanical performances of porous materials, including PUR foams, are dependent on several factors, such as cellular structure (e.g., shape and size of cells) and density of PUR foams [56]. Furthermore, the mechanical performances of PUR composite foams are influenced by the homogenous dispersion of the fillers in the PUR system and the interphase compatibility between the filler surface and the PUR matrix. Figure 6a presents the compressive strength of PUR composite foams containing various contents of BP filler. The compressive strength of PUR composite foams increases initially with the addition of 1 and 2 wt.% of BP filler, and then decreases slightly with the further increase in BP filler up to 5 wt.%. When compared with neat PUR foams, the greatest compressive strength is observed on the addition of 2 wt.% of BP filler—the value of compressive strength (measured parallel) is 272 kPa, which is ~12% higher than that of the neat PUR foams. A similar tendency is observed in the case of compressive strength measured perpendicular to the direction of foam growth, but the compressive strength values are much lower due to the anisotropy of the foam cells [57]. Similarly, the greatest improvement is observed for PUR composite foams with the addition of 2 wt.% BP filler—compressive strength (measured perpendicularly) increases by ~18%. An increase in the formation of a greater number of smaller cells, contributes to a superior mechanical behavior of PUR composite foams; the applied load is encountered by a greater number of cell walls per unit area. This indicates that the addition of 1 and 2 wt.% of BP filler has a reinforcing effect and increases the mechanical performance of PUR composite foams. The obtained results meet the industrial requirements for constructive materials (apparent density > 35 kg m^{-3}, $\sigma_{10\%}$ > 200 kPa) [58]. The addition of BP filler at an optimal level and the appropriate distribution of filler particles in the cell struts strengthened the foam structure, increasing the compressive strength. Increasing the BP filler up to 5 wt.% causes a BP filler agglomeration, which in turn, results in filler–filler interaction, deteriorating the mechanical performances of PUR composite foams [59]. As reported in SEM images (Figure 4), at higher content of BP filler the overall structure of the PUR composite foams becomes distorted and some voids are visible, deteriorating their mechanical properties. To eliminate the effect of density on mechanical performances of PUR composite foams, the specific compressive strength was calculated as a ratio of compressive strength to density. The specific strength of neat PUR foams is 6.5 MPa/kg/m^3 and it increases insignificantly to 6.9 and 6.6 MPa/kg/m^3 for PUR_BP_1 and PUR_BP_2, respectively, and then decreases to 5.1 MPa/kg/m^3 for PUR_BP_5.

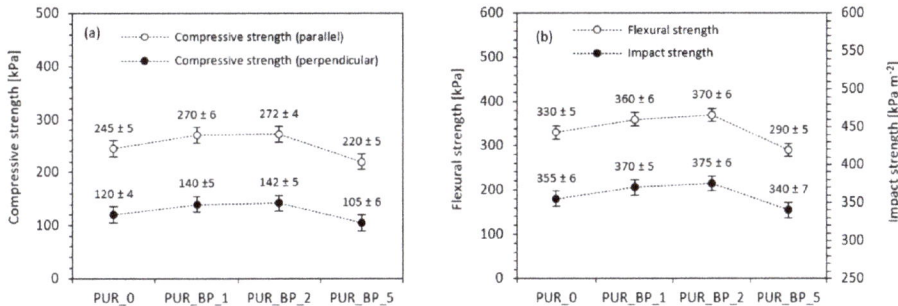

Figure 6. The results of (a) compressive strength and (b) flexural/impact strength of PUR composite foams.

According to the results presented in Figure 6b, the addition of 1 and 2 wt.% of BP filler results in an improvement in flexural and impact strength, although an insignificant deterioration in properties is observed on the addition of 5 wt.% of BP filler. The greatest improvement in flexural and impact strength is observed on the addition of 2 wt.% of the filler—the value of flexural and impact strength increases by ~12 and ~6%, respectively. This result may be connected with the fact that filler particles act as additional stress points for the local stress concentration from which the cracking of the sample begins [60,61]. As the concentration of BP filler increases, the effect is more prominent, leading to the deterioration of the abovementioned properties.

3.8. Thermal Stability of PUR Composite Foams

Thermogravimetric (TGA) and derivative thermogravimetry (DTG) analysis of BP fillers and PUR composite foams are presented in Figure 7a,b, respectively. The resulting parameters are summarized in Table 4.

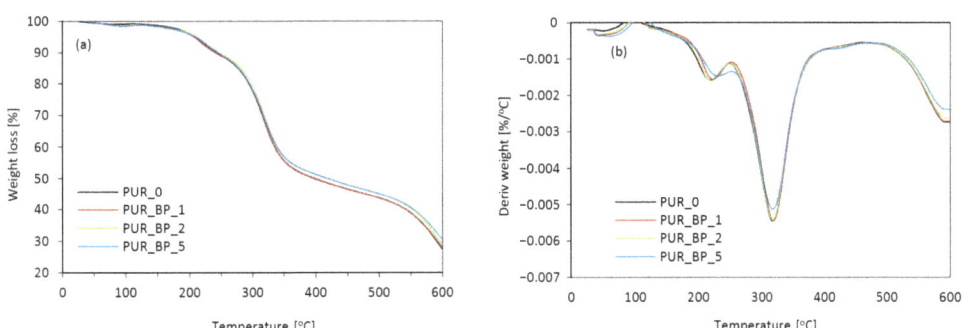

Figure 7. (a) TGA and (b) DTG results obtained for PUR composite foams.

PUR composite foams show three stages of weight loss. The first stage (T_{max1}) occurs around 200–250 °C and refers to the release of volatile compounds of BP filler [62]. When compared with neat PUR_0, T_{max1} increases from 220 °C to 224 and 226 °C for the composite foams with 1 and 2 wt.%

of BP filler, respectively, and then decreases to 217 °C for the composite foams with 5 wt.% of BP filler, due to the heterogeneity of the PUR structure caused by filler agglomeration. T_{max2} occurs in the range of 300–350 °C and refers to the degradation of soft segments of PUR and thermolysis of the residues of organic BP filler [57,63]. The value of T_{max2} of neat PUR foams is 309 °C. Due to the addition of 1 and 2 wt.% of BP filler, the value of T_{max2} increases slightly to 325 and 320 °C, respectively. Such improvement may be due to the fact that fuller particles act as a thermal barrier, which prevents rapid heat transfer and reduces further degradation of PUR composite foams. Such an effect is further enhanced by a greater crosslinking of PUR, due to the reaction between functional groups of BP filler (i.e., hydroxyl groups) and isocyanate groups. The value of T_{max2} decreases slightly for PUR foams with 5 wt.% of BP filler, due to the more open-cell structure. T_{max3} occurs in the range of 500–600 °C and corresponds to the thermal degradation of cellulosic derivatives of BP filler [64,65]. Due to the presence of lignocellulose filler, the value of T_{max3} slightly increases at the addition of BP filler. Moreover, with increasing the content of BP filler, the mass residue (at 600 °C) increases from 29.0 (for neat PUR_0) to 29.1, 30.7, and 31.9% for PUR composite foams with 1, 2, and 5 wt.% of BP filler, respectively. Due to the great thermal stability of PUR composite foams, such developed materials may be successfully used as thermal insulation materials in the building and construction industry.

Table 4. The results of thermogravimetric (TGA) and derivative thermogravimetry (DTG) analysis of PUR composite foams.

Sample	T_{max} (°C)			Residue at 600 °C (wt.%)
	1st Stage	2nd Stage	3rd Stage	
PUR_0	220	309	580	29.0
PUR_BP_1	224	325	586	29.1
PUR_BP_2	226	320	585	30.7
PUR_BP_5	217	302	591	31.9

3.9. Cone Calorimeter Test

The results of fire behavior of PUR composite foams are presented in Table 5 and Figure 8.

Table 5. The results of the cone calorimeter test.

	IT (s)	pHRR (kW m^{-2})	TSR (m^2 m^{-2})	THR (MJ m^{-2})	COY (kg kg^{-1})	CO_2Y (kg kg^{-1})	COY/CO_2Y (-)	LOI (%)
PUR_0	4	260	1500	21.5	0.210	0.240	0.875	20.2
PUR_BP_1	4	170	1200	20.5	0.170	0.200	0.850	20.9
PUR_BP_2	5	155	1100	20.9	0.160	0.195	0.820	21.2
PUR_BP_5	5	190	1450	21.2	0.162	0.190	0.852	20.5

Compared to neat PUR_0, the addition of BP filler has no effect on the ignition time (IT)—in all cases, the value of IT oscillates between 4 and 5 s. The intensity of the flame was determined by measuring the value of heat peak release (pHRR). As presented in Figure 8a, all series of PUR composite foams exhibited one peak of HRR, which refers to the release of low molecular weight compounds of PUR foams, such as isocyanate, amines, or olefins. When compared with neat PUR_0, the pHRR value decreases from 260 kW m^{-2} to 170 and 155 kW m^{-2} for PUR_BP_1 and PUR_BP_2 and then increases to 190 kW m^{-2} on the incorporation of 5 wt.% of BP filler. Such improvement in the fire resistance of PUR composite foams may be connected with the creation of a continuous, distended char layer on the surface of PUR foams with the addition of BP filler, which acts as a physical barrier and effectively limits the mass and heat transfer [66]. Moreover, the BP filler degrades endothermically and during the decomposition releases non-combustible products, which decreases the rate of heat release. Among all series of PUR composite foams, the greatest improvement is observed for PUR composite foams with 1 and 2 wt.% of BP filler—the value of pHRR decreases by ~35 and ~40%, respectively. In the case of

PUR composite foams with 5 wt.% of BP filler, the value of pHRR decreases slightly by ~27%, due to their more open-cell structure when compared with PUR with lower content of the filler. As presented in Figure 8b, the incorporation of each amount of BP filler results in a lower value of total smoke release (TSR). When compared with neat PUR_0, the value of TSR decreases by ~20, ~27, and ~3%, for PUR composite foams with 1, 2, and 5 wt.% of BP filler. This indicates that incorporation of BP filler prevents the heat transfer and protects the PUR structure from further combustion [67]. Moreover, the incorporation of BP fillers decreases the value of total heat release (THR). Compared to neat PUR_0, the addition of 1 and 2 wt.% of BP filler decreases the value of THR from 21.5 MJ m^{-2} (for PUR_0) to 20.5 and 20.9 MJ m^{-2}, respectively. No significant difference is observed for PUR composite foams with 5 wt.% of BP—the value of THR decreases slightly to 21.2 MJ m^{-2}. As presented in Figure 8c,d, the incorporation of BP filler decreases the ratio of carbon monoxide (CO) to carbon dioxide (CO_2), which refers to the foam toxicity. In general, a higher value of the ratio (CO/CO_2) indicates incomplete combustion of PUR composite foams and a greater amount of toxic smoke. The addition of BP filler reduces the value of the ratio, which means that the release of toxic smoke during the PUR foams combustion is reduced. The outer structure of the char residue of PUR composite foams is presented in Figure 9. It can be observed that the addition of 1 wt.% of BP filler has no effect on the morphology of the char residue layer. The difference is observed in the case of PUR composite foams with the addition of 2 and 5 wt.% of BP—the outer surface becomes more dense and the spherical carbon residues are visible in the structure. It can be concluded that BF particles can act as a flame barrier limiting the release of combustible gases.

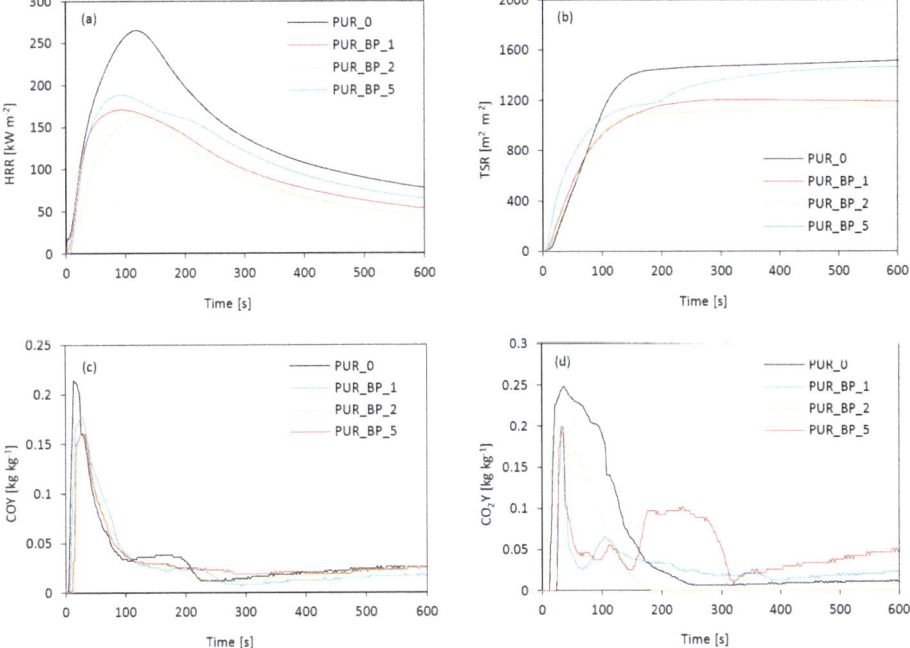

Figure 8. The results of the cone calorimeter test—(a) heat peak release (pHRR), (b) total smoke release (TSR), (c) CO release, and (d) CO_2 release.

The improvement in flame resistance of the PUR composite foams with BP filler has been confirmed by the results of limiting oxygen index (LOI). As presented in Table 5, the addition of BP fillers increases the value of LOI of PUR foams. The greatest improvement was observed for PUR foams with 1 and 2 wt.%—the value of LOI increased from 20.2% (for neat PUR_0) to 20.9 and 21.2%, respectively. A less noticeable improvement was observed for PUR foams with 5 wt.% of BP filler—the value of LOI increased to 20.5%. Similar results have been reported in previous works as well.

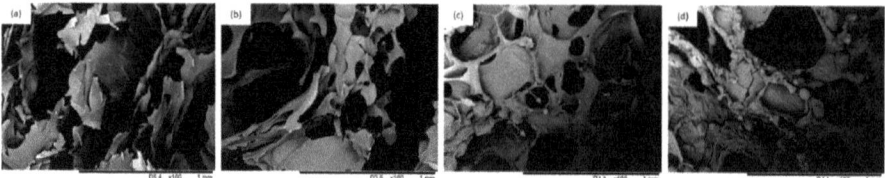

Figure 9. SEM images of char residues after the cone calorimeter test of (**a**) PUR_0, (**b**) PUR_BP_1, (**c**) PUR_BP_2, (**d**) PUR_BP_5.

4. Conclusions

Rigid polyurethane (PUR) foams were reinforced with sugar beet pulp (BP) impregnated with Aminopropylisobutyl-POSS. BP filler was incorporated into PUR at different percentages—1, 2, and 5 wt.%. The influence of BP filler on morphological features, mechanical properties, and thermal stability of PUR composite foams was investigated. The results showed that the greatest improvement in physico-mechanical properties was observed at a lower concentration of BP filler, such as 1 and 2 wt.%. For example, when compared with neat PUR foams, the addition of 2 wt.% of BP resulted in the formation of PUR composite foams with increased compressive strength (~12%), greater flexural strength (~12%), and better impact strength (~6%). Due to the good thermal stability of POSS-impregnated BP filler, the reinforced PUR composite foams were characterized by better thermal stability. Moreover, the addition of each amount of filler resulted in improvements in the fire resistance of PUR foams—the greatest improvement was observed for PUR composite foams with 2 wt.% of BP filler. The results presented in the current study indicate that the addition of a proper amount of POSS-impregnated BP filler may be an effective approach to the synthesis of PUR composite foams with improved physico-mechanical properties.

Author Contributions: Methodology, Investigation, Data Curation, Writing—Original Draft, Writing—Review and Editing, Visualization, S.C.; Methodology, Investigation A.S., A.K. All authors have read and agreed to the published version of the manuscript.

Funding: This research received no external funding.

Conflicts of Interest: The authors declare no conflict of interest.

References

1. Borowicz, M.; Paciorek-Sadowska, J.; Isbrandt, M. Synthesis and application of new bio-polyols based on mustard oil for the production of selected polyurethane materials. *Ind. Crop. Prod.* **2020**, *155*, 112831. [CrossRef]
2. Borowicz, M.; Joanna, P.-S.; Lubczak, J.; Czupryński, B. Biodegradable, Flame-Retardant, and Bio-Based Rigid Polyurethane/Polyisocyanurate Foams for Thermal Insulation Application. *Polymers* **2019**, *11*, 1816. [CrossRef]
3. Kairytė, A.; Kirpluks, M.; Ivdre, A.; Cabulis, U.; Vėjelis, S.; Balčiūnas, G. Paper waste sludge enhanced eco-efficient polyurethane foam composites: Physical-mechanical properties and microstructure. *Polym. Compos.* **2018**, *39*, 1852–1860. [CrossRef]
4. Kuranska, M.; Prociak, A. Porous polyurethane composites with natural fibres. *Compos. Sci. Technol.* **2012**, *72*, 299–304. [CrossRef]
5. Członka, S.; Bertino, M.F.; Strzelec, K. Rigid polyurethane foams reinforced with industrial potato protein. *Polym. Test.* **2018**, *68*, 135–145. [CrossRef]
6. Septevani, A.A.; Evans, D.A.; Annamalai, P.K.; Martin, D.J. The use of cellulose nanocrystals to enhance the thermal insulation properties and sustainability of rigid polyurethane foam. *Ind. Crop. Prod.* **2017**, *107*, 114–121. [CrossRef]

7. Kumar, A.P.; Depan, D.; Tomer, N.S.; Singh, R.P. Nanoscale particles for polymer degradation and stabilization—Trends and future perspectives. *Prog. Polym. Sci.* **2009**, *34*, 479–515. [CrossRef]
8. Mysiukiewicz, O.; Sałasińska, K.; Barczewski, M.; Szulc, J. The influence of oil content within lignocellulosic filler on thermal degradation kinetics and flammability of polylactide composites modified with linseed cake. *Polym. Compos.* **2020**, *41*, 4503–4513. [CrossRef]
9. Matykiewicz, D.; Barczewski, M. On the impact of flax fibers as an internal layer on the properties of basalt-epoxy composites modified with silanized basalt powder. *Compos. Commun.* **2020**, *20*, 100360. [CrossRef]
10. Cichosz, S.; Masek, A. Thermal Behavior of Green Cellulose-Filled Thermoplastic Elastomer Polymer Blends. *Molecules* **2020**, *25*, 1279. [CrossRef]
11. Cichosz, S.; Masek, A. Superiority of Cellulose Non-Solvent Chemical Modification over Solvent-Involving Treatment: Application in Polymer Composite (part II). *Materials* **2020**, *13*, 2901. [CrossRef]
12. Joanna, P.-S.; Czupryński, B.; Borowicz, M.; Liszkowska, J. Rigid polyurethane–polyisocyanurate foams modified with grain fraction of fly ashes. *J. Cell. Plast.* **2020**, *56*, 53–72. [CrossRef]
13. Członka, S.; Strąkowska, A.; Kairytė, A. Effect of walnut shells and silanized walnut shells on the mechanical and thermal properties of rigid polyurethane foams. *Polym. Test.* **2020**, *87*, 106534. [CrossRef]
14. Silva, M.C.; Takahashi, J.A.; Chaussy, D.; Belgacem, M.N.; Silva, G.G. Composites of rigid polyurethane foam and cellulose fiber residue. *J. Appl. Polym. Sci.* **2010**, *117*, 3665–3672. [CrossRef]
15. Hejna, A.; Kopczyńska, M.; Kozłowska, U.; Klein, M.; Kosmela, P.; Piszczyk, Ł. Foamed Polyurethane Composites with Different Types of Ash—Morphological, Mechanical and Thermal Behavior Assessments. *Cell. Polym.* **2016**, *35*, 287–308. [CrossRef]
16. Olszewski, A.; Kosmela, P.; Mielewczyk-Gryn, A.; Piszczyk, Ł. Bio-Based Polyurethane Composites and Hybrid Composites Containing a New Type of Bio-Polyol and Addition of Natural and Synthetic Fibers. *Materials* **2020**, *13*, 2028. [CrossRef]
17. Kurańska, M.; Barczewski, M.; Uram, K.; Lewandowski, K.; Prociak, A.; Michałowski, S. Basalt waste management in the production of highly effective porous polyurethane composites for thermal insulating applications. *Polym. Test.* **2019**, *76*, 90–100. [CrossRef]
18. Li, J.; Jiang, J.; Xu, J.; Xia, H.; Liu, P. Branched polyols based on oleic acid for production of polyurethane foams reinforced with bamboo fiber. *Iran. Polym. J.* **2016**, *25*, 811–822. [CrossRef]
19. Joanna, L.; Borowicz, M.; Joanna, P.-S.; Isbrandt, M.; Czupryński, B.; Moraczewski, K. Assessment of Photodegradation and Biodegradation of RPU/PIR Foams Modified by Natural Compounds of Plant Origin. *Polymers* **2019**, *12*, 33. [CrossRef]
20. Jonjaroen, V.; Ummartyotin, S.; Chittapun, S. Algal cellulose as a reinforcement in rigid polyurethane foam. *Algal Res.* **2020**, *51*, 102057. [CrossRef]
21. Altundoğan, H.S.; Bahar, N.; Mujde, B.; Tümen, F. The use of sulphuric acid-carbonization products of sugar beet pulp in Cr(VI) removal. *J. Hazard. Mater.* **2007**, *144*, 255–264. [CrossRef]
22. Vučurović, V.M.; Razmovski, R.N. Sugar beet pulp as support for Saccharomyces cerivisiae immobilization in bioethanol production. *Ind. Crop. Prod.* **2012**, *39*, 128–134. [CrossRef]
23. Rouilly, A.; Geneau-Sbartaï, C.; Rigal, L. Thermo-mechanical processing of sugar beet pulp. III. Study of extruded films improvement with various plasticizers and cross-linkers. *Bioresour. Technol.* **2009**, *100*, 3076–3081. [CrossRef]
24. Zheng, Z.-Q.; Wang, L.-J.; Li, D.; Huang, Z.-G.; Adhikari, B.; Chen, X.D. Mechanical and Thermal Properties of Polyurethane Foams from Liquefied Sugar Beet Pulp. *Int. J. Food Eng.* **2016**, *12*, 911–919. [CrossRef]
25. Jasiūnas, L.; McKenna, S.T.; Bridžiuvienė, D.; Miknius, L. Mechanical, Thermal Properties and Stability of Rigid Polyurethane Foams Produced with Crude-Glycerol Derived Biomass Biopolyols. *J. Polym. Environ.* **2020**, *28*, 1378–1389. [CrossRef]
26. Członka, S.; Strąkowska, A.; Pospiech, P.; Strzelec, K. Effects of Chemically Treated Eucalyptus Fibers on Mechanical, Thermal and Insulating Properties of Polyurethane Composite Foams. *Materials* **2020**, *13*, 1781. [CrossRef]
27. Cichosz, S.; Masek, A. Drying of the Natural Fibers as A Solvent-Free Way to Improve the Cellulose-Filled Polymer Composite Performance. *Polymers* **2020**, *12*, 484. [CrossRef]
28. Zakaria, S.; Hamzah, H.; Murshidi, J.A.; Deraman, M. Chemical modification on lignocellulosic polymeric oil palm empty fruit bunch for advanced material. *Adv. Polym. Technol.* **2001**, *20*, 289–295. [CrossRef]

29. Wolski, K.; Cichosz, S.; Masek, A. Surface hydrophobisation of lignocellulosic waste for the preparation of biothermoelastoplastic composites. *Eur. Polym. J.* **2019**, *118*, 481–491. [CrossRef]
30. Cichosz, S.; Masek, A. Cellulose Fibers Hydrophobization via a Hybrid Chemical Modification. *Polymers* **2019**, *11*, 1174. [CrossRef]
31. Barczewski, M.; Mysiukiewicz, O.; Lewandowski, K.; Nowak, D.; Matykiewicz, D.; Andrzejewski, J.; Skórczewska, K.; Piasecki, A. Effect of Basalt Powder Surface Treatments on Mechanical and Processing Properties of Polylactide-Based Composites. *Materials* **2020**, *13*, 5436. [CrossRef]
32. Barczewski, M.; Matykiewicz, D.; Szostak, M. The effect of two-step surface treatment by hydrogen peroxide and silanization of flax/cotton fabrics on epoxy-based laminates thermomechanical properties and structure. *J. Mater. Res. Technol.* **2020**, *9*, 13813–13824. [CrossRef]
33. Borysiak, S. Fundamental studies on lignocellulose/polypropylene composites: Effects of wood treatment on the transcrystalline morphology and mechanical properties. *J. Appl. Polym. Sci.* **2013**, *127*, 1309–1322. [CrossRef]
34. Neto, J.S.S.; Lima, R.A.A.; Cavalcanti, D.K.K.; Souza, J.P.B.; Aguiar, R.A.A.; Banea, M. Effect of chemical treatment on the thermal properties of hybrid natural fiber-reinforced composites. *J. Appl. Polym. Sci.* **2019**, *136*, 1–13. [CrossRef]
35. Kabir, M.M.; Wang, H.; Lau, K.T.; Cardona, F. Chemical treatments on plant-based natural fibre reinforced polymer composites: An overview. *Compos. Part B Eng.* **2012**, *43*, 2883–2892. [CrossRef]
36. Liu, Y.; Xie, J.; Wu, N.; Wang, L.; Ma, Y.; Tong, J. Influence of silane treatment on the mechanical, tribological and morphological properties of corn stalk fiber reinforced polymer composites. *Tribol. Int.* **2019**, *131*, 398–405. [CrossRef]
37. Liu, H.; Zheng, S. Polyurethane Networks Nanoreinforced by Polyhedral Oligomeric Silsesquioxane. *Macromol. Rapid Commun.* **2005**, *26*, 196–200. [CrossRef]
38. Hernández, R.; Weksler, J.; Padsalgikar, A.; Runt, J. Microstructural Organization of Three-Phase Polydimethylsiloxane-Based Segmented Polyurethanes. *Macromolecules* **2007**, *40*, 5441–5449. [CrossRef]
39. Liu, L.; Tian, M.; Zhang, W.; Zhang, L.; Mark, J.E. Crystallization and morphology study of polyhedral oligomeric silsesquioxane (POSS)/polysiloxane elastomer composites prepared by melt blending. *Polymer* **2007**, *48*, 3201–3212. [CrossRef]
40. Lee, Y.-J.; Huang, J.-M.; Kuo, S.-W.; Lu, J.-S.; Chang, F.-C. Polyimide and polyhedral oligomeric silsesquioxane nanocomposites for low-dielectric applications. *Polymer* **2005**, *46*, 173–181. [CrossRef]
41. Pellice, S.A.; Fasce, D.P.; Williams, R.J.J. Properties of epoxy networks derived from the reaction of diglycidyl ether of bisphenol A with polyhedral oligomeric silsesquioxanes bearing OH-functionalized organic substituents. *J. Polym. Sci. Part B Polym. Phys.* **2003**, *41*, 1451–1461. [CrossRef]
42. Matějka, L.; Strachota, A.; Pleštil, J.; Whelan, P.; Steinhart, M.; Šlouf, M. Epoxy Networks Reinforced with Polyhedral Oligomeric Silsesquioxanes (POSS). Structure and Morphology. *Macromolecules* **2004**, *37*, 9449–9456. [CrossRef]
43. Członka, S.; Strąkowska, A.; KAIRYTĖ, A. Application of Walnut Shells-Derived Biopolyol in the Synthesis of Rigid Polyurethane Foams. *Materials* **2020**, *13*, 2687. [CrossRef]
44. ISO. *ISO 2555—Plastics-Resins in the Liquid State or as Emulsions or Dispersions—Determination of Apparent Viscosity by the Brookfield Test Method*; International Organization of Standards: Geneva, Switzerland, 2018.
45. ISO. *ISO 845—Cellular Plastics and Rubbers—Determination of Apparent Density*; International Organization of Standards: Geneva, Switzerland, 2006.
46. ISO. ISO 4590:2016—Rigid Cellular Plastics—Determination of the Volume Percentage of Open Cells and of Closed Cells. Available online: https://www.iso.org/standard/60771.html (accessed on 11 November 2020).
47. ISO. *ISO 844—Preview Rigid Cellular Plastics—Determination of Compression Properties*; International Organization of Standards: Geneva, Switzerland, 2014.
48. ISO. *ISO 178-Plastics-Determination of Flexural Properties*; International Organization of Standards: Geneva, Switzerland, 2019.
49. ISO. *ISO 180—Plastics-Determination of Izod Impact Strength*; International Organization of Standards: Geneva, Switzerland, 2019.
50. ISO. ISO 5660-1:2015—Reaction-to-Fire Tests—Heat Release, Smoke Production and Mass Loss Rate—Part 1: Heat Release Rate (Cone Calorimeter Method) and Smoke Production Rate (Dynamic Measurement). Available online: https://www.iso.org/standard/57957.html (accessed on 11 November 2020).

51. Gómez-Fernández, S.; Ugarte, L.; Calvo-Correas, T.; Peña-Rodríguez, C.; Corcuera, M.A.; Eceiza, A. Properties of flexible polyurethane foams containing isocyanate functionalized kraft lignin. *Ind. Crop. Prod.* **2017**, *100*, 51–64. [CrossRef]
52. Lee, L.J.; Zeng, C.; Cao, X.; Han, X.; Shen, J.; Xu, G. Polymer nanocomposite foams. *Compos. Sci. Technol.* **2005**, *65*, 2344–2363. [CrossRef]
53. Sung, G.; Kim, J.H. Influence of filler surface characteristics on morphological, physical, acoustic properties of polyurethane composite foams filled with inorganic fillers. *Compos. Sci. Technol.* **2017**, *146*, 147–154. [CrossRef]
54. Liszkowska, J. The effect of ground coffee on the mechanical and application properties of rigid polyurethane-polyisocyanurate foams (Rapid communication). *Polimery* **2018**, *63*, 305–310. [CrossRef]
55. Kairytė, A.; Kremensas, A.; Balčiūnas, G.; Członka, S.; Strąkowska, A. Closed Cell Rigid Polyurethane Foams Based on Low Functionality Polyols: Research of Dimensional Stability and Standardised Performance Properties. *Materials* **2020**, *13*, 1438. [CrossRef]
56. Zieleniewska, M.; Leszczyński, M.K.; Kurańska, M.; Prociak, A.; Szczepkowski, L.; Krzyżowska, M.; Ryszkowska, J. Preparation and characterisation of rigid polyurethane foams using a rapeseed oil-based polyol. *Ind. Crops Prod.* **2015**, *74*, 887–897. [CrossRef]
57. Mizera, K.; Ryszkowska, J.; Kurańska, M.; Prociak, A. The effect of rapeseed oil-based polyols on the thermal and mechanical properties of ureaurethane elastomers. *Polym. Bull.* **2020**, *77*, 823–846. [CrossRef]
58. Federation of European Rigid Polyurethane Foam Associations. *Thermal Insulation Materials Made of Rigid Polyurethane Foam (PUR/PIR) Properties-Manufacture*; Report N°1; Federation of European Rigid Polyurethane Foam Associations: Brussels, Belgium, 2006.
59. Dolomanova, V.; Rauhe, J.C.M.; Jensen, L.R.; Pyrz, R.; Timmons, A.B. Mechanical properties and morphology of nano-reinforced rigid PU foam. *J. Cell. Plast.* **2011**, *47*, 81–93. [CrossRef]
60. Ciecierska, E.; Jurczyk-Kowalska, M.; Bazarnik, P.; Gloc, M.; Kulesza, M.; Kowalski, M.; Krauze, S.; Lewandowska, M. Flammability, mechanical properties and structure of rigid polyurethane foams with different types of carbon reinforcing materials. *Compos. Struct.* **2016**, *140*, 67–76. [CrossRef]
61. Gu, R.; Konar, S.K.; Sain, M. Preparation and Characterization of Sustainable Polyurethane Foams from Soybean Oils. *J. Am. Oil Chem. Soc.* **2012**, *89*, 2103–2111. [CrossRef]
62. Tian, H.; Wu, J.; Tian, H. Polyether polyol-based rigid polyurethane foams reinforced with soy protein fillers. *J. Vinyl Addit. Technol.* **2018**, *24*, E105–E111. [CrossRef]
63. Kurańska, M.; Polaczek, K.; Auguścik-Królikowska, M.; Prociak, A.; Ryszkowska, J. Open-cell rigid polyurethane bio-foams based on modified used cooking oil. *Polymer* **2020**, *190*, 122164. [CrossRef]
64. Luo, X.; Xiao, Y.; Wu, Q.; Zeng, J. Development of high-performance biodegradable rigid polyurethane foams using all bioresource-based polyols: Lignin and soy oil-derived polyols. *Int. J. Biol. Macromol.* **2018**, *115*, 786–791. [CrossRef]
65. Mahmood, N.; Yuan, Z.; Schmidt, J.; Xu, C. Preparation of bio-based rigid polyurethane foam using hydrolytically depolymerized Kraft lignin via direct replacement or oxypropylation. *Eur. Polym. J.* **2015**, *68*, 1–9. [CrossRef]
66. Chattopadhyay, D.K.; Webster, D.C. Thermal stability and flame retardancy of polyurethanes. *Prog. Polym. Sci.* **2009**, *34*, 1068–1133. [CrossRef]
67. Cheng, J.-J.; Wang, H.; Wang, X.; Li, S.; Zhou, Y.; Zhang, F.; Wang, Y.; Qu, W.; Wang, D.; Pang, X. Effects of flame-retardant ramie fiber on enhancing performance of the rigid polyurethane foams. *Polym. Adv. Technol.* **2019**, *30*, 3091–3098. [CrossRef]

Publisher's Note: MDPI stays neutral with regard to jurisdictional claims in published maps and institutional affiliations.

© 2020 by the authors. Licensee MDPI, Basel, Switzerland. This article is an open access article distributed under the terms and conditions of the Creative Commons Attribution (CC BY) license (http://creativecommons.org/licenses/by/4.0/).

Article

The Impact of Hemp Shives Impregnated with Selected Plant Oils on Mechanical, Thermal, and Insulating Properties of Polyurethane Composite Foams

Sylwia Członka [1,*], Anna Strąkowska [1] and Agnė Kairytė [2]

1. Institute of Polymer & Dye Technology, Lodz University of Technology, 90-924 Lodz, Poland; anna.strakowska@p.lodz.pl
2. Laboratory of Thermal Insulating Materials and Acoustics, Institute of Building Materials, Faculty of Civil Engineering, Vilnius Gediminas Technical University, Linkmenu st. 28, LT-08217 Vilnius, Lithuania; agne.kairyte@vgtu.lt
* Correspondence: sylwia.czlonka@dokt.p.lodz.pl

Received: 4 October 2020; Accepted: 20 October 2020; Published: 22 October 2020

Abstract: Polyurethane (PUR) foams reinforced with 2 wt.% hemp shives (HS) fillers were successfully synthesized. Three different types of HS fillers were evaluated—non-treated HS, HS impregnated with sunflower oil (SO) and HS impregnated with tung oil (TO). The impact of each type of HS fillers on cellular morphology, mechanical performances, thermal stability, and flame retardancy was evaluated. It has been shown that the addition of HS fillers improved the mechanical characteristics of PUR foams. Among all modified series, the greatest improvement was observed after the incorporation of non-treated HS filler—when compared with neat foams, the value of compressive strength increased by ~13%. Moreover, the incorporation of impregnated HS fillers resulted in the improvement of thermal stability and flame retardancy of PUR foams. For example, the addition of both types of impregnated HS fillers significantly decreased the value of heat peak release (pHRR), total smoke release (TSR), and limiting oxygen index (LOI). Moreover, the PUR foams containing impregnated fillers were characterized by improved hydrophobicity and limited water uptake. The obtained results confirmed that the modification of PUR foams with non-treated and impregnated HS fillers may be a successful approach in producing polymeric composites with improved properties.

Keywords: polyurethanes; hemp shives; bio-filler; oil impregnation; mechanical properties

1. Introduction

Recently, the synthesis and development of polyurethane (PUR) composites containing natural fillers have attracted increased attention in industry and academia [1–3]. The application of natural fillers as reinforcing materials in the production of PUR foams has both ecological and economic advantages. Among the natural materials, cellulosic compounds have significant advantages, mostly due to their low density, high stiffness, biodegradability, unlimited availability, and low price. Previous studies have shown that the incorporation of organic and inorganic materials into the polymer matrix may successfully improve the mechanical characteristics of PUR composites [4–6] (Table 1). For example, the basalt waste has been used as a reinforcing filler for the production of rigid PUR foams by Kurańska et al. [7]. Due to the incorporation of 3–40 wt.% of the powdered basalt filler, the resulting PUR composite foams were characterized by improved mechanical performances. Similar results have been reported by Paciorek-Sadowska et al. [8] in the case of PUR composite foams containing 30–60 wt.% rapeseed filler. PUR composites with increased apparent density and enhanced mechanical properties were produced. The improvement of mechanical and thermal performances was also observed after the incorporation of egg-shells [9]. The addition of 20 wt.% egg-shells resulted in a significant improvement of the abovementioned properties. Interesting results were presented by Olszewski et al. [3] in the case of PUR foams containing glass and sisal fibers—the flexural strength, impact strength, and hardness of materials have been improved by the addition of both kinds of fibers. The effect of waste sludge particles on the physical and mechanical properties of PUR foams was studied by Kairyte et al. [10]. The authors reported that the addition of 20 wt.% of the fillers results in the production of the PUR materials with improved characteristics, while the higher content of the filler slightly deteriorates the properties of the foams. Interesting results were presented by de Avila Delucis et al. [11] who synthesized PUR foams reinforced with different ratios (1, 5, and 10 wt.%) of forest-derivatives fillers, e.g., bark, pine trees needles, kraft lignin, and paper sludge. Among the modified samples, the most promising materials were PUR foams reinforced with 1 and 5 wt.% wood, which exhibited improved mechanical and hygroscopic performance.

Table 1. Recent works on filler reinforced polyurethane foams—effect of different fillers on mechanical properties of polyurethane foams.

Filler Used	Percentage of Filler	Results
Kenaf fibre	20–50 wt.%	Improvement of mechanical properties [12]
Pulp fibre	0–5 wt.%	Deterioration of mechanical properties, improvement of thermal stability [13]
Rice husk ash	0–5 wt.%	Improvement of mechanical properties and flame-retardancy, deterioration of thermal conductivity [14]
Cellulose microfibres	0–2 wt.%	Improvement of mechanical properties [15]
Cellulose nanocrystals	1–8 wt.%	Improvement of mechanical properties [16]
Egg shell waste	20 wt.%	Improvement of mechanical properties and thermal stability [17]
Potato protein waste	0.1–5 wt.%	Deterioration of mechanical properties and thermal stability with increasing filler content [18]
Buffing dust waste	0.1–5 wt.%	Deterioration of mechanical properties and thermal stability with increasing filler content [19]
Keratin feathers	0.1–1.5 wt.%	Mechanical properties and thermal stability decrease with increasing filler content [20]
Forest based wastes	10 wt.%	Deterioration of mechanical properties and thermal conductivity, improvement of flame-retardancy [11]
Ground coffee	2.5–15 wt.%	No significant influence on the mechanical and thermal properties, reduced brittleness and aging process [21]
Jute fibre	0.5–4 wt.%	Deterioration of mechanical properties [22]
Ramie fiber	0.2–0.8 wt.%	Improvement of mechanical properties, thermal stability, and flame-retardancy [23]
Rapeseed cake	30–60 wt.%	Improvement of mechanical properties, thermal stability, and flame-retardancy [8]
Wood flour	0–15 wt.%	Deterioration of mechanical properties, improvement of thermal conductivity and thermal stability [24]
Coir fibre	2.5 wt.%	Improvement of mechanical properties [25]
Fly ashes	5–35 wt.%	Improvement of mechanical properties and fire resistance [26]
Cinnamon extract, green coffee extract, cocoa extract	10 wt.%	Improvement of susceptibility to biodegradation [27]
Soy protein	2.4–9.6 wt.%	Improvement of mechanical properties, deterioration of thermal stability [28]

Among different organic fillers, the chemical composition of hemp shives have great potential as sustainable reinforcements for novel polyurethane composite foams. The basic unit of hemp shives is composed of cellulose microfibrils, which are combined by an interphase mixture of different pectins, hemicellulose, and other low-molecular polysaccharides [29]. The hydrogen bonds between different chemical components provide stiffness and mechanical strength of hemp shives. For example, hemicellulose determines the thermal degradation and moisture absorption, while the lignin content determines the UV degradation of the hemp shives [30,31]. Hemp shives offer several advantages, such as sufficient reactive functional groups, high carbon content, compatibility with diverse industrial chemicals, good stability and mechanical properties due to the presence of aromatic rings, and good rheological and viscoelastic properties, making it a potential candidate to be used as reinforcing material in polymer composites.

Nevertheless, the application of cellulosic materials as reinforcement of polymeric composite materials presents some limitations. As for mechanical strength, chemical or physical treatment of cellulose surface may improve the mechanical and thermal properties of polymer composites [32–34]. In previous works, a chemical modification, such as acetylation [35], alkalization [36], benzoylation [37] of cellulosic compounds, has been reported. For example, Du et al. have reported an improvement of interfacial compatibility between polyimide matrix and wood fibers treated by 3-Aminopropyltriethoxysilane [38]. Such reinforced composites were characterized by improved abrasive and tensile properties. Alkali-treated coir fibers were developed by Valášek et al. [39]. Due to the improved interphase adhesive, epoxy composites reinforced with alkali-treated coir fibers exhibited improved mechanical properties. Similar results have been shown in the case of epoxy composites reinforced with palm fibers chemically treated with sodium hydroxide [40]. Improvement of wear characteristics and mechanical performances of composites was observed due to the addition of the fibers. Chemical treatments of oil palm fibers, such as latex coating, acetylation, or acrylonitrile grafting have been evaluated by Sreekala et al. [41]. The authors have shown that phenol formaldehyde composites containing modified fibers were characterized by better flexural characteristics and improved impact resistance.

Many previous works have studied the impact of natural fillers on the mechanical and thermal characteristics of polymeric composites; however, no studies have been devoted to the examination of the polyurethane foam composites reinforced with physically-treated hemp shives. Keeping in view the advantageous properties of hemp shives, it seems logical to use hemp shives as a reinforcing filler for new bio-based polyurethane composite foams. The preparation of novel materials from hemp shives products may improve the mechanical properties of the polyurethane materials as well as possibly solve the problem of their waste disposal. Therefore, the impact of hemp shives impregnated with sunflower oil and tung oil on morphological, mechanical, and thermal properties of polyurethane foam composites was examined.

2. Materials and Methods

2.1. Materials

PUR foams were synthesized using polyether polyol (Stapanpol PS-2352) and polymeric diphenylmethane diisocyanate (Purocyn B). As catalysts, Kosmos 75 and Kosmos 33 (potassium octoate and potassium acetate, respectively) were used. Silicone surfactant (Tegostab B8513) was used for stabilizing the foam's structure and the mixture of pentane and cyclopentane (50:50 $v/v\%$) was selected as a blowing agent in forming cellular structure. Hemp shives, sunflower oil, and tung oil were obtained from a local company.

2.2. Methods

2.2.1. Impregnation of Hemp Shives (HS) with Sunflower Oil and Tung Oil

Hemp shives were milled and wetted with selected oil (sunflower/tung oil). The mixture was thoroughly mixed and poured into cups. Subsequently, the cups with the mixture were put into the vacuuming dish and the vacuuming process proceeded until 0.01 MPa of pressure was achieved. Then, the green handle of the vacuum dish was screwed and the vacuum was left for another 30 min. A total of 10 cycles were done for the mixtures and, after that, all mixtures were thermally treated at 70 °C for 24 h. After the thermal treatment, the mixtures were left to cool down at 23 ± 5 °C temperature and 50 ± 5% humidity conditions.

2.2.2. Synthesis of PUR Foams

PUR foams were produced by a one-shot method according to the procedure reported in the previous works. In brief, the synthesis of PUR foams modified with the addition of HS was as follows (Figure 1): To form a polyol premix, the calculated amounts of polyol (Stepanpol), catalysts (Kosmos 75 and Kosmos 33), blowing agent (the mixture of pentane and cyclopentane), and surfactant (Tegostab) were placed in a plastic cup and intensively mixed at 1500 rpm by a mechanical stirrer for 60 s. Then, the previously impregnated HS fillers were added to the cup and mixed for another 60 s to form a homogenous dispersion. A calculated amount of isocyanate (Purocyn) was added to the reaction mixture and thoroughly mixed for 10 s. The free rise PUR composite foam was left at room temperature for 24 h to provide complete curing of composites. A schematic procedure for the synthesis of PUR foams is shown in Figure 2. PUR foams were synthesized following the formulations presented in Table 2.

Figure 1. Hemp shives (HS) fillers used as reinforcing fillers: (**a**) non-treated hemp shives (HS), (**b**) hemp shives impregnated with sunflower oil (HS/SO), and (**c**) hemp shives impregnated with tung oil (HS/TO).

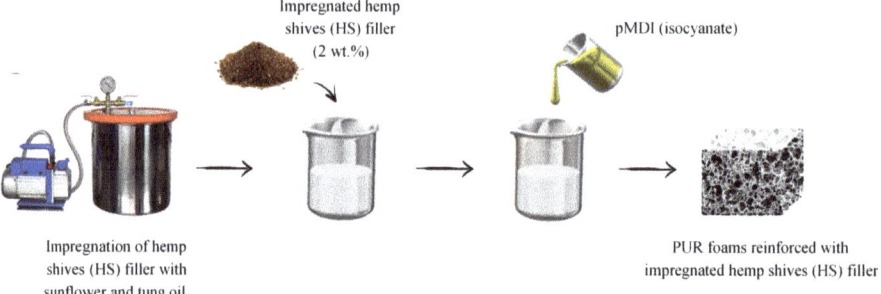

Figure 2. Synthesis of PUR foams reinforced with HS fillers.

Table 2. Composition of PUR composite foams.

Component	PUR_0	PUR_HS	PUR_HS/SO	PUR_HS/TO
	Parts by Weight (wt.%)			
STEPANPOL PS-2352	100	100	100	100
PUROCYN B	160	160	160	160
Kosmos 75	6	6	6	6
Kosmos 33	0.8	0.8	0.8	0.8
Tegostab B8513	2.5	2.5	2.5	2.5
Water	0.5	0.5	0.5	0.5
Pentane/cyclopentane	11	11	11	11
Hemp shives (HS)	0	2	0	0
Hemp shives/sunflower oil (HS/SO)	0	0	2	0
Hemp shives/tung oil (HS/TO)	0	0	0	2

2.2.3. Sample Characterization

The dynamic viscosity of polyol premixes was examined following ISO 2555 [42] using Viscometer DVII+. (Viscometer DVII+, Brookfield, Berlin, Germany). The cellular structure of PUR foams was evaluated using a scanning electron microscope using JSM-5500 LV (JEOL JSM 5500 LV, JEOL Ltd., Peabody, MA, USA). The cell sizes of PUR foams was determined by ImageJ software (Java 1.8.0, Media Cybernetics Inc., Rockville, MD, USA). The average pore diameters, pore size distribution, and the closed-cell content were identified based on SEM micrographs using the binarization threshold—an average of 400 individual measurements was reported. The apparent density of PUR foams was calculated as the ratio between the weight and volume of the samples according to ISO 845. The number of closed-cells was evaluated according to the ISO 4590 standard. Thermal conductivity (λ) of PUR foams was measured at 25 °C by using LaserComp 50. The mechanical performances of PUR foams were performed using a Zwick Z100 Testing Machine (Zwick/Roell Group, Germany). Compressive strength was examined parallel to the foam rise direction according to the ISO 844 [43] standard. Flexural and impact strength of PUR foams were evaluated according to the ISO 178 [44] and ISO 180 [45] standards. The dynamic–mechanical characteristic (DMA) was performed using an ARES rheometer (ARES, TA Instruments, New Castle, DE, USA) under the selected parameters (applied deformation of 0.1% and a frequency of 1 Hz, temperature range of 0–250 °C). The thermogravimetric analysis (TGA) test was performed in the function of temperature (0–600 °C) using an STA 449 F1 Jupiter Analyzer (Netzsch Group, Selb, Germany). The fire behavior of PUR foams was evaluated using a cone calorimeter apparatus according to ISO 5660 in S.Z.T.K. "TAPS"—Maciej Kowalski Company (Lodz, Poland).

3. Results and Discussion

3.1. Topography and an Average Size of HS Fillers

The external morphology of HS, and impregnated HS fillers (HS/SO and HS/TO), is presented in Figure 3. Comparing non-treated HS and impregnated HS fillers, it is clear that the oil impregnation affects an external morphology of the filler. After the impregnation with sunflower and tung oils, the fillers possess a similar structure; however, the particles tend to agglomerate, forming the bigger clusters of filler particles. The size of HS particles ranges from 400 to 800 nm, while, after the impregnation with sunflower oil, the average size of particles increases and it ranges from 3 to 5 μm. A similar relationship is observed for HS impregnated with tung oil—the average diameter ranges from 3 nm to 6 μm. Moreover, the addition of HS fillers increases the viscosity of the PUR systems (Table 3). Among all modified systems, the greatest viscosity is observed for PUR systems containing impregnated HS fillers. The viscosity increases rapidly from 840 (for neat PUR system) to 1800 and 2200 mPa·s after the addition of HS fillers impregnated with sunflower and tung oil, respectively. This result is not surprising, considering a high tendency of the filler to agglomeration and formation of coarse domains, as presented in Figure 3.

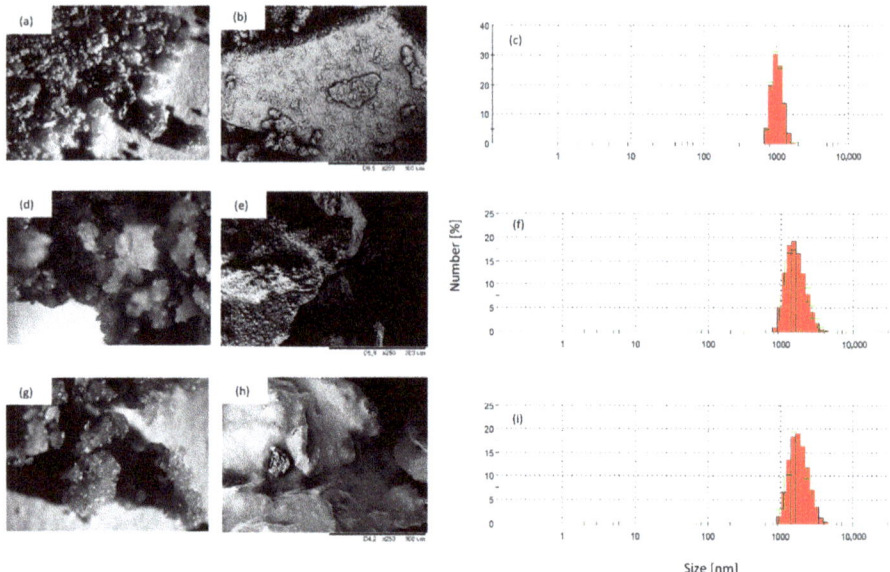

Figure 3. Topography and particle size of (a–c)—HS filler, (d–f)—HS/SO filler and (g–i)—HS/TO filler.

Table 3. The impact of HS fillers on viscosity and processing times of PUR systems.

	Dynamic Viscosity η (mPa·s)			Processing Times (s)		
	0.5 RPM	50 RPM	100 RPM	Cream Time (s)	Free Rise Time (s)	Tack-Free Time (s)
PUR	850 ± 10	410 ± 8	330 ± 9	42 ± 4	280 ± 9	350 ± 12
PUR_HS	1100 ± 10	980 ± 10	420 ± 12	50 ± 2	320 ± 9	345 ± 10
PUR_HS/SO	1800 ± 11	1300 ± 10	750 ± 10	60 ± 1	355 ± 8	330 ± 8
PUR_HS/TO	2200 ± 11	1550 ± 10	850 ± 10	66 ± 2	370 ± 8	320 ± 8

3.2. Reactivity of PUR Foam Formulations

The reactivity of PUR systems was investigated by measuring the start time, free rise time, and tack-free time during the foaming process (Table 3). Incorporation of HS, HS/SO, and HS/TO decreased the reactivity of the PUR systems. When compared to neat PUR_0, the foaming reaction rate was lower as HS fillers were introduced, probably due to the steric hindrance effects of hydroxyl groups (-OH) of HS filler. On the other hand, the presence of hydroxyl groups of HS filler can affect the proper stoichiometry of PUR synthesis due to the reaction between hydroxyl groups of HS filler and isocyanate groups. Consequently, the higher number of isocyanate groups is consumed and the reduced amount of carbon dioxide is produced, slowing down the foaming behavior of PUR systems. The extended processing times may be also connected with increased viscosity of the PUR dispersion containing HS fillers, which affect the expansion of PUR systems, extending the free rise time of PUR foams. Comparing PUR systems containing HS fillers, the highest values of processing times are observed for PUR_HS/TO, which may be connected with the higher viscosity of PUR systems containing HS/TO and the presence of bigger aggregates of HS particles. Similar results have been also found in previous works [46].

3.3. Morphology, Apparent Density and Thermal Conductivity of PUR Foams

Figure 4 presents the SEM images of PUR foams containing HS fillers without and with oil impregnation. The morphology of neat PUR_0 is smooth and regular. With the incorporation of HS fillers, the cellular shape is more irregular with the formation of a higher number of open cells. It can be seen that, with the incorporation of impregnated HS/SO and HS/TO, some agglomerates of the fillers are visible in the cell struts. The alteration in cellular morphology may be connected with increased viscosity of PUR systems containing HS fillers. As a result, the formation and expansion of air bubbles are hindered, which results in the creation of a more heterogeneous structure of PUR foams. When compared to neat PUR_0, the overall shape of cells becomes more irregular after the incorporation of HS fillers. All series of modified PUR foams possess a poor structure with a higher number of open cells. This effect is more prominent in the case of PUR foams containing both types of impregnated HS fillers. Similar dependence was also observed in previous works and was connected with the attachment of the filler particles to the cell walls leading to the rupturing and collapsing of foam's cells and ultimately to the weakening of the modified foam's structure [47]. The addition of HS fillers affects average cell size and this effect is more prominent after the incorporation of impregnated HS fillers (Figure 4). In general, neat PUR_0 possess an average cell size of 450 μm. The addition of HS fillers results in the formation of more inhomogeneous PUR foams with an average size in the range of 380–620 μm for each series of modified foams. Moreover, the addition of HS fillers decreases the content of closed-cell—the value decreases from 91.4% (for neat PUR_0) to 89.2, 88.6, and 85.6% for PUR_HS, PUR_HS/SO, and PUR_HS/TO, respectively. Previous studies have shown that the opening of cells due to the incorporation of organic filler may be connected with poor interphase adhesion between polyurethane matrix and filler surface, which results in disruption of the foaming process and more defective morphology of modified PUR foams [48]. Such an explanation may be found in our study as well. According to SEM results, some particles of HS fillers are localized in empty pores and they are not completely built in the foams' struts. This confirms the poor compatibility between the filler and PUR matrix, leading to the cell collapsing and formation of PUR foams with open-pore structure [47]. Moreover, with the addition of HS fillers, the gas may form additional nucleation sites on the surface of HS filler particles, providing heterogeneous centers for the formation of air bubbles, which in turn increase the cell number of the PUR foam structure. As reported previously, the viscosity of PUR systems containing HS fillers is increased, and the expansion of the cells is reduced. Because of this, the HS fillers react with isocyanate groups, forming an interpenetrating cross-linked network, which in turn disturbs the gas release, reducing the size of cells.

The cellular structure affects the apparent density of PUR foams (Table 4). Due to the incorporation of HS, HS/SO, and HS/TO, the value of apparent density increases by 6, 16, and 17%, respectively. A greater apparent density of PUR foams containing HS fillers should be attributed to the increased viscosity and limited expansion of modified PUR systems. Moreover, the apparent density of modified PUR foams is further enhanced by the molecular weight of HS fillers.

Table 4. Selected properties of PUR foams containing HS fillers.

Sample	Closed-Cell Content (%)	APPARENT Density (kg m^{-3})	Thermal Conductivity (W m^{-1} K^{-1})	Water Uptake (%)	Contact Angle (°)
PUR_0	91.4 ± 0.5	37.2 ± 0.6	0.025 ± 0.001	21.5 ± 0.6	123 ± 1
PUR_HS	89.2 ± 0.4	40.6 ± 0.7	0.026 ± 0.001	23.8 ± 0.5	120 ± 1
PUR_HS/SO	88.6 ± 0.4	43.1 ± 0.6	0.030 ± 0.001	19.2 ± 0.6	129 ± 1
PUR_HS/TO	85.6 ± 0.4	43.5 ± 0.6	0.031 ± 0.001	20.1 ± 0.5	130 ± 1

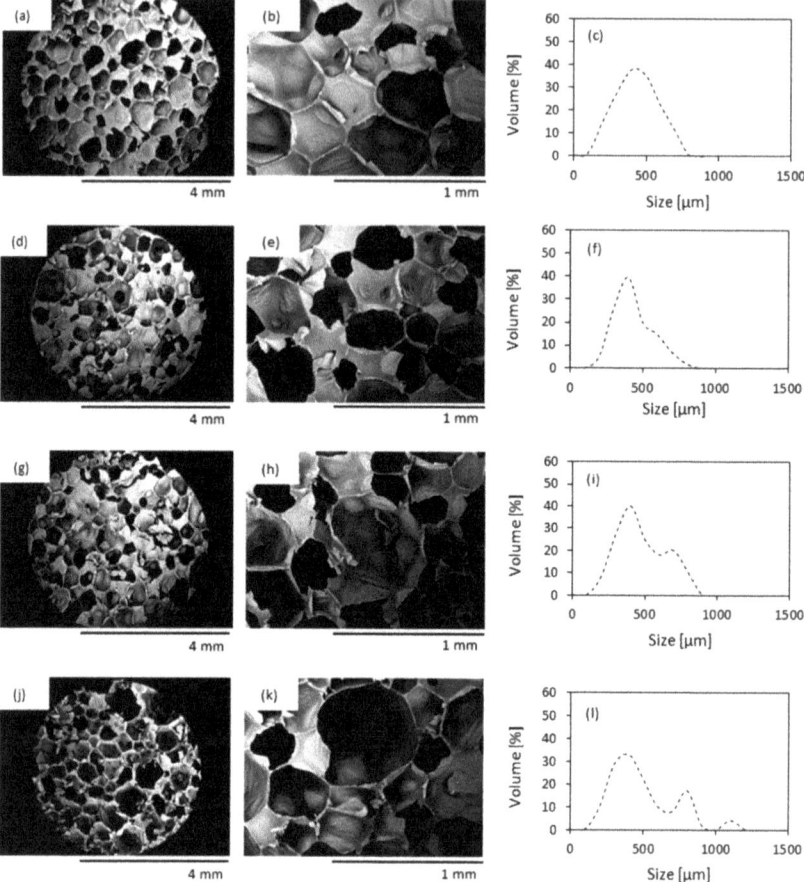

Figure 4. Optical image, SEM image and cell size distribution of (**a–c**) PUR_0, (**d–f**) PUR_HS, (**g–i**) PUR_HS/SO and (**j–l**) PUR_HS/TO.

Thermal conductivity (λ) is an important parameter that defines the thermal insulation properties of PUR foams [10]. The value of λ measured for neat PUR_0 is 0.025 W m^{-1} K^{-1} (Table 4). The addition of HS filler has no significant influence on the value of λ; however, the incorporation of impregnated HS/SO and HS/TO increases the value of λ by about 20 and 24%, respectively. In general, the value of λ involves the thermal conductivity of the gas captured in the foam cells (λ_{gas}), solid backbone of the foams (λ_{solid}), heat transfer between foam cells (λ_{solid}), and gas convection ($\lambda_{convection}$). With the addition of HS fillers, a greater number of filler particles are built in the polymer matrix, thus the value of λ_{solid} increases. As mentioned previously, due to the increased viscosity of modified PUR foams, the functional groups of filler particles are involved in the reaction with isocyanate groups. This results in an increased crosslinking degree (Figure 5) of PUR molecular chains and formation of PUR foams with smaller cells and a greater apparent density, which additionally increases the value of λ_{solid}. Besides this, all series of PUR foams are in line with commercial requirements for commercial thermal insulation boards [49].

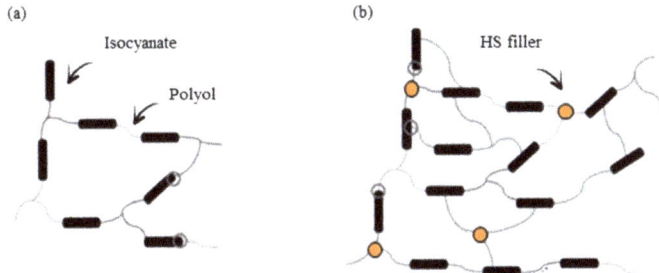

Figure 5. Crosslinking structure of (**a**) neat PUR foams and (**b**) PUR foams containing HS fillers.

Another important parameter determining the further use of PUR foams as construction materials is water uptake. Previous studies reported that the cellular structure of PUR foams and the hydrophobic nature of the filler affects the water uptake of porous materials [11,50,51]. The results of the water uptake of PUR foams containing HS fillers are presented in Table 4. When compared with neat PUR_0, the addition of HS filler increases the water uptake of PUR_HS—the value increases from 21.5 to 23.8%. Increased absorption of water may be connected with a less uniform structure of PUR_HS and a greater number of open cells. Moreover, the particles of HS filler tend to agglomerate, creating a "pathway" that facilitates water penetration into the foam structure [11,50–52]. Surprisingly, an opposite effect has been observed for PUR foams containing impregnated HS/SO and HS/TO—the water uptake slightly decreases by ~6% due to the hydrophobic character of impregnated RC fillers, which was also confirmed by a decreased value of the contact angle (θ) (Table 4).

3.4. Mechanical Characteristics of PUR Foams

The mechanical properties of porous materials mostly depend on their cellular structure [11]. The compressive strength–strain graph measured during the external loading is presented in Figure 6a. All samples are characterized by analog plots—the linear region refers to the elastic response of PUR foams, while the second region (which presents a plateau region) refers to the plastic deformation and cell's rupture. When compared with neat PUR_0, in the case of foams containing HS fillers, the transition from elastic to plateau region is more abrupt, while the elongation at break decreases, indicating the more rigid structure of modified foams. A similar trend has been found in previous studies as well [53,54].

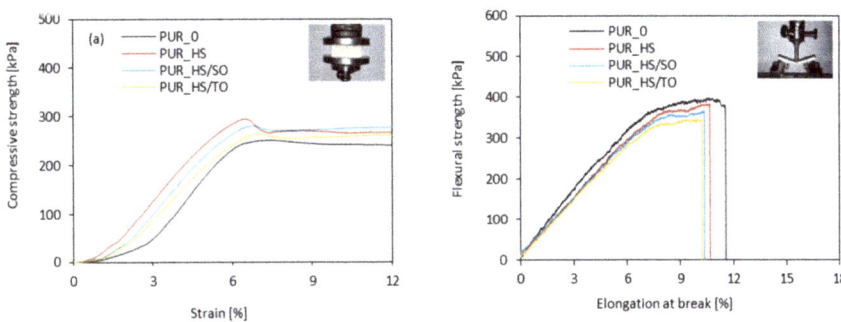

Figure 6. Mechanical behavior of PUR foams' (**a**) compressive and (**b**) flexural behavior.

Depending on the type of HS filler, the compression strength increases by ~12, ~6, and ~8% for PUR_HS, PUR_HS/SO, and PUR_HS/TO, respectively (Figure 7a). Previous studies have shown that the value of compression strength depends on the crosslinking density of PUR foams, which refers to the content of hard segments (urethane groups) [55,56]. It may be concluded that the incorporation of HS fillers affects the crosslinking density of PUR foams. After the incorporation of cellulosic HS fillers, active groups of HS (i.e., hydroxyl groups of cellulose and lignin) are involved in the reaction with isocyanate groups, increasing the number of urethane groups and creating a more dense structure of PUR foams. An increased number of urethane groups provides additional crosslinking points, increasing the number of hard segments and enhancing the mechanical performances of PUR foams.

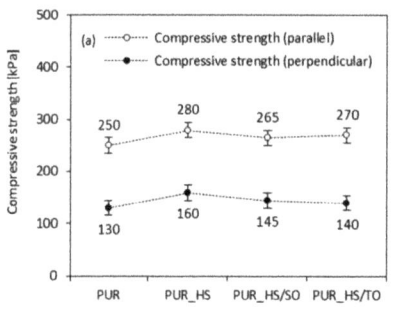 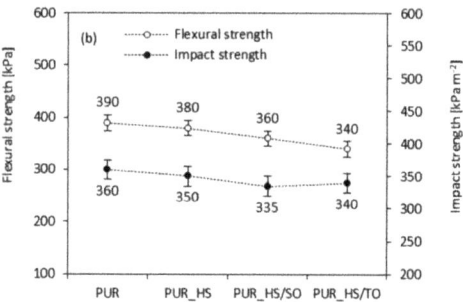

Figure 7. Mechanical characteristics of PUR foams containing HS fillers: (**a**) compressive strength, (**b**) flexural and impact strength.

A different trend is observed in the case of measuring the flexural and impact strength (Figure 7b). The addition of HS fillers affects both parameters, decreasing their values. Depending on the HS filler type, the value of flexural strength decreases by 3–13%, while the value of impact strength decreases by 3–7%. The higher content of hard segments makes the PUR foams harder but more brittle. Thus, the values of flexural and impact strength decrease when the HS fillers are added. Besides, some aggregates of HS particles, which are localized in the PUR structure, can act as stress concentrations, promoting the cracking of the samples and deteriorating the mechanical performances of PUR foams. As presented on the stress–elongation graph (Figure 6b), all modified PUR foams exhibit a similar mechanical performance, which involves elastic and plastic deformation; however, the value of elongation at break decreases when the HS fillers are added. The aggregates of HS particles act as additional defects, promoting the cracking of the sample under an external force.

3.5. Dynamic–Mechanical Properties of PUR Foams

The dynamic–mechanical properties of PUR foams are presented in Figure 8. The modification of PUR foams with HS fillers affects the glass transition temperature (T_g), determined as a maximum peak of the curve loss tangent (tanδ) in the function of the temperature. When compared with neat PUR_0, the addition of HS fillers increases the value of T_g. Among all series of PUR foams, the highest value of T_g is observed for PUR foams containing impregnated HS fillers—the value of T_g increases from 128 to 148 °C for PUR_HS/SO and PUR_HS/TO. The results confirm that the addition of impregnated HS fillers results in the formation of PUR foams with a greater cross-linking density, limiting the mobility of polymer chains. Besides, HS filler particles that are built into the foam structure can act as additional blockages, increasing the amount of energy that is required to achieve the T_g. Moreover, when compared to neat PUR_0, the addition of each type of HS filler results in the improvement of the storage modulus of PUR foams. The greatest improvement is observed for PUR foams containing impregnated HS fillers. This may be connected with an increased viscosity of PUR systems containing HS/SO and HS/TO, which limits the mobility of polymer chains, leading to the formation of PUR foams with higher stiffness. In addition, the filler particles can act as reinforcing centers, effectively transferring the external stresses from particles to the PUR matrix. Similar results have been reported in previous works as well [57,58].

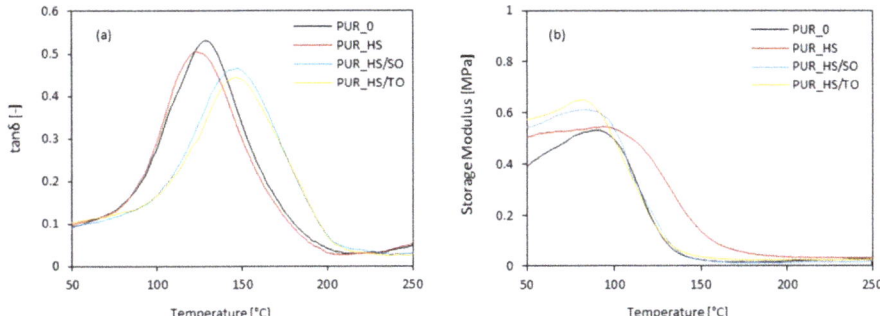

Figure 8. The dynamic–mechanical properties of PUR foams containing HS fillers: (**a**)—tanδ and (**b**)—storage modulus results.

3.6. Thermogravimetric Analysis (TGA) of PUR Foams

Figure 9a,b present the TG/DTG curves of HS fillers. The first loss of mass occurs at ~100 °C and refers to the release of water accommodated in the fillers. A second loss occurs in the range of 300–400 °C and is mostly connected with the decomposition of the cellulosic derivatives—cellulose and lignin, respectively [34,59]. Similar TG curves have been observed in the case of other cellulosic fillers [60].

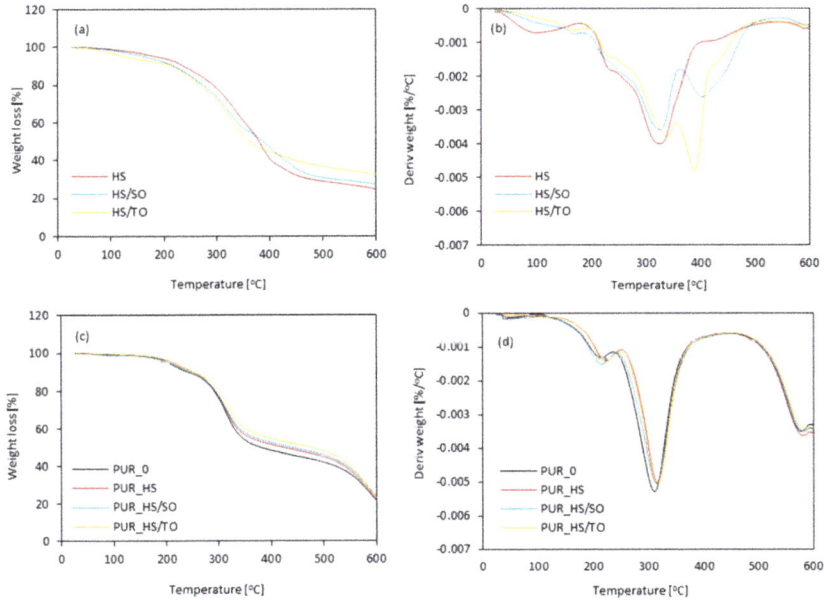

Figure 9. TG and DTG results obtained for (**a**,**b**) HS fillers, and (**c**,**d**) PUR foams.

Figure 9c,d present TG and DTG curves of PUR foams. The resulting data are depicted in Table 5. Only slight shifts in the level of each curve are observed, which correspond to the weight loss. However, a meaningful change in the position of the main peak connected with the decomposition temperature of PUR can be observed. When compared with neat PUR_0, the addition of HS fillers results in the production of PUR foams with higher values of $T_{2\%}$, which refers to the release of volatile products presented in the HS fillers. Previous studies have reported that cellulosic fillers tend to degrade in the range of lower temperatures, thus a higher value of $T_{2\%}$ should be attributed to the partial crosslinking between functional groups of HS fillers and isocyanate groups [61]. The second degradation of mass loss occurs in the range of 309–326 °C and refers to the thermal decomposition of hard segments—urethane bonds [62,63]. A slower degradation of mass is observed when HS fillers are added and confirms a higher crosslinking degree of modified PUR foams. The third stage, which occurs at nearly 590 °C, refers to the degradation of cellulose and lignin. The thermal degradation mechanism of PUR foams modified with HS fillers seems similar to PUR_0. The addition of HS fillers results in a slight increase in mass loss of PUR foams, because of the presence of lignocellulosic compounds and incomplete miscibility of soft and hard segments of PUR [64]. Moreover, the rigid structure of cellulosic HS fillers can cause them to act as additional cross-linker centers between PUR matrices, improving the heat resistance of modified PUR foams—the value of mass residue at 600 °C increases from 23.8% (for PUR_0) to 26.2 and 26.5% for PUR_HS/SO and PUR_HS/TO, respectively. An analog trend has been shown previously in the case of PUR foams containing another type of cellulosic filler [65,66]. For example, Tian et al. [61] have stated that the reduced thermal decomposition PUR foams enhanced with soy-protein filler refer to the higher degree of crosslinking of PUR foams, due to the reaction between functional groups of soy-protein and PUR systems. More cross-linked structure of PUR foams limits the number of volatile products that are generated and released during the thermal degradation process, reducing the thermal decomposition of PUR foams. A similar explanation may be found in our study as well.

Table 5. The results of TGA and DTG analysis of PUR foams.

Sample	T_{max} (°C)			Residue at 600 °C (wt.%)
	1st Stage	2nd Stage	3rd Stage	
PUR_0	218	309	581	23.8
PUR_HS	217	318	584	23.1
PUR_HS/SO	220	325	586	26.2
PUR_HS/TO	222	326	589	26.5

3.7. Flammability of PUR Foams

The addition of RC fillers significantly affects the flame retardancy of PUR foams. The results of the cone calorimeter test are presented in Figure 10 and Table 6.

When compared to neat PUR_0, the addition of HS fillers increases the ignition time (IT) and this trend is more prominent in the case of PUR foams containing impregnated HS/SO and HS/TO fillers—the value of IT increases from 3 s to 5 and 6 s, respectively. The results presented in Figure 10a indicate that the value of heat peak release (pHRR) increases by ~8% when the non-treated HS filler is added. An opposite effect is observed for PUR foams containing impregnated RC fillers—the value of pHRR decreases by ~30 and 20% for PUR foams containing HS/SO and HS/TO, respectively. Such an improvement may be connected with the formation of a char layer, which effectively reduces the release of combustible gases. In general, the pHRR values of modified PUR foams are in the line with the regulations, which determines the accessible value of pHRR as 300 kW m^{-2} [67]. Below this value, the materials are approved for use as insulating materials for building construction. Moreover, the addition of impregnated HS fillers decreases the value of total smoke release (TSR) (Figure 10b). Compared to neat PUR_0, the value of TSR decreases by 15 and 27% for PUR_HS/SO and PUR_HS/TO, respectively. Moreover, due to the incorporation of HS fillers, the values of COY and CO_2Y slightly decrease (Figure 10c,d). This may be connected with the fact that HS filler particles act as a physical barrier for flame spread, reducing the heat transfer through the PUR sample and decreasing the intensity of the flame.

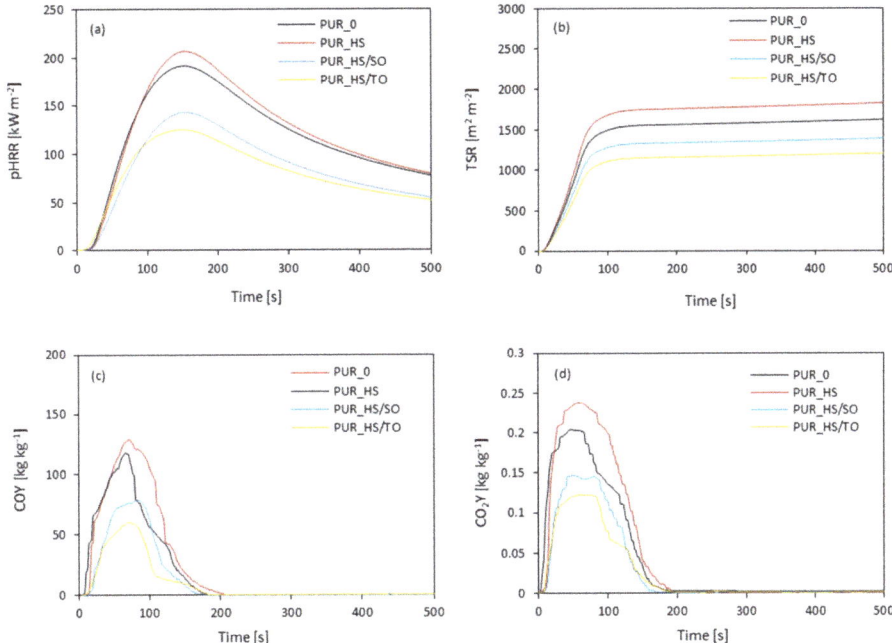

Figure 10. Cone calorimeter results: (a) pHRR, (b) TSR, (c) COY and (d) CO_2Y.

Table 6. Flammability parameter measured for PUR foams.

	IT [s]	pHRR [kW m^{-2}]	THR [MJ m^{-2}]	TSR [m^2 m^{-2}]	COY [kg kg^{-1}]	CO_2Y [kg kg^{-1}]
PUR_0	4	268	21.0	1550	0.204	0.204
PUR_HS	3	288	22.5	1740	0.238	0.238
PUR_HS/SO	5	172	20.2	1315	0.147	0.147
PUR_HS/TO	6	211	19.4	1135	0.122	0.122

The results of limiting oxygen index (LOI) are presented in Figure 11a. Similarly to the results of the cone calorimeter test (pHRR and TSR), the addition of non-treated HS filler decreases the LOI value from 20.2 to 19.8%. Increased value of LOI is observed for PUR foams containing impregnated HS fillers—the LOI increases to 20.8 and 21.6%, for PUR_HS/SO and PUR_HS/TO, respectively. Chan et al. [68] have reported similar results in the case of PUR foams enhanced with ramie fibers in the amount of 0.2–0.8 wt.%. It has been shown that the carbonization of the fibers during the combustion process results in the formation of pores, preventing the propagation of the flame through the PUR matrix and increasing the value of LOI.

The self-extinguish capacity was measured according to the UL-94 standard, which corresponds to the burning time measured after continuous ignitions of the samples. According to the UL-94 standard, neat PUR_0 was not classified under test standard, while PUR foam containing RC fillers achieved a V-0 rating. Moreover, the total time of burning was evaluated. According to the results presented in Figure 11b, the flame keeps longer in the case of neat PUR_0, for which the total time of burning was 32 s. The total time of burning decreases when the HS fillers were added. The most visible effect is observed in the case of PUR foams containing impregnated HS fillers—the value decreases to 22 and 21 s for PUR_HS/SO and PUR_HS/TO, respectively. The results of UL-94 are in agreement with the results of LOI. In summary, the highest standards among UL-94 and LOI tests are obtained for PUR foams containing impregnated HS fillers.

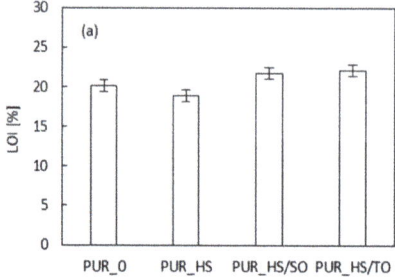 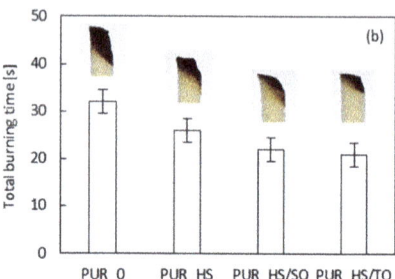

Figure 11. (a) LOI and (b) total burning time of PUR foams.

4. Conclusions

PUR foams were reinforced with 2 wt.% non-treated and impregnated HS fillers were successfully synthesized. It has been shown that each type of HS filler affects the morphology and further mechanical, thermal, and insulating properties of PUR foams. It has been shown that the addition of HS fillers improved the mechanical characteristics of PUR foams. Among all modified series, the greatest improvement was observed after the incorporation of non-treated HS filler—when compared with neat foams, the value of compressive strength increased by ~13%. Moreover, the incorporation of impregnated HS fillers resulted in the improvement of thermal stability and flame retardancy of PUR foams. For example, the addition of both types of impregnated HS fillers significantly decreased the value of heat peak release (pHRR), total smoke release (TSR), and limiting oxygen index (LOI). Moreover, the PUR foams containing impregnated fillers were characterized by improved hydrophobicity and limited water uptake. The obtained results confirmed that the modification of PUR foams with non-treated and impregnated HS fillers may be a successful approach in producing polymeric composites with improved properties.

Author Contributions: S.C.: Methodology, Investigation, Data Curation, Writing—Original Draft, Writing—Review and Editing, Visualization; A.S.: Methodology, Investigation; A.K.: Methodology, Investigation. All authors have read and agreed to the published version of the manuscript.

Funding: This research received no external funding.

Conflicts of Interest: The authors declare no conflict of interest.

References

1. Silva, M.C.; Takahashi, J.A.; Chaussy, D.; Belgacem, M.N.; Silva, G.G. Composites of rigid polyurethane foam and cellulose fiber residue. *J. Appl. Polym. Sci.* **2010**, *117*, 3665–3672. [CrossRef]
2. Hayati, A.N.; Evans, D.A.C.; Laycock, B.; Martin, D.J.; Annamalai, P.K. A simple methodology for improving the performance and sustainability of rigid polyurethane foam by incorporating industrial lignin. *Ind. Crops Prod.* **2018**, *117*, 149–158. [CrossRef]
3. Olszewski, A.; Kosmela, P.; Mielewczyk-Gryń, A.; Piszczyk, Ł. Bio-based polyurethane composites and hybrid composites containing a new type of bio-polyol and addition of natural and synthetic fibers. *Materials (Basel)* **2020**, *13*, 2028. [CrossRef] [PubMed]
4. Barczewski, M.; Kurańska, M.; Sałasińska, K.; Michałowski, S.; Prociak, A.; Uram, K.; Lewandowski, K. Rigid polyurethane foams modified with thermoset polyester-glass fiber composite waste. *Polym. Test.* **2020**, *81*, 106190. [CrossRef]
5. Matykiewicz, D.; Barczewski, M. On the impact of flax fibers as an internal layer on the properties of basalt-epoxy composites modified with silanized basalt powder. *Compos. Commun.* **2020**, *20*, 100360. [CrossRef]
6. Barczewski, M.; Mysiukiewicz, O.; Matykiewicz, D.; Kloziński, A.; Andrzejewski, J.; Piasecki, A. Synergistic effect of different basalt fillers and annealing on the structure and properties of polylactide composites. *Polym. Test.* **2020**, *89*, 106628. [CrossRef]

7. Kurańska, M.; Barczewski, M.; Uram, K.; Lewandowski, K.; Prociak, A.; Michałowski, S. Basalt waste management in the production of highly effective porous polyurethane composites for thermal insulating applications. *Polym. Test.* **2019**, *76*, 90–100. [CrossRef]
8. Paciorek-Sadowska, J.; Borowicz, M.; Isbrandt, M.; Czupryński, B.; Apiecionek, Ł. The use of waste from the production of rapeseed oil for obtaining of new polyurethane composites. *Polymers (Basel)* **2019**, *11*, 1431. [CrossRef]
9. Leszczyńska, M.; Ryszkowska, J.; Szczepkowski, L.; Kurańska, M.; Prociak, A.; Leszczyński, M.K.; Gloc, M.; Antos-Bielska, M.; Mizera, K. Cooperative effect of rapeseed oil-based polyol and egg shells on the structure and properties of rigid polyurethane foams. *Polym. Test.* **2020**, *90*, 106696. [CrossRef]
10. Kairytė, A.; Kirpluks, M.; Ivdre, A.; Cabulis, U.; Vėjelis, S.; Balčiūnas, G. Paper waste sludge enhanced eco-efficient polyurethane foam composites: Physical-mechanical properties and microstructure. *Polym. Compos.* **2018**, *39*, 1852–1860. [CrossRef]
11. De Avila Delucis, R.; Magalhães, W.L.E.; Petzhold, C.L.; Amico, S.C. Forest-based resources as fillers in biobased polyurethane foams. *J. Appl. Polym. Sci.* **2018**, *135*, 1–7. [CrossRef]
12. El-Shekeil, Y.A.; Sapuan, S.M.; Abdan, K.; Zainudin, E.S. Influence of fiber content on the mechanical and thermal properties of Kenaf fiber reinforced thermoplastic polyurethane composites. *Mater. Des.* **2012**, *40*, 299–303. [CrossRef]
13. Xue, B.-L.; Wen, J.-L.; Sun, R.-C. Lignin-Based Rigid Polyurethane Foam Reinforced with Pulp Fiber: Synthesis and Characterization. *ACS Sustain. Chem. Eng.* **2014**, *2*, 1474–1480. [CrossRef]
14. Ribeiro da Silva, V.; Mosiewicki, M.A.; Yoshida, M.I.; Coelho da Silva, M.; Stefani, P.M.; Marcovich, N.E. Polyurethane foams based on modified tung oil and reinforced with rice husk ash II: Mechanical characterization. *Polym. Test.* **2013**, *32*, 665–672. [CrossRef]
15. Zhu, M.; Bandyopadhyay-Ghosh, S.; Khazabi, M.; Cai, H.; Correa, C.; Sain, M. Reinforcement of soy polyol-based rigid polyurethane foams by cellulose microfibers and nanoclays. *J. Appl. Polym. Sci.* **2011**, *124*. [CrossRef]
16. Zhou, X.; Sain, M.M. Semi-rigid biopolyurethane foams based on palm-oil polyol and reinforced with cellulose nanocrystals. *Compos. Part A Appl. Sci. Manuf.* **2016**, *83*, 56–62. [CrossRef]
17. Zieleniewska, M.; Leszczyński, M.K.; Szczepkowski, L.; Bryśkiewicz, A.; Krzyżowska, M.; Bień, K.; Ryszkowska, J. Development and applicational evaluation of the rigid polyurethane foam composites with egg shell waste. *Polym. Degrad. Stab.* **2016**, *132*, 78–86. [CrossRef]
18. Członka, S.; Bertino, M.F.; Strzelec, K. Rigid polyurethane foams reinforced with industrial potato protein. *Polym. Test.* **2018**, *68*, 135–145. [CrossRef]
19. Członka, S.; Bertino, M.F.; Strzelec, K.; Strąkowska, A.; Masłowski, M. Rigid polyurethane foams reinforced with solid waste generated in leather industry. *Polym. Test.* **2018**, *69*, 225–237. [CrossRef]
20. Członka, S.; Sienkiewicz, N.; Strąkowska, A.; Strzelec, K. Keratin feathers as a filler for rigid polyurethane foams on the basis of soybean oil polyol. *Polym. Test.* **2018**, *72*, 32–45. [CrossRef]
21. Soares, B.; Gama, N.; Freire, C.S.R.; Barros-Timmons, A.; Brandão, I.; Silva, R.; Neto, C.P.; Ferreira, A. Spent coffee grounds as a renewable source for ecopolyols production. *J. Chem. Technol. Biotechnol.* **2015**, *90*, 1480–1488. [CrossRef]
22. Huang, G.; Wang, P. Effects of preparation conditions on properties of rigid polyurethane foam composites based on liquefied bagasse and jute fibre. *Polym. Test.* **2017**, *60*, 266–273. [CrossRef]
23. Nam, S.; Netravali, A.N. Characterization of ramie fiber/soy protein concentrate (SPC) resin interface. *J. Adhes. Sci. Technol.* **2004**, *18*, 1063–1076. [CrossRef]
24. Mosiewicki, M.A.; Casado, U.; Marcovich, N.E.; Aranguren, M.I. Vegetable oil based-polymers reinforced with wood flour. *Mol. Cryst. Liq. Cryst.* **2008**, *484*, 509–516. [CrossRef]
25. Shan, C.W.; Idris, M.I.; Ghazali, M.I. Study of Flexible Polyurethane Foams Reinforced with Coir Fibres and Tyre Particles. *Int. J. Appl. Phys. Math.* **2012**, *2*, 123–130. [CrossRef]
26. Paciorek-Sadowska, J.; Czupryński, B.; Borowicz, M.; Liszkowska, J. Rigid polyurethane–polyisocyanurate foams modified with grain fraction of fly ashes. *J. Cell. Plast.* **2020**, *56*, 53–72. [CrossRef]
27. Liszkowska, J.; Borowicz, M.; Paciorek-Sadowska, J.; Isbrandt, M.; Czupryński, B.; Moraczewski, K. Assessment of photodegradation and biodegradation of RPU/PIR foams modified by natural compounds of plant origin. *Polymers* **2020**, *12*, 33. [CrossRef]

28. Zhang, S.; Xiang, A.; Tian, H.; Rajulu, A.V. Water-Blown Castor Oil-Based Polyurethane Foams with Soy Protein as a Reactive Reinforcing Filler. *J. Polym. Environ.* **2018**, *26*, 15–22. [CrossRef]
29. Števulová, N.; Terpáková, E.; Čigášová, J.; Junák, J.; Kidalová, L. Chemically treated hemp shives as a suitable organic filler for lightweight composites preparing. *Procedia Eng.* **2012**, *42*, 948–954. [CrossRef]
30. Terzopoulou, Z.N.; Papageorgiou, G.Z.; Papadopoulou, E.; Athanassiadou, E.; Reinders, M.; Bikiaris, D.N. Development and study of fully biodegradable composite materials based on poly(butylene succinate) and hemp fibers or hemp shives. *Polym. Compos.* **2016**, *37*, 407–421. [CrossRef]
31. Brazdausks, P.; Puke, M.; Rizhikovs, J.; Pubule, J. Evaluation of cellulose content in hemp shives after salt catalyzed hydrolysis. *Energy Procedia* **2017**, *128*, 297–301. [CrossRef]
32. Zakaria, S.; Hamzah, H.; Murshidi, J.A.; Deraman, M. Chemical modification on lignocellulosic polymeric oil palm empty fruit bunch for advanced material. *Adv. Polym. Technol.* **2001**, *20*, 289–295. [CrossRef]
33. Wolski, K.; Cichosz, S.; Masek, A. Surface hydrophobisation of lignocellulosic waste for the preparation of biothermoelastoplastic composites. *Eur. Polym. J.* **2019**, *118*, 481–491. [CrossRef]
34. Cichosz, S.; Masek, A. Cellulose fibers hydrophobization via a hybrid chemical modification. *Polymers* **2019**, *11*, 1174. [CrossRef]
35. Borysiak, S. Fundamental studies on lignocellulose/polypropylene composites: Effects of wood treatment on the transcrystalline morphology and mechanical properties. *J. Appl. Polym. Sci.* **2013**, *127*, 1309–1322. [CrossRef]
36. Neto, J.S.S.; Lima, R.A.A.; Cavalcanti, D.K.K.; Souza, J.P.B.; Aguiar, R.A.A.; Banea, M.D. Effect of chemical treatment on the thermal properties of hybrid natural fiber-reinforced composites. *J. Appl. Polym. Sci.* **2019**, *136*, 1–13. [CrossRef]
37. Kabir, M.M.; Wang, H.; Lau, K.T.; Cardona, F. Chemical treatments on plant-based natural fibre reinforced polymer composites: An overview. *Compos. Part B Eng.* **2012**, *43*, 2883–2892. [CrossRef]
38. Gang, D. The influence of surface treatment on the tensile and tribological properties of wood fiber-reinforced polyimide composite. *Surf. Interface Anal.* **2018**, *50*, 304–310. [CrossRef]
39. Valášek, P.; D'Amato, R.; Müller, M.; Ruggiero, A. Mechanical properties and abrasive wear of white/brown coir epoxy composites. *Compos. Part B Eng.* **2018**, *146*, 88–97. [CrossRef]
40. Shalwan, A.; Yousif, B.F. Influence of date palm fibre and graphite filler on mechanical and wear characteristics of epoxy composites. *Mater. Des.* **2014**, *59*, 264–273. [CrossRef]
41. Sreekala, M.S.; Kumaran, M.G.; Joseph, S.; Jacob, M. Oil Palm Fibre Reinforced Phenol Formaldehyde Composites: Influence of Fibre Surface Modifications on the Mechanical Performance. *Appl. Compos. Mater.* **2000**, *7*, 295–329. [CrossRef]
42. *Plastics—Resins in the Liquid State or as Emulsions or Dispersions—Determination of Apparent Viscosity by the Brookfield Test Method*; ISO 2555; ISO: Geneva, Switzerland, 1989.
43. *Preview Rigid Cellular Plastics—Determination of Compression Properties*; ISO 844; ISO: Geneva, Switzerland, 2014.
44. *Plastics—Determination of Flexural Properties*; ISO 178; ISO: Geneva, Switzerland, 2019.
45. *Plastics—Determination of Izod Impact Strength*; ISO 180; ISO: Geneva, Switzerland, 2019.
46. Lee, L.J.; Zeng, C.; Cao, X.; Han, X.; Shen, J.; Xu, G. Polymer nanocomposite foams. *Compos. Sci. Technol.* **2005**, *65*, 2344–2363. [CrossRef]
47. Sung, G.; Kim, J.H. Influence of filler surface characteristics on morphological, physical, acoustic properties of polyurethane composite foams filled with inorganic fillers. *Compos. Sci. Technol.* **2017**, *146*, 147–154. [CrossRef]
48. Gu, R.; Khazabi, M.; Sain, M. Fiber reinforced soy-based polyurethane spray foam insulation. Part 2: Thermal and mechanical properties. *BioResources* **2011**, *6*, 3775–3790.
49. Kairytė, A.; Kizinievič, O.; Kizinievič, V.; Kremensas, A. Synthesis of biomass-derived bottom waste ash based rigid biopolyurethane composite foams: Rheological behaviour, structure and performance characteristics. *Compos. Part A Appl. Sci. Manuf.* **2019**, *117*, 193–201. [CrossRef]
50. Aranberri, I.; Montes, S.; Wesołowska, E.; Rekondo, A.; Wrześniewska-Tosik, K.; Grande, H.-J. Improved Thermal Insulating Properties of Renewable Polyol Based Polyurethane Foams Reinforced with Chicken Feathers. *Polymers (Basel)* **2019**, *11*, 2002. [CrossRef] [PubMed]

51. Sair, S.; Oushabi, A.; Kammouni, A.; Tanane, O.; Abboud, Y.; El Bouari, A. Mechanical and thermal conductivity properties of hemp fiber reinforced polyurethane composites. *Case Stud. Constr. Mater.* **2018**, *8*, 203–212. [CrossRef]
52. Joanna, P.S.; Bogusław, C.; Joanna, L. Application of waste products from agricultural-food industry for production of rigid polyurethane-polyisocyanurate foams. *J. Porous Mater.* **2011**, *18*, 631–638. [CrossRef]
53. Mosiewicki, M.A.; Dell'Arciprete, G.A.; Aranguren, M.I.; Marcovich, N.E. Polyurethane foams obtained from castor oil-based polyol and filled with wood flour. *J. Compos. Mater.* **2009**, *43*, 3057–3072. [CrossRef]
54. Finlay, K.A.; Gawryla, M.D.; Schiraldi, D.A. Effects of fiber reinforcement on clay aerogel composites. *Materials (Basel)* **2015**, *8*, 5440–5451. [CrossRef]
55. Ciecierska, E.; Jurczyk-Kowalska, M.; Bazarnik, P.; Gloc, M.; Kulesza, M.; Kowalski, M.; Krauze, S.; Lewandowska, M. Flammability, mechanical properties and structure of rigid polyurethane foams with different types of carbon reinforcing materials. *Compos. Struct.* **2016**, *140*, 67–76. [CrossRef]
56. Gu, R.; Konar, S.; Sain, M. Preparation and characterization of sustainable polyurethane foams from soybean oils. *J. Am. Oil Chem. Soc.* **2012**, *89*, 2103–2111. [CrossRef]
57. Ye, L.; Meng, X.Y.; Ji, X.; Li, Z.M.; Tang, J.H. Synthesis and characterization of expandable graphite-poly(methyl methacrylate) composite particles and their application to flame retardation of rigid polyurethane foams. *Polym. Degrad. Stab.* **2009**, *94*, 971–979. [CrossRef]
58. Gama, N.V.; Silva, R.; Mohseni, F.; Davarpanah, A.; Amaral, V.S.; Ferreira, A.; Barros-Timmons, A. Enhancement of physical and reaction to fire properties of crude glycerol polyurethane foams filled with expanded graphite. *Polym. Test.* **2018**, *69*, 199–207. [CrossRef]
59. Cichosz, S.; Masek, A. Superiority of cellulose non-solvent chemical modification over solvent-involving treatment: Solution for green chemistry (Part I). *Materials* **2020**, *13*, 2552. [CrossRef]
60. Cichosz, S.; Masek, A. Thermal Behavior of Green Cellulose-Filled Thermoplastic Elastomer Polymer Blends. *Molecules* **2020**, *25*, 1279. [CrossRef]
61. Tian, H.; Wu, J.; Xiang, A. Polyether polyol-based rigid polyurethane foams reinforced with soy protein fillers. *J. Vinyl Addit. Technol.* **2018**, *24*, E105–E111. [CrossRef]
62. Mizera, K.; Ryszkowska, J.; Kurańska, M.; Prociak, A. The effect of rapeseed oil-based polyols on the thermal and mechanical properties of ureaurethane elastomers. *Polym. Bull.* **2020**, *77*, 823–846. [CrossRef]
63. Kurańska, M.; Polaczek, K.; Auguścik-Królikowska, M.; Prociak, A.; Ryszkowska, J. Open-cell rigid polyurethane bio-foams based on modified used cooking oil. *Polymer (Guildf)* **2020**, *190*, 1–7. [CrossRef]
64. Gómez-Fernández, S.; Ugarte, L.; Calvo-Correas, T.; Peña-Rodríguez, C.; Corcuera, M.A.; Eceiza, A. Properties of flexible polyurethane foams containing isocyanate functionalized kraft lignin. *Ind. Crops Prod.* **2017**, *100*, 51–64. [CrossRef]
65. Luo, X.; Xiao, Y.; Wu, Q.; Zeng, J. Development of high-performance biodegradable rigid polyurethane foams using all bioresource-based polyols: Lignin and soy oil-derived polyols. *Int. J. Biol. Macromol.* **2018**, *115*, 786–791. [CrossRef]
66. Mahmood, N.; Yuan, Z.; Schmidt, J.; Xu, C. Preparation of bio-based rigid polyurethane foam using hydrolytically depolymerized Kraft lignin via direct replacement or oxypropylation. *Eur. Polym. J.* **2015**, *68*, 1–9. [CrossRef]
67. Qian, L.; Feng, F.; Tang, S. Bi-phase flame-retardant effect of hexa-phenoxy-cyclotriphosphazene on rigid polyurethane foams containing expandable graphite. *Polymer (Guildf)* **2014**, *55*, 95–101. [CrossRef]
68. Cheng, J.; Wang, H.; Wang, X.; Li, S.; Zhou, Y.; Zhang, F.; Wang, Y.; Qu, W.; Wang, D.; Pang, X. Effects of flame-retardant ramie fiber on enhancing performance of the rigid polyurethane foams. *Polym. Adv. Technol.* **2019**, *30*, 3091–3098. [CrossRef]

Publisher's Note: MDPI stays neutral with regard to jurisdictional claims in published maps and institutional affiliations.

© 2020 by the authors. Licensee MDPI, Basel, Switzerland. This article is an open access article distributed under the terms and conditions of the Creative Commons Attribution (CC BY) license (http://creativecommons.org/licenses/by/4.0/).

Article

The Eco-Friendly Biochar and Valuable Bio-Oil from *Caragana korshinskii*: Pyrolysis Preparation, Characterization, and Adsorption Applications

Tongtong Wang [1,2], Hongtao Liu [1,2], Cuihua Duan [1,2], Rui Xu [1,2], Zhiqin Zhang [1,2], Diao She [1,3] and Jiyong Zheng [1,3,*]

1. State Key Laboratory of Soil Erosion and Dryland Farming on the Loess Plateau, Northwest A & F University, Yangling 712100, China; tongtwang@163.com (T.W.); liuhongtao@nwafu.edu.cn (H.L.); chduan@nwafu.edu.cn (C.D.); 2019055450@nwafu.edu.cn (R.X.); zhangzhiqin@nwafu.edu.cn (Z.Z.); diaoshe888@163.com (D.S.)
2. College of Natural Resources and Environment, Northwest A & F University, Yangling 712100, China
3. Institute of Soil and Water Conservation, Chinese Academy of Sciences and Ministry of Water Resources, Yangling 712100, China
* Correspondence: zhjy@ms.iswc.ac.cn; Tel.: +86-150-2928-2338

Received: 11 July 2020; Accepted: 30 July 2020; Published: 31 July 2020

Abstract: Carbonization of biomass can prepare carbon materials with excellent properties. In order to explore the comprehensive utilization and recycling of *Caragana korshinskii* biomass, 15 kinds of *Caragana korshinskii* biochar (CB) were prepared by controlling the oxygen-limited pyrolysis process. Moreover, we pay attention to the dynamic changes of microstructure of CB and the by-products. The physicochemical properties of CB were characterized by Scanning Electron Microscope (SEM), BET-specific surface area (BET-SSA), X-ray photoelectron spectroscopy (XPS), X-ray diffraction (XRD), Fourier Transform Infrared (FTIR), and Gas chromatography-mass spectrometry (GC-MS). The optimal preparation technology was evaluated by batch adsorption application experiment of NO_3^-, and the pyrolysis mechanism was explored. The results showed that the pyrolysis temperature is the most important factor in the properties of CB. With the increase of temperature, the content of C, pH, mesoporous structure, BET-SSA of CB increased, the cation exchange capacity (CEC) decreased and then increased, but the yield and the content of O and N decreased. The CEC, pH, and BET-SSA of CB under each pyrolysis process were 16.64–81.4 cmol·kg^{-1}, 6.65–8.99, and 13.52–133.49 m^2·g^{-1}, respectively. CB contains abundant functional groups and mesoporous structure. As the pyrolysis temperature and time increases, the bond valence structure of C 1s, Ca 2p, and O 1s is more stable, and the phase structure of $CaCO_3$ is more obvious, where the aromaticity increases, and the polarity decreases. The CB prepared at 650 °C for 3 h presented the best adsorption performance, and the maximum theoretical adsorption capacity for NO_3^- reached 120.65 mg·g^{-1}. The Langmuir model and pseudo-second-order model can well describe the isothermal and kinetics adsorption process of NO_3^-, respectively. Compared with other cellulose and lignin-based biomass materials, CB showed efficient adsorption performance of NO_3^- without complicated modification condition. The by-products contain bio-soil and tail gas, which are potential source of liquid fuel and chemical raw materials. Especially, the bio-oil of CB contains α-D-glucopyranose, which can be used in medical tests and medicines.

Keywords: pyrolysis process; *Caragana korshinskii* biochar; physicochemical properties; adsorption characteristics; nitrate nitrogen; bio-oil

1. Introduction

The *Caragana korshinskii* Kom. (1909) is a legume shrub that is widely distributed across desert habitats with gravely, sandy, and saline soils in arid and semi-arid areas of Asia and Africa [1]. Many scholars believe that it may be an ideal biomass energy crop with high heating value, strong drought tolerance, and high sprouting capacity [2]. As a pioneer species of arid areas in northern China, the *Caragana korshinskii* is essential for windbreak function, sand fixation, and water and soil conservation [3]. The planting area of *Caragana korshinskii* is conservatively estimated more than 1.6 million km^2 [4]. However, because of the lack of corresponding development technologies and products worldwide, the utilization rate of *Caragana korshinskii* resources is less than 40% and even a nearly null rate in some areas. Other countries or regions also produce a large number of *Caragana korshinskii* yearly, which has been regarded as a new forestry wastes [5]. They represent not only a waste of energy but also a threat to the local environment [6,7]. Therefore, developing the new technology and product and enhancing the efficient use of *Caragana korshinskii* resources are necessary and urgent. It cannot only help local farmers solve waste problems and promote environmental sustainability [8], but also produce positive economic benefits in arid areas of Asia and Africa which are usually relatively underdeveloped areas.

The high temperature pyrolysis is generally the most common and important carbonization method [9]. Under limited oxygen supply or anoxic conditions and at a certain temperature (<700 °C), biomass resources can be carbonized into biochar, a carbon-rich, stable, and highly aromatic solid product [10–13]. At present, because of its large specific surface area, high stability, strong adsorption capacity, porous structure, as well as its wide range of raw materials, and low cost, biochar was widely used in agricultural and environmental protection fields such as pollution remediation, modified materials, soil improvement, etc., [14,15]. The carbonization process is mainly controlled by parameters such as pyrolysis temperature, time, and heating rate. It was a general consensus that pyrolysis process had great influence on the physical and chemical properties of biochar. Chen et al. [16] and Fan et al. [17] set different pyrolysis temperatures to study the properties of biochar, and found that the pyrolysis temperature had a vital impact on the carbonization process of raw materials, and direct influence on the yield, element content, surface functional group, and pore size distribution. Higher pyrolysis temperatures had a positive effect on the adsorption applications. Xue et al. [18] observed that the biochar yield rate ranged from 89.7% to 51.2% at temperature from 300 to 600 °C. The surface characteristics and micromorphology of biochar changed with increasing pyrolysis temperature and time, but the heating rate has little effect on the properties of biochar [19]. Biomass pyrolysis usually also produces by-products, bio-oil, and tail gas. High temperature heating of the biomass in the pyrolysis process caused irregular thermal movement in the internal molecular structure of the biomass and eventually decomposition into small molecule polymer [20]. Both biochar and by-products can be generally well-obtained at conventional heating rate of 10 °C·min^{-1} [21].

However, some reports on biomass pyrolysis are not systematic. Most scholars pay more attention to yield data and basic properties than the changes of bio-oil and tail gas during pyrolysis process. The research on the effects of microscopic indexes such as XPS, XRD, and FTIR is also not sufficient and deep enough [22]. In order to better promote the application of biomass pyrolysis technology, it is meaningful to systematically study the impact of pyrolysis process on its properties [23,24]. Moreover, nitrate nitrogen (NO_3^-) is the main form of nitrogen, which plays an important role in plant nutrition, microbial transformation, and natural nitrogen cycle [25]. But the nitrogen leaching from farmland will lead to many problems such as groundwater pollution and water eutrophication [26]. Fortunately, the adsorption of NO_3^- by biochar is the most effective method to remove nitrogen because of high efficiency, simple equipment, and reliable operation [27]. Indeed, NO_3^- was selected as the object for adsorption application, which has always been a research hotspot [28]. Li et al. [25] found that biochar has limited adsorption of NO_3^- from switchgrass and water oak at variable pyrolysis condition. Even if the biochar is modified by $FeCl_3$ impregnated, the maximum adsorption capacity for nitrogen is only 14 mg·g^{-1} [29]. Therefore, *Caragana korshinskii* was used as the raw material to prepare 15 kinds of biochar by oxygen-limited pyrolysis with controlled pyrolysis temperature and time. Some characterization methods were used to investigate the effects of different pyrolysis processes on the changes of biochar properties. At the same time, the by-products are systematically analyzed for chemical composition, and the pyrolysis mechanism is explored. The adsorption characteristics of NO_3^- in aqueous solution were studied by batch adsorption experiment. This work will provide a theoretical and practical basis for the efficient utilization of *Caragana korshinskii* resource and development technology and products about the eco-friendly biochar and valuable by-products.

2. Materials and Methods

2.1. Production of Biochar and By-Products

The *Caragana korshinskii* samples (20 years old) were collected from the Shanghuang Village, Guyuan City, Ningxia Hui Autonomous Region, China. The ratios of cellulose, hemicellulose, and lignin in the biomass of *Caragana korshinskii* are approximately 45%, 20%, and 35% respectively. The biomass was cut lengthwise into 10-cm pieces, washed with distilled water several times, and dried. Subsequently, a box was filled with the samples, sealed with a cover, and placed into a GF11Q-B box atmosphere muffle furnace (Nanjing Boyuntong Instrument Technology Co., Ltd., Nanjing, China). The pyrolysis temperatures were set to 450, 500, 550, 600, and 650 °C. The pyrolysis time (maintain the constant temperature) was set to 2, 3, and 4 h, and the constant heating rate to 10 °C·min^{-1}. Nitrogen gas was used to protect the anaerobic environment. A three-stage temperature-programmed pyrolysis method was used to uniformly cool the samples to room temperature. The bio-oil and tail gas were collected by a condensing reflux device (Figure 1). Subsequently, the *Caragana korshinskii* biochar (CB) was crushed into small pieces of approximately 2 cm by a pulveriser (Zhejiang Fengli Pulverization Equipment Co., Ltd., Shaoxing, China).

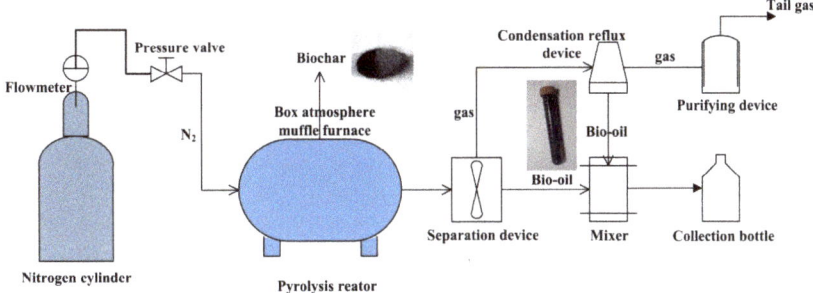

Figure 1. Schematic for pyrolysis process of *Caragana korshinskii* biomass.

2.2. Characterisation and Test

2.2.1. Basic Properties

The element contents of C, H, O, and N were measured using a Vario EL cube elemental analyzer (Element Analysensysteme Co., Ltd., Berlin, Germany) with argon as the carrier gas. The pH of the CB samples were determined by mixing 1.00 g of biochar and 20.0 mL of deionized water for 1 h in a rotary shaker. After that, the mixture was still at room temperature for 1 h, and its pH was measured by a PHS-3C pH meter (Rex INESA Scientific Instrument Co., Ltd., Shanghai, China). According to Li et al. [21], the barium chloride-sulfuric acid-forced exchange method was used to determine the CEC of CB, and the yield of CB is equal to the ratio of the mass of biochar produced by pyrolysis to the mass of original dried *Caragana korshinskii*.

2.2.2. Microscopic Indexes

CB samples of approximately 0.5 g were degassed for 3 h at 125 °C. The surface morphology of the CB samples was analyzed in a JSM-6510LV scanning electron microscope (JEOL Ltd., Tokyo, Japan) at a 20 kV. The BET-SSA and pore size analysis were determined using N_2 as the adsorbate at 77 K and relative pressure of 0.05–0.20, for which a V-Sorb 2800P BET-SSA and pore size analyzer (Gold APP Instrument Co., Ltd., Beijing, China) was used. All CB samples were scanned with an Escalab 250 Xi X-ray photoelectron spectrometer (Thermo Fisher Scientific Co., Ltd., Waltham, MA, USA) to investigate the chemical state of the main elements and the functional groups on the surface. The chemical bonds were identified by published references and the National Institute of Standards and Technology database (https://srdata.nist.gov/xps/). The mineral species of the CB samples were identified using a D/max2400 X-ray powder diffractometer (Rigaku Co., Ltd., Wilmington, NC, USA) at a 0.02 scan step size, 2 deg·min^{-1} scan speed, 0.15 receiving slit width, 30–40 kV, and 30–40 mA. The surface functional groups of CB samples were measured with KBr pellet methods by infrared spectra analysis using a Vertex70 FTIR spectrometer (Bruker Co., Ltd., Billerica, USA) for 16 scans over a range of 400–4000 cm^{-1} with a resolution of 2 cm^{-1}.

2.2.3. By-Product Analysis

The gas chromatography tandem mass spectrometer (GC-MS, Agilent 7890A/7000B GC-QQQ, Santa Clara, CA, USA) was used to analyze the components of bio-oil and tail gas. GC-MS analysis conditions are: Shunt injection with a ratio of 25:1, HP-5 (30 m × 0.25 mm × 0.25 m) column, the temperature of vaporization chamber at 260 °C, high purity helium as carrier gas, and flow rate 1.0 mL/min. The column temperature is programmed: the initial temperature is 50 °C and kept for 5 min, the temperature is raised to 100 °C at 2 °C/min, and then 180 °C at 1 °C/min, finally 260 °C at 1 °C/min and kept for 5 min. MS electron energy was 70 eV, filament current 80 A, ion source temperature 230 °C.

2.3. Batch Experiments

The biochar sorption capacity was evaluated by batch experiments. KNO_3 was used to prepare standard NO_3^- solutions at different concentrations. Accurately weighted samples of 0.1000 g, including above 15 kinds of CB, were placed in 250 mL conical flasks and mixed with 50 ml of the NO_3^- solution. The concentration of NO_3^- in the solution is 50 mg·L^{-1}. Moreover, the background electrolyte was 0.01 mol·L^{-1} KCl. The conical flasks were sealed with a plastic film and placed into an air bath constant temperature oscillator at 25 ± 1 °C, which was shaken at 150 rpm for 3 h. The obtained suspension was filtered with a 0.45 μm microfiltration membrane. The concentrations of NO_3^- in the filtrate was considered the equilibrium concentrations of the solutions, and measured using a continuous flow auto analyzer (Bran+luebbe, Analyzer 3-AA3, Hamburg, Germany). All experiments were performed in triplicate with appropriate blanks (biochar and H_2O), and statistical methods were used to analyse the mean values. The adsorption capacity (Q_e, mg·g^{-1}) and the removal efficiency (RE, %) were calculated according to the following equations:

$$Q_e = (C_0 - C_e)\frac{V}{m} \qquad (1)$$

$$RE = 1 - \frac{C_e}{C_0} \times 100\% \qquad (2)$$

Throughout this paper, Q_e is the adsorption capacity at equilibrium, and C_e and C_0 are the equilibrium and initial NO_3^- concentrations (mg·L^{-1}), respectively. V is the volume of the solution (L), and m is the mass of added biochar (g). The following steps were used for batch experiments using optimal CB (alternative biochar). (1) Isothermal adsorption: the initial concentration gradients of NO_3^- were set to 0, 2, 5, 10, 20, 30, 40, 50, 60, 70, 80, 90 and 100 mg·L^{-1}; (2) adsorption kinetics: the adsorption times of NO_3^- were set to 5, 10, 15, 30, 60, 180, 360, 540, 720, 1080, and 1440 min, other settings are the same.

2.4. Adsorption Models and Data Analysis

To better investigate the characteristics of the adsorption, four isotherms models (Langmuir, Freundlich, Temkin, and Dubinin-Radushkevich) [11] and four kinetic models (pseudo-first-order, pseudo-second-order, Elovich, and intraparticle diffusion model) [30] were used, as shown in Table 1. In this work, data statistics were calculated using Excel 2016 Pro (Microsoft, Redmond, WA, USA) and SPSS 20.0 (IBM Corporation, Armonk, NY, USA). One-way Analysis of Variance (ANOVA) using SPSS 20.0 was conducted to compare the means of the measured values at $p < 0.05$ for each treatment. Isothermal adsorption and adsorption kinetic curves were fitted and plotted using OriginPro 8.5.1 SR2 (OriginLab, Northampton, MA, USA). In addition, XRD was performed using Jade 6.5 (Materials data Inc., Livermore, CA, USA) to determine the mineral composition of biochar. Wide XPS scans were analyzed using Avantage v5.979 (Thermo Fisher Scientific Co., Ltd., Waltham, MA, USA) to determine the surface elements content. Narrow scans of C1s were fitted using Avantage v5.979 to analyze and quantify the existing state of C.

Table 1. The list of kinetic and isotherm models.

Models	Expression	Parameters
Langmuir	$Q_e = \frac{aQ_mC_e}{1+aC_e}$, $R_L = \frac{1}{1+aC_0}$	Q_m, a
Freundlich	$Q_e = K_F C_e^{1/n}$	$K_F, 1/n$
Temkin	$Q_e = B\ln A + B\ln C_e$	A, B
Dubinin-Radushkevich (D-R) model	$\ln Q_e = \ln Q_0 - \beta\varepsilon^2$, $\varepsilon = RT\ln(1+\frac{1}{C_e})$ $E = \frac{1}{(2\beta)^{0.5}}$	$Q_0, \beta, \varepsilon, E$
Pseudo-first-order	$Q_t = Q_e(1 - \exp(-\frac{k_1}{2.303}t))$	Q_e, k_1
Pseudo-second-order	$Q_t = \frac{k_2 Q_e^2 t}{1+k_2 Q_e t}$, $h = k_2 Q_e^2$	Q_e, k_2, h
Elovich model	$Q_t = \frac{1}{b}\ln(a'bt + 1)$	a', b
Intraparticle diffusion	$Q_t = k_i t^{0.5} + C$	C, k_i

Where Q_e and Q_t are the adsorbed capacity (mg·g^{-1}) at an equilibrium concentration (C_e, mg·g^{-1}) and a given time of NO_3^- in solution, respectively. Q_m (mg·g^{-1}) denotes the maximum adsorption capacity. a and K_F are the Langmuir (mg·L^{-1}) and Freundlich (mg·g^{-1}) constants, respectively. $1/n$ is the heterogeneity factor. A is the equilibrium binding constant (mg·L^{-1}), and B is the Temkin constant related to the adsorption heat. β is the D-R model coefficient (mol^2·J^{-2}), Q_0 is the maximum unit adsorption capacity (mmol·g^{-1}), ε is the Polanyi adsorption potential, R is the ideal gas constant 8.314 J·(mol·K)$^{-1}$, T is the absolute temperature, and E is the adsorption free energy (J·mol^{-1}). k_1, k_2, and k_i are the rate constants for the pseudo-first-order (min^{-1}), pseudo-second-order (g·mg^{-1}·min^{-1}), and the intraparticle diffusion (mg·g^{-1}·h$^{-1/2}$) rate constant, respectively. h is the initial adsorption rate (mg·g^{-1}·min^{-1}). a' is the initial sorption ratio (g·mg^{-1}·min^{-1}), and b is the desorption constant (g·mg^{-1}), which is related to the extent of surface coverage and activation energy for chemisorption. C is a constant indicating the number of boundary layers of the adsorbent.

3. Results and Discussion

3.1. Basic Physicochemical Properties

The physicochemical properties of CB in all processes are shown in Table 2. With the increase of pyrolysis temperature, the content of C in biochar increases continuously, while the content of O and N decreases gradually, which is consistent with many research results [31,32]. This is because they are precipitated in the form of small molecule organic matter and water during the pyrolysis process [18,33]. As an important index of elemental analysis, the O/C has been used to determine the aromatic structure and composition of polymer, and the N/C can reflect the metabolic decomposition of microorganisms, and the (O + N)/C can be used to characterize the polarity of adsorbent. It can be clearly seen from Table 2 that O/C, N/C, and (O + N)/C decrease obviously with the increase of temperature, which is mainly due to the intensification of dehydrogenation and deoxidation reactions during the pyrolysis process. But, at the same temperature, O/C, N/C, and (O + N)/C change irregularly with the pyrolysis time. The change in O/C shows that the increase of temperature can promote the fracture of weak chemical bonds and the formation of condensation products, so the CB prepared at high temperature is more stable. In addition, O/C can also characterize the CEC of CB, and the larger the O/C, the larger the CEC [34]. In this work, the N/C of 15 kinds of CB is not high, which indicates that it is easy to decompose and will not consume the available nitrogen when applied to the soil [25]. The results in (O + N)/C shows that the oxygen-containing functional groups such as hydroxyl, carboxyl, and carbonyl groups of CB may be burned out with the increase of temperature, that is, the polar functional groups decreases and the hydrophobicity of biochar increases.

Table 2. Basic physical and chemical properties of biochar produced from *Caragana korshinskii*.

biochar	C (%)	O (%)	N (%)	O/C	N/C	(O + N)/C	pH	CEC (cmol·kg^{-1})	Yield (%)
CB450(2h)	77.99	16.18	3.10	0.21	0.04	0.25	6.65	62.51	33.12
CB500(2h)	78.25	15.6	2.05	0.2	0.03	0.23	8.55	66.12	33.00
CB550(2h)	77.9	13.52	1.26	0.17	0.02	0.19	8.80	35.00	28.93
CB600(2h)	82.17	12.52	1.10	0.15	0.01	0.17	8.96	16.64	25.94
CB650(2h)	85.1	10.04	0.84	0.12	0.01	0.13	8.93	46.44	27.24
CB450(3h)	65.73	22.9	2.79	0.35	0.04	0.39	8.04	55.72	33.23
CB500(3h)	74.55	18.73	2.41	0.25	0.03	0.28	7.72	74.42	31.80
CB550(3h)	77.02	16.84	2.13	0.22	0.03	0.25	8.84	31.97	31.43
CB600(3h)	79.12	13.19	1.47	0.17	0.02	0.19	8.97	20.48	28.20
CB650(3h)	85.31	10.59	0.72	0.12	0.01	0.13	8.99	38.93	23.08
CB450(4h)	74.35	18.17	3.28	0.24	0.04	0.29	7.51	81.40	30.31
CB500(4h)	72.13	18.83	2.95	0.26	0.04	0.30	8.74	44.62	29.78
CB550(4h)	81.42	11.52	2.14	0.14	0.03	0.17	8.99	32.91	28.65
CB600(4h)	79.79	12.92	1.5	0.16	0.02	0.18	8.68	30.66	27.56
CB650(4h)	81.04	10.86	0.62	0.13	0.01	0.14	8.97	17.75	27.38

The pH of CB increases with the increase of pyrolysis temperature. For pyrolysis time, the average pH is 8.38 for 2 h, 8.51 for 3 h, and 8.58 for 4 h. With the increase of time, the pH of the obtained CB increases. The pH of the CB under each pyrolysis conditions was 6.65–8.99, which was close to other lignin-based biochar [35]. Singh et al. [36] found that the biochar pH ranged from 6.93 to 10.26. This is due to the gradual separation of alkali salts from the biomass structure and the increase of alkaline functional groups on the surface of biochar [37], which is mutually supportive of the findings by Cantrell et al. [38]. Cantrell et al. [38] believed that the inorganic minerals (alkaline salts) in biochar were the main reason for the alkaline pH of biochar, and the oxygen-containing functional groups on the surface of biochar may also have certain contribution to the pH. Shaaban et al. [34] also believed that the biochar at high temperature had less acidic volatiles and more ash than low temperature, so the pH of biochar was higher. In addition, prolonging the pyrolysis time will help the further development of biochar.

The average statistics of 2 h and 3 h showed that the CEC of CB decreased first and then increased as the increase of temperature from 450 to 650 °C. The CEC of CB for 4 h decreased with the increase of temperature. For the pyrolysis time, the average CEC was 45.34 cmol·kg^{-1} for 2 h, 44.31 cmol·kg^{-1} for 3 h, and 41.47 cmol·kg^{-1} for 4 h. The CEC of CB under each pyrolysis conditions is 16.64–81.4 cmol·kg^{-1}. In the pyrolysis process, the loss rate of some acidic functional groups of biomass is the fastest at low temperature but slows down after 450 °C. The general trend is to decrease with the increase of temperature. Guo et al. [39] also believed this phenomenon of the CEC decrease. But, Liang et al. [40]

found that CEC increased with the increase of temperature by studying the pyrolysis process of pine wood. In this work, the CEC only increased at 650 °C.

On average, the yield of CB was 32.22% at 450 °C, followed by 31.53% at 500 °C, 29.67% at 550 °C, 27.23% at 600 °C, and 25.90% at 650 °C. This shows that even if the temperature rises slightly at 50 °C, the yield will obviously decrease by 2–8%, especially in the low temperature region. It may be because pyrolysis at lower temperatures promotes the degradation of organic components of raw materials and produces various volatile substances and high-boiling substances, while at high temperatures, high-boiling and non-volatile substances decompose slowly [34,41]. Pyrolysis time also has a significant effect on the yield of CB. The average yield of CB was 29.65% at 2 h, and then decreased, 29.55% at 3 h, and 28.74% at 4 h. It can be obviously concluded that with the increase of pyrolysis time, the yield of CB decreases, and the decrease range of yield is about 1.5% for every 1 h of increase. Compared with the pyrolysis temperature, the effect of pyrolysis time is not significant, which is mainly because the heating rate for preparing CB is constant in this work, that is, the time stays for a long time, but the temperature cannot reach, and the decomposition of high-boiling and non-volatile substances is slow. This is consistent with the results reported by Xue et al. [18].

3.2. Biochar Characterization

3.2.1. SEM

The SEM images of the different processes biochar are shown in Figure 2 and Figure S1. It can be seen that CB has obvious pore structure under different carbonization conditions, but the number and size of pores are different. From Figure 2A–E, the microscopic morphology of CB shows flaky irregular polygons with uneven distribution of some broken particles, and there are more loose pores and thin sheets of tubular structure. With the increase of pyrolysis temperature, the surface roughness further increases, and a large number of micropores begin to appear, similar to honeycomb-shaped porous bodies. These honeycomb-shaped micropore structures vary in size and shape, and are closely arranged. From 450 to 650 °C, the more obvious the surface pores texture is, the rougher the surface roughness is, the better the micropore structure is, the thinner the pore wall becomes, and the more the number of pores is. There are the similar rules in Figure 2F–J and Figure 2K–O. These pores are an important basis for biochar to have larger surface area and stronger adsorption performance [18,42]. Comparing Figure 2A,F,K, it is found that the images are very similar, which shows that pyrolysis time has little effect on the pore structure of CB. Further EDS elementary analysis (Figure S1) confirmed that CB not only has higher amounts of elemental C and O, but also Si (0.14%), Cl (3.49%), K (0.3%), Ca (1.19%), and other elements.

3.2.2. BET Specific Surface Area and Pore Size Analysis

The effects of different pyrolysis processes on BET-SSA, pore volume, and pore size of CB are detailed in Figure 3. As shown in Figure 3a, the BET-SSA of CB prepared ranges from 13.52–133.49 $m^2 \cdot g^{-1}$. With the increase of pyrolysis temperature, the BET-SSA of CB increases. It showed that even if the temperature rises by 50 °C, the BET-SSA will increase significantly in the range of 21–61%, especially in the low temperature regions. At the same temperature, the effect of pyrolysis time on the BET-SSA increases first, then decreases, or remains unchanged. The BET-SSA at 650 °C/3 h (Table S1) reached the maximum (133.49 $m^2 \cdot g^{-1}$), which was an ideal alternative biochar. Because of the increase of pyrolysis temperature, the carbon content and the adsorption of hydrophobic pollutants increases while the others decrease, including the oxygen content, the O/C, the hydrophilicity, and polarity of biochar and the affinity for water molecules, resulting in larger specific surface area along with stronger adsorption effect [18,43]. It can be clearly seen from Figure 3b,c that the pore volume increases with the increase of pyrolysis temperature, which has the same trend as the BET-SSA. The average pore size decreases with the increase of pyrolysis temperature. It implies that the large pores on the surface can become mesopores and then develop into microporous structures. At the same temperature, pyrolysis time has little effect on the pore volume and pore size of CB. It is because as the temperature increases, the micropore structure develops, the average pore diameter decreases, the pore wall becomes thinner, the pore number and volume increase [44], and the specific surface area increases.

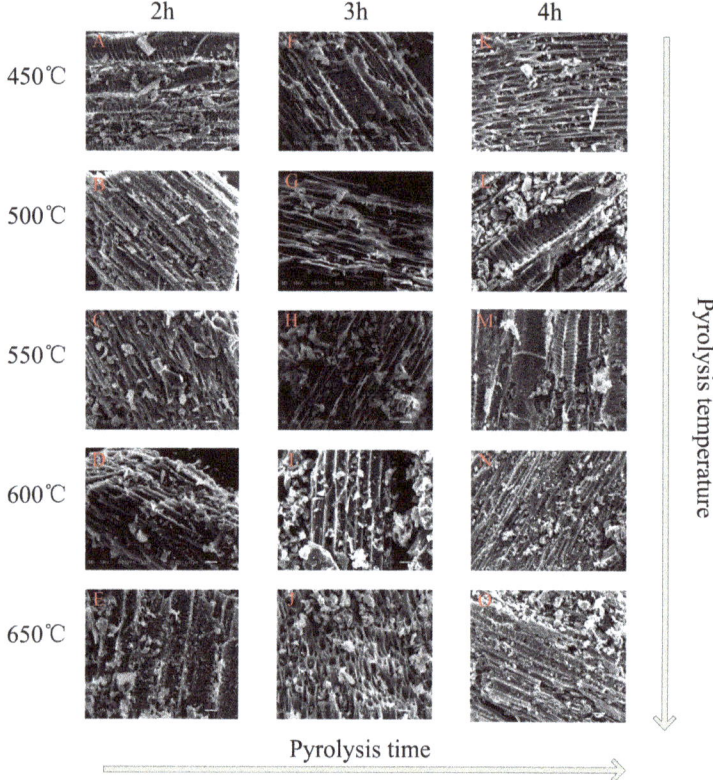

Figure 2. The scanning electron microscope (SEM) images of CB at 1000 times. (**A**) 450 °C/2 h, (**B**) 500 °C/2 h, (**C**) 550 °C/2 h, (**D**) 600 °C/2 h, (**E**) 650 °C/2 h, (**F**) 450 °C/3 h, (**G**) 500 °C/3 h, (**H**) 550 °C/3 h, (**I**) 600 °C/3 h, (**J**) 650 °C/3 h, (**K**) 450 °C/4 h, (**L**) 500 °C/4 h, (**M**) 550 °C/4 h, (**N**) 600 °C/4 h, (**O**) 650 °C/4 h.

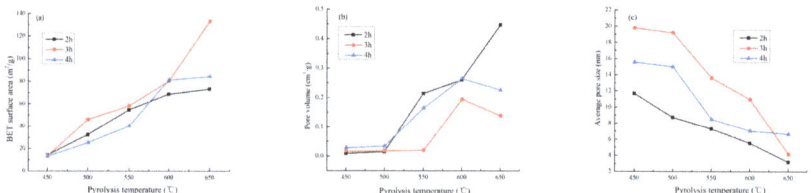

Figure 3. Effects of different pyrolysis processes on BET surface area and pore size analysis of *Caragana korshinskii* biochar (CB). (**a**) BET surface area, (**b**) pore volume, (**c**) average pore size.

According to the definition of the International Union of Pure and Applied Chemistry [45], when a pore size is between 2 and 50 nm, it is called mesopores. It means that the pore structure of CB is mainly mesopores (Figure 3c and Figure S2). Pyrolysis temperature has a great influence on the BET-SSA and pore structure of CB, followed by pyrolysis time, which is basically consistent with the results of SEM image observation. Many reports have found that the BET-SSA of biochar decreases at 700 °C [17,25], which is attributed to the evolution of the volatile phenol bubbles, the structural sequence, the decrease in the number of micropores, and the increase in the number of macropores. In view of this change trend, the pyrolysis temperature of 700 °C was not set in this work.

3.2.3. XPS and XRD

The surface of CB was investigated using XPS (Figure 4), and the results indicated six main elements, namely C, O, N, Si, K, and Ca, and two minor elements, namely Fe and Cl. Their average mass percentages were 74.55%, 18.73%, 0.41%, 3.92%, 0.72%, 0.45%, 0.42%, and 0.30%, respectively. The photoelectron lines with binding energy (E_B) of approximately 100.1, 284.8, 292.9, 346.6, 398.4, and 531.8 eV were attributed to Si 2p, C 1s, K 2p, Ca 2p, N 1s, and O 1s, respectively (Figure 4a). Detailed XPS data for the C 1s, Ca 2p and Si 2p core level spectrum with a peak fitting of its envelope are presented in Figure 4b–d. The C1s XPS spectrum can be curve fitted into four peak components at approximately 284.4 eV (C–C), 285.4 eV (C–O), 287.8 eV (C=O), and 289.9 eV (C-OOR). In addition, according to the relevant literature, there may be a peak at approximately 293.2 eV (CO_3^{2-}) attributed to the high cellulose, hemicelluloses, and lignin contents of *Caragana korshinskii*. With the increase of temperature, the bond valence structure of each element slowly appears and tends to be stable, and the peak value becomes higher with the increase of temperature.

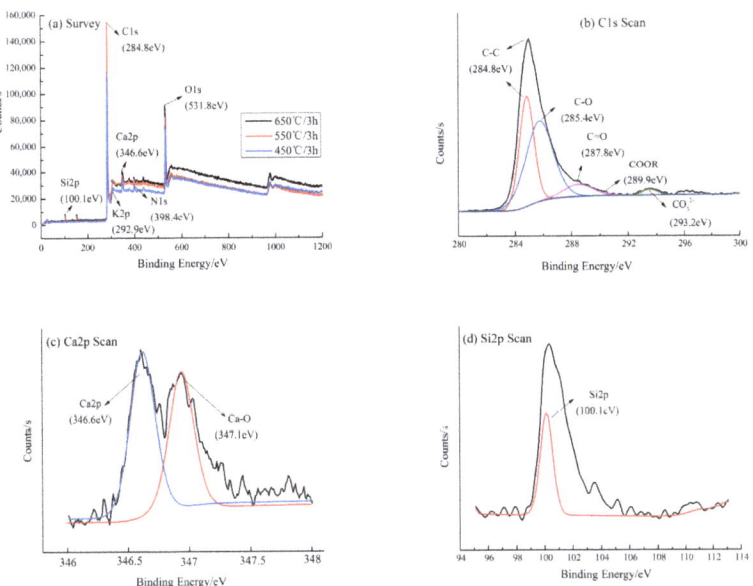

Figure 4. X-ray photoelectron spectra of CB. (a) A typical survey scan under different pyrolysis processes (taking 3 h as an example); (b) C1s binding energy at 650 °C/3 h; (c) Ca2p binding energy at 650 °C/3 h; (d) Si2p binding energy at 650 °C/3 h.

The XRD analysis (Figure 5) revealed that the principal diffraction peaks at low 2θ angles were sharp and symmetric, indicating the presence of mineral crystals. According to the Jade 6.5 PDF cards [46], the two strong peaks at 26.6 and 29.5° suggest the presence of SiO_2 and $CaCO_3$, respectively. Other strong peaks which suggested CaO, KCl, and $CaHPO_4·2H_2O$ were observed on the CB's surface. Moreover, there was a small peak of $(Mg_{0.3}Ca_{0.97})(CO_3)$ with a highly crystalline structure. As the pyrolysis temperature increases, the crystallization of CaO, KCl, $CaHPO_4·2H_2O$ and $MgCa(CO_3)$ gradually transforms, and the peaks of the XRD pattern decrease, or tend to be flat. The peaks of SiO_2

and $CaCO_3$ become more obvious, indicating that pyrolysis temperature increase may be beneficial to the particle growth of inorganic deposits and thus promote the formation of crystallization [34,38]. It means that the cellulose and lignin components in CB gradually degrade, and release a lot of ash. The ash was mainly composed of mineral components such as silicate, potassium salt, and calcium carbonate. Fan et al [17] believed the calcium oxalate decomposed when the temperature raised above 500 °C, the ash content decreased, and calcium carbonate crystals gradually formed. Thus, the XRD results are in good agreement with the XPS analysis (Figure 4c,d). At the same temperature, pyrolysis time has little effect on the XPS and XRD of CB (Figures 4a and 5).

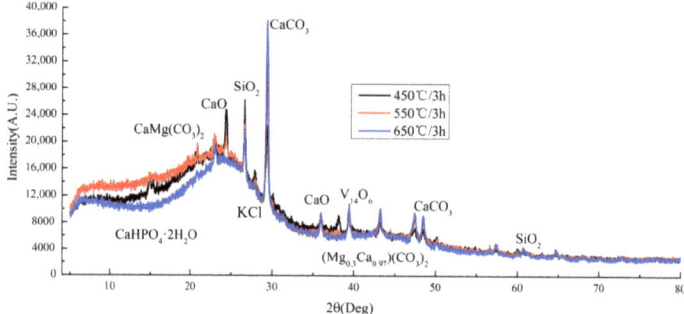

Figure 5. XRD patterns of CB biochar under different pyrolysis processes (taking 3 h as an example).

3.2.4. FTIR

The FTIR of CB is shown in Figure 6. It can be clearly seen that CB has about six identical peak positions, which are located near 500–900, 1300–1400, 1600, 2300, 2800, and 3400 cm^{-1}. Viewing the spectrum data table [3,6,11,17,18,24], it was found that CB has -OH vibration absorption peak (3418 cm^{-1}), and the C-H in alkanes are mainly vibration absorption peaks of methyl and methylene groups (2854–2966 cm^{-1}), the fatty C-H and C=O groups (2280 cm^{-1}), aromatic acids -COOH groups (1697 cm^{-1}), amide stretching vibration -C=O group (1650 cm^{-1}), NH_4^+ group (1396 cm^{-1}), the pyridine and indole of aromatic compounds (500–900 cm^{-1}) groups. It shows that CB has rich functional groups on its surface. Qualitatively, the peak positions of the 15 kinds of CB are roughly the same, and the peak curves are similar, indicating that the functional group types are basically the same. There are certain differences in the surface functional groups of biochar prepared at different temperatures. With the temperature increases, the total content of surface functional groups of biochar decreases, which is consistent with the related research results [19].

Comparing Figure 6a–c, we found that with the increase of temperature, the absorption peak near 3420 cm^{-1} gradually weakens, indicating that the -OH group decreases, which is probably due to the gradual breakage of -OH group by hydrogen-bonded and the separation of bound water [18]. At 2941 cm^{-1}, the absorption peak of C-H vibration in alkanes decreases, which means the loss of alkyl group of CB and the aromatization degree of CB gradually increases with the increase of temperature. Among them, the methylene group around 2927 cm^{-1} is gradually degraded or changed. For the fatty C-H and C=O groups of 2280 cm^{-1}, the absorption peak gradually increases, especially in Figure 6b, which reached the maximum at 650 °C/3 h. With the increase of pyrolysis temperature, the biochar components have experienced the transition of excessive char, amorphous char, composite char, and chaotic char in turn [27]. The aromatic acid -COOH group (1697 cm^{-1}) and amide stretching vibration -CON- group (1650 cm^{-1}) gradually decrease, and even disappear in some scientific reports [32,41]. In the characteristic zone of benzene rings around 1400 cm^{-1}, the absorption peak gradually increased, indicating that the aromatization degree was increased. On the whole, the functional groups of biochar with different pyrolysis processes have certain differences, mainly showing that with the increase of pyrolysis temperature, the content of basic functional groups (benzene ring and pyridine) increases, while the content of acidic functional groups (phenolic hydroxyl and carboxyl) decreases, this meaning the aromaticity increases and the polarity decreases. At the same time, new functional groups appeared around 500–900 cm^{-1}, which means that an aromatic ring is formed, and the aromatization degree is further enhanced. In addition, we found that the pyrolysis time has basically no effect on the biochar surface functional groups, which is basically consistent with many research results [47,48].

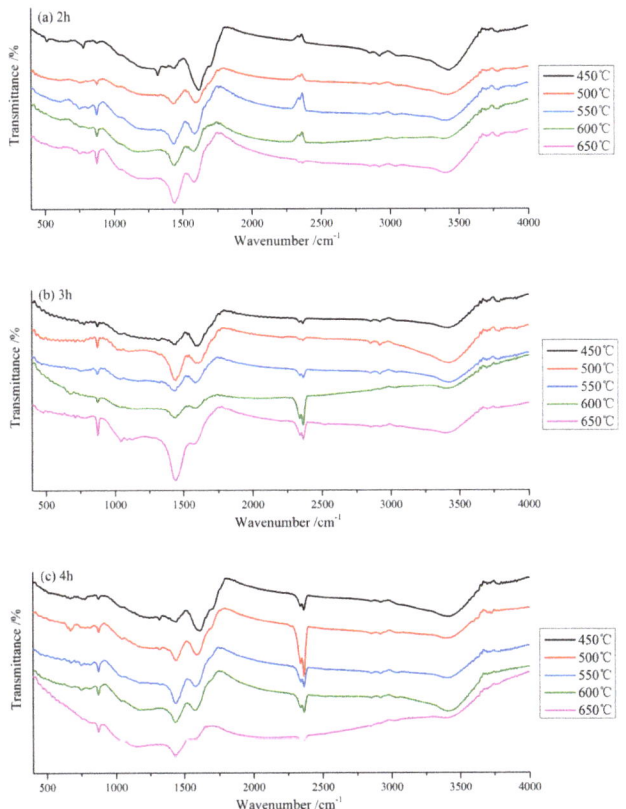

Figure 6. FTIR spectrograms of CB at different pyrolysis processes. (**a**) 2 h; (**b**) 3 h; (**c**) 4 h.

3.3. Adsorption Application

3.3.1. Screening for Optimal Preparation Conditions

The effects of CB prepared under different pyrolysis processes on the adsorption capacity and removal rate of NO_3^- are shown in Figure 7. It can be seen that the adsorption capacity of NO_3^- by CB prepared at 650 °C and 3 h was the largest, reaching 19.67 mg·g^{-1}, and the corresponding RE reached 78.67% (NO_3^- concentration 50 mg·L^{-1}). It is mutually verified by BET-SSA data of CB at 650 °C for 3 h. With the increase of pyrolysis temperature and time, the adsorption capacity and RE of NO_3^- by CB showed a rising trend, but the curve presented twists and turns. On average, regardless of the preparation process, the adsorption capacity of CB for NO_3^- is above 18.8 mg·g^{-1}, and the removal rate is above 76%, which implies that the *Caragana korshinskii* is an excellent and low-cost biomass raw material. Therefore, CB prepared at 650 °C for 3 h was the optimal biochar and the most appropriate preparation conditions.

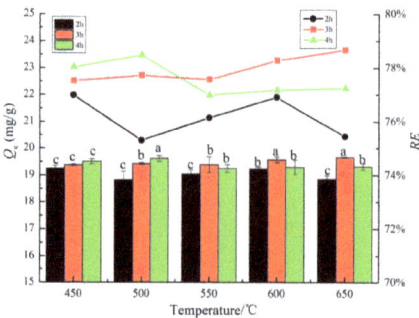

Figure 7. Effects of different pyrolysis process on CB adsorption capacity and removal rate for NO_3^-.

3.3.2. Isotherms Adsorption

As shown in Figure 8, with increasing initial NO_3^- concentration, the adsorption capacity of CB gradually increased until saturation. At low initial concentrations, the adsorption curve is steep, and the adsorption capacity increases rapidly in a straight line. At high initial concentrations, the adsorption increased slowly and eventually stabilized to reach equilibrium. The initial concentration of the solution is essential to overcome the mass transfer resistance between the liquid and the solid phase. So, higher initial concentration of the solution improves the adsorption capacity of the biochar [49]. The related parameters calculated from the four isotherms models are listed in Table 3. The isotherm models reproduced the adsorption data well. However, the Langmuir models for NO_3^- (highest R^2 value of 0.9364) matched the experimental data better than the other models. These results suggest that the adsorption process is similar to a monolayer adsorption for NO_3^- and mainly physical adsorption [4]. But the Freundlich models presented high fitting decision coefficients, which may also imply that the process involved multilayer complex and associated with chemical interactions [11,15]. In the Langmuir model (Table 3), the theoretical maximum adsorption capacity (Q_m) of NO_3^- were 120.65 mg·g^{-1}. The Q_m values and the actual maximum adsorption capacity obtained by the experiments were significantly different. The difference is attributed to the limitations and errors of the model, which are reflected on the isothermal adsorption curve [4]. The R_L values of NO_3^- are between 0 and 1, thus indicating a favorable adsorption process. In the Freundlich model, K_F is related to the adsorption capacity of the adsorbent, which indicates that the adsorption capacity and strength was stable. In the Temkin model, the R^2 values is bigger than 0.8, thus suggesting that the adsorption process includes chemisorption. In the D-R model, E represents the chemical adsorption at 8–16 kJ·mol^{-1} and physical adsorption below 8 kJ·mol^{-1} [30]. This indicates that the adsorption of NO_3^- (E = 2.52 kJ·mol^{-1}) by CB is dominated by physical adsorption. The Q_0 is the maximum unit adsorption capacity, which is associated with the ordinate of last point (Figure 8).

Table 3. Isothermal adsorption fitting parameters of CB for NO_3^-.

Models		Langmuir			Freundlich			Temkin			D-R Model		
	a	Q_m (mg·g^{-1})	R_L	R^2	K_F	n	R^2	A	B	R^2	Q_0 (mmol·g^{-1})	E (kJ·mol^{-1})	R^2
CB (650 °C/3 h)	0.02	120.65	0.0008–0.97	0.9364	2.45	1.20	0.9355	0.77	10.21	0.8506	33.04	2.52	0.8568

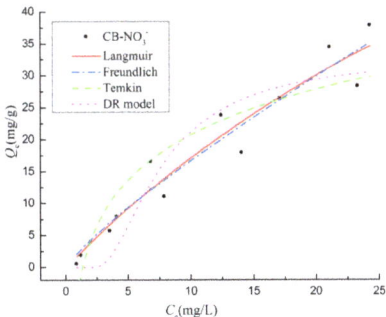

Figure 8. The isotherm adsorption of CB biochar for NO_3^- (650 °C/3 h).

3.3.3. Adsorption Kinetics

The adsorption kinetics fitting curves are presented in Figure 9, and the kinetic parameters are summarized in Table 4. With increasing adsorption time, the adsorption rate of NO_3^- was fast and large at the beginning, then the curve slowed down and increased slightly, and finally reached adsorption equilibrium. For example, the adsorption capacity reached 92% of the saturated adsorption capacity at 5 min, and gradually tended to equilibrium. The correlation coefficients of the pseudo-second-order model ($R^2 > 0.99$) were higher than other models, and the fitted equilibrium adsorption capacities were closer to the experimental values. These results suggest that the adsorption of NO_3^- were better described by the pseudo-second-order kinetic model [11]. The adsorption process is a composite adsorption reaction, which includes surface adsorption, external liquid film diffusion, and intraparticle diffusion. This phenomenon can be similarly verified in many reports [26,28,43]. In the pseudo-second-order model (Table 4), h represents the initial adsorption rate, and its value is 0.38 mg·g^{-1}·min^{-1}. The Q_e is very similar in the pseudo-first-order and pseudo-second-order model, both of which are related to the adsorption capacity at an equilibrium concentration. The R^2 values from the Elovich model is larger than 0.9, which indicates that the CB surface adsorption energy is uniformly distributed during the entire adsorption process. For the intraparticle diffusion model, the fitted straight line does not cross the origin, which indicates that particle diffusion is not the only rate-limiting factor [30]. There are other processes (surface adsorption, liquid film diffusion, etc.,) that jointly control the adsorption reaction rate [15]. The C are 15.23, which implies the boundary layer of CB has a greater effect on the adsorption processes.

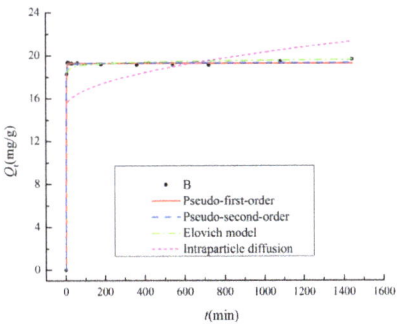

Figure 9. The adsorption kinetics of CB biochar for NO_3^- (650 °C/3 h).

Table 4. Adsorption kinetics fitting parameters of CB for NO_3^-.

Models	Pseudo-First-Order			Pseudo-Second-Order				Elovich			Intraparticle Diffusion		
	Q_e (mg·g^{-1})	k_1	R^2	Q_e (mg·g^{-1})	k_2	H (mg·g^{-1}·min^{-1})	R^2	a'	b	R^2	k_i	C	R^2
CB (650 °C/3 h)	19.28	1.37	0.9991	19.45	0.001	0.38	0.9998	4.68	5.70	0.9959	0.16	15.23	0.06

In conclusion, relevant reports on the adsorption of NO_3^- by biochar are summarized in Table 5. Compared to other cellulose- and lignin-based biomass raw materials, the adsorption capacity of NO_3^- by the CB in this study was larger. Even if compared with commercial activated carbon (1.22 mg·g^{-1}) [50], the adsorption performance of CB also has certain advantages. Therefore, in view of the preparation temperature and time, the CB can be developed and popularized because of low cost and efficient adsorption performance without complicated modification conditions.

Table 5. Comparison of sorption capacity of CB with selected biochars derived from other materials for NO_3^-.

Raw Materials	Temperature (°C) [1]	Time (h) [1]	Maximum Adsorption Capacity (mg·g^{-1}) [2]	Literature
Wheat straw	300	1	1.10	[50]
Mustard straw	300	1	1.30	[50]
Ponderosa pine wood	650	0.3	6.20	[51]
Switchgrass	650	0.3	1.84	[51]
Corn stover	650	0.3	6.25	[51]
Sugar beet bagasse	700	1.5	9.14	[52]
Rice husk	600	-	2.1	[53]
Date palm	700	4	8.37	[54]
Fe-impregnated corn stalk	550	0.5	15.41	[29]
$ZnCl_2$ activated Sugar beet bagasse	700	1.5	27.55	[52]
Amine-grafted corn cob	100	24	49.9	[55]
Amine-grafted coconut copra	100	48	59.2	[55]
Caragana korshinskii	650	3	120.65	This work

[1] Pyrolysis temperature and time. [2] the maximum adsorption capacity in the table is the maximum of the actual adsorption capacity or the theoretical maximum adsorption capacity calculated by Langmuir model.

3.4. By-Products and Process Analysis

The by-products of *Caragana korshinskii* biomass include bio-oil and tail gas in pyrolysis process. All analytical components of bio-oil and tail gas analyzed by GC-MS are listed in Tables S2 and S3, respectively. After searching in the NIST05 standard mass spectrometry library and consulting relevant data [56], a total of 83 chromatographic peaks were identified in the bio-oil, including 73 compounds in this work. There are 41 chromatographic peaks in the liquid collected by tail gas condensation, including 40 compounds. According to the classification of these organic compounds (Table S2), there are 29 kinds of aliphatic compounds with a relative content of 32.80%, including alkanes, olefins, alcohols, fatty acids, ketones, aldehydes, carbohydrate, etc., 28 aromatic compounds with a relative content of 56.64%, and 16 heterocyclic compounds with a relative content of 10.56%. In general, the content of aromatic compounds in bio-oil is more than the content of aliphatic compounds, and the ratio is 1.73:1. The analysis results in the liquid collected by tail gas condensation (Table S3) are similar to those of bio-oil. The main components of bio-oil analyzed by GC-MS are listed in Table 6 through the screening conditions that the quality matching of GC-MS is greater than 60%. It is worth noting that the content of these compounds accounts for 68.93% of the total, where the higher content is 1,2-cyclopentanedione and 3-methyl-. The content of C6-C15 hydrocarbon compounds in bio-oil is close to 10%, which means that it is of great significance as a potential bio-diesel. The content of ketones and heterocyclic compounds in bio-oils exceeds 20%, which can be used as important chemical raw materials. Benzene and its derivatives and naphthalene are intermediates in the synthesis of spices, plastics, dyes, and drugs [57]. The bio-oil also contains a certain amount of high-value chemicals, such as Guaiacol (Phenol, 2-methoxy-) and α-D-glucose (1,4:3,6-Dianhydro-α-D-glucopyranose). The Guaiacol (relative content 5.08%) is an important fine chemical intermediate and widely used in the synthesis of medicines, spices and dyes [58]. The α-D-glucopyranose (relative content 2.63%), which is mainly used in medical liver function test or medicine, is rarely synthesized artificially considering its cost. Interestingly, the bio-oil contains α-D-glucopyranose and thus can be a valuable product obtained by pyrolysis biomass, which usually exists in the form of black-brown organic liquid mixture. Discussing with some related literature reports [56–58], the bio-oil of *Caragana korshinskii* biomass has the advantages of wide raw material sources, abundant reserves, renewable, high energy density, and easy transportation and storage. Khuenkaeo et al. [57] and Lu et al. [58] also believed the bio-oil is a potential source of liquid fuel and chemical raw materials. Therefore, biomass pyrolysis is an important technical method for solving fossil energy problems in the future, and many scholars have reached consensus on this point [56–58].

Table 6. The main components of bio-oil analyzed by GC-MS.

Number	Library (ID)	CAS	Quality (%)	Peak Area (%)
1	2-Furanmethanol	000098-00-0	90	2.40
2	Pyridine, 3-methyl-	000108-99-6	92	1.22
3	2,6-Lutidine	000108-48-5	70	0.39
4	2-Cyclopenten-1-one, 2-methyl-	001120-73-6	80	1.73
5	Pyridine, 3,5-dimethyl-	000591-22-0	90	0.80
6	Pyridine, 2,3-dimethyl-	000583-61-9	94	0.50
7	2-Cyclopenten-1-one, 3-methyl-	002758-18-1	91	2.49
8	Phenol	000108-95-2	91	4.24
9	3-Methylpyridazine	001632-76-4	64	0.22
10	Pyridine, 3-methoxy-	007295-76-3	80	0.54
11	1,2-Cyclopentanedione, 3-methyl-	000765-70-8	95	6.70
12	2-Cyclopenten-1-one, 2,3-dimethyl-	001121-05-7	90	1.18
13	Phenol, 2-methyl-	000095-48-7	97	2.12
14	p-Cresol	000106-44-5	95	5.51
15	Phenol, 2-methoxy-	000090-05-1	94	5.08
16	Phenol, 2,6-dimethyl-	000576-26-1	76	0.12
17	Maltol	000118-71-8	62	0.17
18	2-Cyclopenten-1-one, 3-ethyl-2-hydroxy-	021835-01-8	97	1.67
19	Phenol, 2,4-dimethyl-	000105-67-9	95	2.92
20	Phenol, 4-ethyl-	000123-07-9	93	0.78
21	Naphthalene	000091-20-3	87	1.39
22	Creosol	000093-51-6	97	3.37
23	1,4:3,6-Dianhydro-.alpha.-D-glucopyranose	1000098-14-8	95	2.63
24	Phenol, 4-ethyl-3-methyl-	001123-94-0	64	0.75
25	1,2-Benzenediol, 3-methoxy-	000934-00-9	94	2.33
26	4-Methoxybenzene-1,2-diol	003934-97-2	62	1.27
27	Phenol, 4-ethyl-2-methoxy-	002785-89-9	90	2.70
28	Naphthalene, 2-methyl-	000091-57-6	97	2.33
29	Phenol, 2,6-dimethoxy-	000091-10-1	96	6.54
30	Naphthalene, 1,3-dimethyl-	000575-41-7	94	0.69
31	Naphthalene, 1,5-dimethyl-	000571-61-9	95	0.61
32	3,5-Dimethoxy-4-hydroxytoluene	006638-05-7	96	3.51

A very interesting process about the change of tail gas is recorded in detail as follows: a slight wood burning odor begins to form at 280 °C, and green smoke is generated at 300 °C, accompanied by an irritating smoke smell. A lot of brown smoke is generated at 350 °C, and the tail gas devices start to collect the bio-oil. The tail gas turns into white smoke at 400 °C followed by a stable smoke flow and a large amount of bio-oil. When the tail gas color becomes lighter at 450 °C, the bio-oil tends to be stable. At 550 °C, the smoke gradually decreases until there is no smoke. No irritating smoke can be smelled at 600 °C. The yield of bio-oil increased with the rise of temperature, and increased greatly from 350 to 450 °C, but slightly from 450 to 550 °C, then gradually decreased to 0 from 550 °C to 600 °C. This result shows that some components in bio-oil further crack to generate small molecules of non-condensable gas, which reduces the yield of bio-oil and further increases the output of non-condensable gas [18]. Up to now, we may not have found any relevant reports on the tail gas color change of pyrolysis biomass.

Therefore, we speculate that the possible mechanism is as follows. The chemical structure of biomass is extremely complex, and its main components are cellulose, hemicellulose, and lignin. Among them, the pyrolysis mechanism of cellulose has been widely studied and a preliminary theoretical system has been formed [59]. However, hemicellulose and lignin are relatively lacking on their pyrolysis mechanism because of their complex structure and many research branches. In general, during the pyrolysis process, water is evaporated at 105 °C, and the main pyrolysis temperature of cellulose is 240–375 °C, which usually produces dehydrated fiber and L-glucose. Hemicellulose softens and decomposes at 200–295 °C and usually produces volatile products. Lignin decomposes at 280–500 °C, usually producing carbon and various organic compounds. From the perspective of matter and energy, heat is first transferred to the biomass surface and then the inside. The pyrolysis process proceeds layer by layer from the inside to the outside, and the heated part of the biomass is rapidly decomposed into biochar and volatiles [60]. Among them, the volatile matter is composed of condensable gas and non-condensable gas, and the condensable gas is rapidly condensed to obtain bio-oil [57]. With the transfer of heat, the organic matter inside the biomass is heated to continue pyrolysis, and part of the primary pyrolysis product undergoes a second pyrolysis or multiple pyrolysis until it is stable.

4. Conclusions

This work used *Caragana korshinskii* biomass as a raw material to prepare 15 kinds of biochar by controlling the oxygen-limited pyrolysis process. On the whole, the effect of pyrolysis temperature on the CB properties is the most important, which is greater than pyrolysis time. The CEC, pH, and BET specific surface area of CB under each pyrolysis process were 16.64–81.4 cmol·kg^{-1}, 6.65–8.99, and 13.52–133.49 m^2·g^{-1}, respectively. With the increase of pyrolysis temperature and time, the bond valence structure of C 1s, Ca 2p, and O 1s is more stable, and the phase structure of CaCO$_3$ is more obvious. The surface of CB biochar contains abundant functional groups, such as hydroxyl, carboxyl, ester carbonyl, NH$_4$$^+$, pyridine, etc. With the increase of pyrolysis temperature, the content of basic functional groups (benzene ring and pyridine) and the aromaticity increases, while the content of acidic functional groups (phenolic hydroxyl and carboxyl) decreases and the polarity decrease. The CB prepared at 650 °C for 3 h presented the best adsorption performance, and the maximum theoretical adsorption capacity for NO$_3$$^-$ reached 120.65 mg·g^{-1}. The Langmuir model and pseudo-second-order model can well describe the isothermal and kinetics process of NO$_3$$^-$, respectively. This indicated that the beneficial adsorption process via monolayer was affected by surface adsorption, intraparticle diffusion and liquid film diffusion, and the physical adsorption is the more important mechanism. The by-products contain bio-soil and tail gas, which are potential source of liquid fuel and chemical raw materials. The bio-oil of CB contains α-D-glucopyranose, which can be used in medical tests and medicines. The color and flow rate of the tail gas also change correspondingly with the increase of temperature during the pyrolysis process. Therefore, the CB and bio-oil from *Caragana korshinskii* biomass will serve arid areas of Asia and Africa as innovative products, and CB is worth to be further developed and popularized because of its low cost and efficient adsorption performance without complicated modification condition.

Supplementary Materials: The following are available online at http://www.mdpi.com/1996-1944/13/15/3391/s1, Table S1: The surface characteristics of alternative CB biochar and Caragana korshinskii, Table S2: The all components of bio-oil analyzed by GC-MS, Table S3: The all components of the liquid collected by tail gas condensation analyzed, Figure S1: Scanning electron microscope (SEM) images of Energy Dispersive Spectrometer (EDS) spectra of CB at 650 °C/3 h, Figure S2: BJH (Barren-Joyner-Halenda)-adsorption-pore size distribution of CB at 650 °C/3 h.

Author Contributions: T.W. designed and performed all the experimental work, and wrote the draft of the manuscript. H.L., C.D., and R.X. conducted the characterizations. Z.Z. performed model analysis. J.Z. and D.S. supervised the research and reviewed the manuscript. All authors have read and agreed to the published version of the manuscript.

Funding: We acknowledge funding from the National Key Research and Development Plan of China (2017YFC0504504), the Key Research and Development Plan of Ningxia Hui Autonomous Region (2020BCF01001), and the Science and Technology Service Network Initiative of the Chinese Academy of Sciences (KFJ-STS-QYZD-177). We also thank the Opening Fund of Chongqing Key Laboratory of Environmental Materials and Remediation Technology (CEK1805).

Acknowledgments: The authors are thankful to Xiaoyan Zhou for the kind help in the experiment.

Conflicts of Interest: The authors declare no conflicts of interest.

References

1. Liu, W.; Yu, Z.; Zhu, Q.; Zhou, X.; Peng, C. Assessment of biomass utilization potential of Caragana korshinskii and its effect on carbon sequestration on the Northern Shaanxi Loess Plateau, China. *Land Degrad. Dev.* **2019**, *31*, 53–64. [CrossRef]
2. Long, Y.; Liang, F.; Zhang, J.; Xue, M.; Zhang, T.; Pei, X. Identification of drought response genes by digital gene expression (DGE) analysis in Caragana korshinskii Kom. *Gene* **2020**, *725*, 144170. [CrossRef]
3. Wang, T.; Stewart, C.E.; Sun, C.; Wang, Y.; Zheng, J. Effects of biochar addition on evaporation in the five typical Loess Plateau soils. *CATENA* **2018**, *162*, 29–39. [CrossRef]
4. Cui, Q.; Jiao, G.; Zheng, J.; Wang, T.; Wu, G.; Li, G. Synthesis of a novel magnetic Caragana korshinskii biochar/Mg–Al layered double hydroxide composite and its strong adsorption of phosphate in aqueous solutions. *RSC Adv.* **2019**, *9*, 18641–18651. [CrossRef]
5. Dou, Y.; Yang, Y.; An, S. Above-Ground Biomass Models of Caragana korshinskii and Sophora viciifolia in the Loess Plateau, China. *Sustainability* **2019**, *11*, 1674. [CrossRef]
6. Zheng, X.; Shi, T.; Song, W.; Xu, L.; Dong, J. Biochar of distillers' grains anaerobic digestion residue: Influence of pyrolysis conditions on its characteristics and ammonium adsorptive optimization. *Waste Manag. Res.* **2020**, *38*, 86–97. [CrossRef]
7. Alcantara, J.C.; Gonzalez, I.; Pareta, M.M.; Vilaseca, F. Biocomposites from Rice Straw Nanofibers: Morphology, Thermal and Mechanical Properties. *Materials* **2020**, *13*, 2138. [CrossRef]
8. Zhang, Z.S.; Li, X.R.; Liu, L.C.; Jia, R.L.; Zhang, J.G.; Wang, T. Distribution, biomass, and dynamics of roots in a revegetated stand of Caragana korshinskii in the Tengger Desert, northwestern China. *J. Plant Res.* **2009**, *122*, 109–119. [CrossRef]
9. Hoffmann, V.; Jung, D.; Zimmermann, J.; Correa, C.R.; Elleuch, A.; Halouani, K.; Kruse, A. Conductive Carbon Materials from the Hydrothermal Carbonization of Vineyard Residues for the Application in Electrochemical Double-Layer Capacitors (EDLCs) and Direct Carbon Fuel Cells (DCFCs). *Materials* **2019**, *12*, 1703. [CrossRef]
10. Lehmann, J.; Rillig, M.C.; Thies, J.; Masiello, C.A.; Hockaday, W.C.; Crowley, D. Biochar effects on soil biota—A review. *Soil Biol. Biochem.* **2011**, *43*, 1812–1836. [CrossRef]
11. Cui, Q.; Xu, J.; Wang, W.; Tan, L.; Cui, Y.; Wang, T.; Li, G.; She, D.; Zheng, J. Phosphorus recovery by core-shell γ-Al_2O_3/Fe_3O_4 biochar composite from aqueous phosphate solutions. *Sci. Total Environ.* **2020**, *729*, 138892. [CrossRef] [PubMed]
12. Wang, T.; Stewart, C.E.; Ma, J.; Zheng, J.; Zhang, X. Applicability of five models to simulate water infiltration into soil with added biochar. *J. Arid Land* **2017**, *9*, 701–711. [CrossRef]
13. Zhao, N.; Yang, X.; Zhang, J.; Zhu, L.; Lv, Y. Adsorption Mechanisms of Dodecylbenzene Sulfonic Acid by Corn Straw and Poplar Leaf Biochars. *Materials* **2017**, *10*, 1119. [CrossRef]
14. Irfan, M.; Lin, Q.; Yue, Y.; Ruan, X.; Chen, Q.; Zhao, X.; Dong, X. Co-production of Biochar, Bio-oil, and Syngas from Tamarix chinensis Biomass under Three Different Pyrolysis Temperatures. *Bioresources* **2016**, *11*, 8929–8940. [CrossRef]
15. Inyang, M.I.; Gao, B.; Yao, Y.; Xue, Y.; Zimmerman, A.; Mosa, A.; Pullammanappallil, P.; Ok, Y.S.; Cao, X. A review of biochar as a low-cost adsorbent for aqueous heavy metal removal. *Crit. Rev. Environ. Sci. Technol.* **2015**, *46*, 406–433. [CrossRef]

16. Chen, Z.; Liu, T.; Tang, J.; Zheng, Z.; Wang, H.; Shao, Q.; Chen, G.; Li, Z.; Chen, Y.; Zhu, J.; et al. Characteristics and mechanisms of cadmium adsorption from aqueous solution using lotus seedpod-derived biochar at two pyrolytic temperatures. *Environ. Sci. Pollut. Res. Int.* **2018**, *25*, 11854–11866. [CrossRef] [PubMed]
17. Fan, L.; Zhou, X.; Liu, Q.; Wan, Y.; Cai, J.; Chen, W.; Chen, F.; Ji, L.; Cheng, L.; Luo, H. Properties of Eupatorium adenophora Spreng (Crofton Weed) Biochar Produced at Different Pyrolysis Temperatures. *Environ. Eng. Sci.* **2019**, *36*, 937–946. [CrossRef]
18. Xue, Y.; Wang, C.; Hu, Z.; Zhou, Y.; Xiao, Y.; Wang, T. Pyrolysis of sewage sludge by electromagnetic induction: Biochar properties and application in adsorption removal of Pb(II), Cd(II) from aqueous solution. *Waste Manag.* **2019**, *89*, 48–56. [CrossRef]
19. Rodriguez Correa, C.; Kruse, A. Biobased Functional Carbon Materials: Production, Characterization, and Applications—A Review. *Materials* **2018**, *11*, 1568. [CrossRef]
20. Zhang, X.; Zhu, J.; Wu, C.; Wu, Q.; Liu, K.; Jiang, K. Preparation and Properties of Wood Tar-based Rejuvenated Asphalt. *Materials* **2020**, *13*, 1123. [CrossRef]
21. Cui, Y.; Wang, W.; Chang, J. Study on the Product Characteristics of Pyrolysis Lignin with Calcium Salt Additives. *Materials* **2019**, *12*, 1609. [CrossRef] [PubMed]
22. Mo, H.; Qiu, J.; Yang, C.; Zang, L.; Sakai, E. Preparation and characterization of magnetic polyporous biochar for cellulase immobilization by physical adsorption. *Cellulose* **2020**, *27*, 4963–4973. [CrossRef]
23. Li, S.; Yao, Y.; Zhao, T.; Wang, M.; Wu, F. Biochars preparation from waste sludge and composts under different carbonization conditions and their Pb(II) adsorption behaviors. *Water Sci. Technol.* **2019**, *80*, 1063–1075. [CrossRef]
24. Wang, Z.; Li, J.; Zhang, G.; Zhi, Y.; Yang, D.; Lai, X.; Ren, T. Characterization of Acid-Aged Biochar and its Ammonium Adsorption in an Aqueous Solution. *Materials* **2020**, *13*, 2270. [CrossRef] [PubMed]
25. Li, S.; Barreto, V.; Li, R.; Chen, G.; Hsieh, Y.P. Nitrogen retention of biochar derived from different feedstocks at variable pyrolysis temperatures. *J. Anal. Appl. Pyrolysis* **2018**, *133*, 136–146. [CrossRef]
26. Dai, Y.; Wang, W.; Lu, L.; Yan, L.; Yu, D. Utilization of biochar for the removal of nitrogen and phosphorus. *J. Clean. Prod.* **2020**, *257*, 120573. [CrossRef]
27. Luo, L.; Wang, G.; Shi, G.; Zhang, M.; Zhang, J.; He, J.; Xiao, Y.; Tian, D.; Zhang, Y.; Deng, S.; et al. The characterization of biochars derived from rice straw and swine manure, and their potential and risk in N and P removal from water. *J. Environ. Manag.* **2019**, *245*, 1–7. [CrossRef]
28. Jiang, Y.H.; Li, A.Y.; Deng, H.; Ye, C.H.; Wu, Y.Q.; Linmu, Y.D.; Hang, H.L. Characteristics of nitrogen and phosphorus adsorption by Mg-loaded biochar from different feedstocks. *Bioresour. Technol.* **2019**, *276*, 183–189. [CrossRef]
29. Min, L.; Zhongsheng, Z.; Zhe, L.; Haitao, W. Removal of nitrogen and phosphorus pollutants from water by FeCl3- impregnated biochar. *Ecol. Eng.* **2020**, *149*, 105792. [CrossRef]
30. Tümsek, F.; Avcı, Ö. Investigation of Kinetics and Isotherm Models for the Acid Orange 95 Adsorption from Aqueous Solution onto Natural Minerals. *J. Chem. Eng. Data* **2013**, *58*, 551–559. [CrossRef]
31. Tan, X.F.; Liu, Y.G.; Gu, Y.L.; Xu, Y.; Zeng, G.M.; Hu, X.J.; Liu, S.B.; Wang, X.; Liu, S.M.; Li, J. Biochar-based nano-composites for the decontamination of wastewater: A review. *Bioresour. Technol.* **2016**, *212*, 318–333. [CrossRef]
32. Tsai, W.-T.; Huang, C.-N.; Chen, H.-R.; Cheng, H.-Y. Pyrolytic Conversion of Horse Manure into Biochar and Its Thermochemical and Physical Properties. *Waste Biomass Valorization* **2015**, *6*, 975–981. [CrossRef]
33. Shen, Z.; Jin, F.; Wang, F.; McMillan, O.; Al-Tabbaa, A. Sorption of lead by Salisbury biochar produced from British broadleaf hardwood. *Bioresour. Technol.* **2015**, *193*, 553–556. [CrossRef] [PubMed]
34. Shaaban, A.; Se, S.-M.; Dimin, M.F.; Juoi, J.M.; Husin, M.H.M.; Mitan, N.M.M. Influence of heating temperature and holding time on biochars derived from rubber wood sawdust via slow pyrolysis. *J. Anal. Appl. Pyrolysis* **2014**, *107*, 31–39. [CrossRef]
35. Yao, Y.; Gao, B.; Zhang, M.; Inyang, M.; Zimmerman, A.R. Effect of biochar amendment on sorption and leaching of nitrate, ammonium, and phosphate in a sandy soil. *Chemosphere* **2012**, *89*, 1467–1471. [CrossRef]
36. Singh, B.; Singh, B.P.; Cowie, A.L. Characterisation and evaluation of biochars for their application as a soil amendment. *Aust. J. Soil Res.* **2010**, *48*, 516–525. [CrossRef]

37. Ahmad, Z.; Gao, B.; Mosa, A.; Yu, H.; Yin, X.; Bashir, A.; Ghoveisi, H.; Wang, S. Removal of Cu(II), Cd(II) and Pb(II) ions from aqueous solutions by biochars derived from potassium-rich biomass. *J. Clean. Prod.* **2018**, *180*, 437–449. [CrossRef]
38. Cantrell, K.B.; Hunt, P.G.; Uchimiya, M.; Novak, J.M.; Ro, K.S. Impact of pyrolysis temperature and manure source on physicochemical characteristics of biochar. *Bioresour. Technol.* **2012**, *107*, 419–428. [CrossRef]
39. Guo, Y.; Rockstraw, D.A. Activated carbons prepared from rice hull by one-step phosphoric acid activation. *Microporous Mesoporous Mater.* **2007**, *100*, 12–19. [CrossRef]
40. Liang, B.; Lehmann, J.; Solomon, D.; Kinyangi, J.; Grossman, J.; O'Neill, B.; Skjemstad, J.O.; Thies, J.; Luizao, F.J.; Petersen, J.; et al. Black Carbon increases cation exchange capacity in soils. *Soil Sci. Soc. Am. J.* **2006**, *70*, 1719–1730. [CrossRef]
41. Chen, T.; Zhang, Y.; Wang, H.; Lu, W.; Zhou, Z.; Zhang, Y.; Ren, L. Influence of pyrolysis temperature on characteristics and heavy metal adsorptive performance of biochar derived from municipal sewage sludge. *Bioresour. Technol.* **2014**, *164*, 47–54. [CrossRef]
42. Goliszek, M.; Podkoscielna, B.; Sevastyanova, O.; Gawdzik, B.; Chabros, A. The Influence of Lignin Diversity on the Structural and Thermal Properties of Polymeric Microspheres Derived from Lignin, Styrene, and/or Divinylbenzene. *Materials* **2019**, *12*, 2847. [CrossRef]
43. Tong, Y.; McNamara, P.J.; Mayer, B.K. Adsorption of organic micropollutants onto biochar: A review of relevant kinetics, mechanisms and equilibrium. *Environ. Sci. -Water Res. Technol.* **2019**, *5*, 821–838. [CrossRef]
44. Butt, M.T.Z.; Preuss, K.; Titirici, M.-M.; Rehman, H.U.; Briscoe, J. Biomass-Derived Nitrogen-Doped Carbon Aerogel Counter Electrodes for Dye Sensitized Solar Cells. *Materials* **2018**, *11*, 1171. [CrossRef] [PubMed]
45. Hverett, D.H. IUPAC manual of symbols and terminology for physicochemical quantities and units. *Pure Appl. Chem.* **1972**, *31*, 579–638.
46. Tan, L.; Ma, Z.; Yang, K.; Cui, Q.; Wang, K.; Wang, T.; Wu, G.L.; Zheng, J. Effect of three artificial aging techniques on physicochemical properties and Pb adsorption capacities of different biochars. *Sci. Total Environ.* **2019**, *699*, 134223. [CrossRef] [PubMed]
47. Angin, D. Effect of pyrolysis temperature and heating rate on biochar obtained from pyrolysis of safflower seed press cake. *Bioresour. Technol.* **2013**, *128*, 593–597. [CrossRef]
48. Cimo, G.; Kucerik, J.; Berns, A.E.; Schaumann, G.E.; Alonzo, G.; Conte, P. Effect of Heating Time and Temperature on the Chemical Characteristics of Biochar from Poultry Manure. *J. Agric. Food Chem.* **2014**, *62*, 1912–1918. [CrossRef]
49. Kalinke, C.; Mangrich, A.S.; Marcolino-Junior, L.H.; Bergamini, M.F. Biochar prepared from castor oil cake at different temperatures: A voltammetric study applied for Pb(2^+), Cd(2^+) and Cu(2^+) ions preconcentration. *J. Hazard. Mater.* **2016**, *318*, 526–532. [CrossRef]
50. Mishra, P.C.; Patel, R.K. Use of agricultural waste for the removal of nitrate-nitrogen from aqueous medium. *J. Environ. Manag.* **2009**, *90*, 519 522. [CrossRef]
51. Chintala, R.; Mollinedo, J.; Schumacher, T.E.; Papiernik, S.K.; Malo, D.D.; Clay, D.E.; Kumar, S.; Gulbrandson, D.W. Nitrate sorption and desorption in biochars from fast pyrolysis. *Microporous Mesoporous Mater.* **2013**, *179*, 250–257. [CrossRef]
52. Demiral, H.; Gunduzoglu, G. Removal of nitrate from aqueous solutions by activated carbon prepared from sugar beet bagasse. *Bioresour. Technol.* **2010**, *101*, 1675–1680. [CrossRef] [PubMed]
53. Pratiwi, E.P.A.; Hillary, A.K.; Fukuda, T.; Shinogi, Y. The effects of rice husk char on ammonium, nitrate and phosphate retention and leaching in loamy soil. *Geoderma* **2016**, *277*, 61–68. [CrossRef]
54. Alsewaileh, A.S.; Usman, A.R.; Al-Wabel, M.I. Effects of pyrolysis temperature on nitrate-nitrogen ($NO_3{-}$-N) and bromate ($BrO3-$) adsorption onto date palm biochar. *J. Environ. Manag.* **2019**, *237*, 289–296. [CrossRef]
55. Kalaruban, M.; Loganathan, P.; Shim, W.G.; Kandasamy, J.; Ngo, H.H.; Vigneswaran, S. Enhanced removal of nitrate from water using amine-grafted agricultural wastes. *Sci. Total Environ.* **2016**, *565*, 503–510. [CrossRef]
56. Cui, Y.; Hou, X.; Wang, W.; Chang, J. Synthesis and Characterization of Bio-Oil Phenol Formaldehyde Resin Used to Fabricate Phenolic Based Materials. *Materials* **2017**, *10*, 668. [CrossRef]
57. Khuenkaeo, N.; Tippayawong, N. Production and characterization of bio-oil and biochar from ablative pyrolysis of lignocellulosic biomass residues. *Chem. Eng. Commun.* **2020**, *207*, 153–160. [CrossRef]
58. Lu, H.R.; El Hanandeh, A. Life cycle perspective of bio-oil and biochar production from hardwood biomass; what is the optimum mix and what to do with it? *J. Clean. Prod.* **2019**, *212*, 173–189. [CrossRef]

59. Zhixia, Z.; Jing, W.; Jun, M.; Wenfu, C. Study of Biochar Pyrolysis Mechanism and Production Technology. *Appl. Mech. Mater.* **2015**, *709*, 364–369.
60. Fakayode, O.A.; Aboagarib, E.A.A.; Zhou, C.; Ma, H. Co-pyrolysis of lignocellulosic and macroalgae biomasses for the production of biochar—A review. *Bioresour. Technol.* **2020**, *297*, 122408. [CrossRef]

© 2020 by the authors. Licensee MDPI, Basel, Switzerland. This article is an open access article distributed under the terms and conditions of the Creative Commons Attribution (CC BY) license (http://creativecommons.org/licenses/by/4.0/).

Article

Rigid Polyurethane Foams Based on Bio-Polyol and Additionally Reinforced with Silanized and Acetylated Walnut Shells for the Synthesis of Environmentally Friendly Insulating Materials

Sylwia Członka * and Anna Strąkowska

Institute of Polymer and Dye Technology, Faculty of Chemistry, Lodz University of Technology, Stefanowskiego 12/16, 90-924 Lodz, Poland; anna.strakowska@p.lodz.pl
* Correspondence: sylwia.czlonka@dokt.p.lodz.pl

Received: 24 June 2020; Accepted: 20 July 2020; Published: 22 July 2020

Abstract: Rigid polyurethane (PUR) foams produced from walnut shells-derived polyol (20 wt.%) were successfully reinforced with 2 wt.% of non-treated, acetylated, and silanized walnut shells (WS). The impact of non-treated and chemically-treated WS on the morphology, mechanical, and thermal characteristics of PUR composites was determined. The morphological analysis confirmed that the addition of WS fillers promoted a reduction in cell size, compared to pure PUR foams. Among all the modified PUR foams, the greatest improvement of mechanical characteristics was observed for PUR foams with the addition of silanized WS—the compressive, flexural, and impact strength were enhanced by 21, 16, and 13%, respectively. The addition of non-treated and chemically-treated WS improved the thermomechanical stability of PUR foams. The results of the dynamic mechanical analysis confirmed an increase in glass transition temperature and storage modulus of PUR foams after the incorporation of chemically-treated WS. The addition of non-treated and chemically-treated WS did not affect the insulating properties of PUR foams, and the thermal conductivity value did not show any significant improvement and deterioration due to the addition of WS fillers.

Keywords: rigid polyurethane foams; lignocellulosic materials; filler; chemical treatment; mechanical characteristics

---

1. Introduction

Nowadays, polyurethanes (PUR) are used in many areas, such as thermal insulation materials, automotive elements, construction parts, medical devices as well as in the production of elastomers, adhesives, and foams [1]. PUR are synthesized by a reaction between hydroxyl groups (–OH) and isocyanate groups (–NCO). Due to the non-biodegradability and high toxicity, both petrochemical-based compounds have a negative impact on the environment [2,3]. Due to this, bio-based polyols from sustainable raw materials, such as plant oils, have attracted great interest in the synthesis of environmentally-friendly PUR foams [4–7]. Bio-polyols derived from certain types of cellulosic sources, such as spent coffee [8], cassava residue [9] or jute fibers [10] have been examined for the synthesis of PUR materials. In our previous work, PUR foams were prepared using the bio-polyol derived from lignocellulosic walnut shells (WS) [11]. It has been shown that due to the rich organic nature (~50% of lignin, ~24% of cellulose and ~24% of hemicellulose) [12,13], WS can be successfully applied for the production of bio-polyols for the synthesis of PUR foams. Therefore, liquefaction of WS has been conducted using a mixture of glycerine and polyethylene glycol (PEG-400) in the presence of acid catalyst (sulphuric acid). The impact of WS-based polyol on the mechanical and thermal characteristics of PUR foams have been examined. The results showed that it is possible to convert these lignocellulosic residues into polyol and produce PUR foams with the properties somewhat similar to those of commercial foams, although with higher thermal conductivity.

Unfortunately, a common phenomenon observed after the synthesis of PUR foams from bio-polyols is the deterioration of mechanical parameters of these materials [11]. Many previous studies have shown that the incorporation of a selected amount of fillers with both, organic or inorganic nature, helps to improve the mechanical characteristic of polymeric composites [14,15]. Therefore, to improve the mechanical characteristic of PUR foams, an additional fillers in the form of nanoparticles or fibers are incorporated. For example, PUR foams modified with a ground waste of bulk moulding composites were successfully synthesized by Barczewski et. al. [16]. When compared with unmodified PUR foams, the addition of inorganic filler resulted in the formation of composites with improved thermo-mechanical performances and better fire retardancy. Zhou et al. [17] have synthesized PUR composites from palm oil-based polyol. The bio-composites were additionally modified with the addition of selected amounts of cellulose nanocrystals. The resulting composites exhibited improved mechanical performance and better dimensional stability under the selected temperatures. Rigid PUR composites synthesized from bio-polyol and additionally enhanced with paper waste sludge (PWS) particles were synthesized by Kairyte et. al. [18]. The addition of 5 wt.% of PWS resulted in the production of PUR foams with better water vapor resistance, higher density, and improved mechanical characteristics, such as compressive strength and elastic modulus. Paciorek-Sadowska et al. [19] synthesized PUR foams with rapeseed cake as a natural filler. The addition of 30-60 wt.% of rapeseed cake filler resulted in the production of PUR composites with higher density, greater mechanical performances, and improved fire resistance.

Besides, many lignocellulosic fillers can be successfully used as reinforcing materials in polymer matrices [20–27], it was reported, that their hydrophilic nature may limit their further application. Due to this, the surface modification of the fillers seems to be a sufficient step before the application of natural fillers as reinforcing materials for polymeric composites [28–31]. Previous studies have proved that chemical modifications of the filler surface, such as acetylation [32], alkalization [33], benzoylation [34], and silanization [35] may successfully improve the adhesion between the filler and the polymeric matrix. Although many published studies are devoted to the examination of PUR composites, just a few research has been done on the surface treatment of WS and the impact of acetylated and silanized WS on selected properties of PUR foams. Therefore, in this study, the effect of acetylated and silanized WS on the morphological, mechanical, insulating, and thermal characteristics of the PUR composites was investigated.

2. Experimental Section

2.1. Chemicals and Materials

Commercial polyester polyol with a brand name STEPANPOL PS-2352 and aromatic diisocyanate with a brand name PUROCYN B were purchased from Purinova Sp. z o.o (Bydgoszcz, Poland). Silicone surfactant with a brand name TEGOSTAB B8513 and PUR metal catalysts-Kosmos 33 (potassium acetate) and Kosmos 75 (octoate catalyst) were provided by Evonik Industries AG (Essen, Germany). Cyclopentane (blowing agent) and pentane (blowing agent) were purchased from Merck KGaA (Darmstadt, Germany). Sodium hydroxide (anhydrous), acetic acid (≥99.9%), ethanol (≥99.9%), sulfuric acid (purity 95–98%) were provided by Sigma-Aldrich Corporation (Saint Louis, MO, USA). Triphenylsilanol was provided by abcr GmbH Company (Karlsruhe, Germany). Walnut shells were kindly provided by a Polish local company (Lodz, Poland).

2.2. Methods

2.2.1. Pre-Treatment of WS with an Alkali Solution

WS filler was pretreated with 10% NaOH solution. A calculated amount of WS filler was soaked in the NaOH solution for 1 h. After that, the solution was neutralized with 1% acetic acid. Such obtained alkali-treated filler was washed with ultrapure water to pH of 7 and dried in an oven (24 h, 80 °C).

2.2.2. Silanization of WS

Alkali-treated WS filler was treated with a 5% triphenylsilanol solution. The calculated amount of WS filler was soaked in a solution of triphenylsilanol in ethanol, maintaining the ratio of WS filler to a solution at the level of 1:20 (by weight). After 3 h, the triphenylsilanol solution was evaporated, and the silane-treated WS filler was separated from the solvent. The silane-treated WS was washed with ultrapure water and dried in an oven (24 h, 80 °C).

2.2.3. Acetylation of WS

Alkali-treated WS was soaked in a mixture of acetic acid and acetic anhydride (1:1 v/v). To promote the reaction, a few drops of sulfuric acid were dropped into the mixture and the solution was intensively stirred for 30 min. Lastly, the WS was removed from the mixture, thoroughly washed with ultrapure water to pH of 7. The acetylated WS filler was placed in an oven (120 °C) and dried, until a constant weight of the WS filler was obtained.

2.2.4. PUR Foams Preparation

The selected amounts of polyester polyol, WS-based polyol, surfactant, catalysts, flame retardant, and blowing agent were placed in a cylindrical form and mixed with mechanical stirring (1000 rpm, 30 s). The obtained mixture was modified with the addition of 2 wt.% of non-treated, silanized, or acetylated-WS, and thoroughly mixed at 1000 rpm for 60 s. After the complete dispersion of WS filler, an isocyanate component was poured into the mixture and stirred for another 30 s. The resulting PUR foams were expanded freely in open forms. Before further characterization, PUR foams were conditioned at a standardized temperature of 25 °C and humidity of 50% for 24 h. The detailed formulations presenting the weight ratio of components are shown in Table 1. The schematic procedure of PUR foams synthesis is given in Figure 1.

Table 1. The weight ratio of components used for polyurethanes (PUR) foam synthesis.

Component	PUR_0	PUR_WS_NT	PUR_WS_A	PUR_WS_S
Amount, parts by STEPANPOL PS-2352 weight [pbw]				
STEPANPOL PS-2352	80	80	80	80
WS-based polyol	20	20	20	20
PUROCYN B	160	160	160	160
Kosmos 75	6	6	6	6
Kosmos 33	0.8	0.8	0.8	0.8
Tegostab B8513	2.5	2.5	2.5	2.5
Water	0.5	0.5	0.5	0.5
Pentane/cyclopentane	11	11	11	11
Non-treated WS	0	2	0	0
Acetylated WS	0	0	2	0
Silanized WS	0	0	0	2

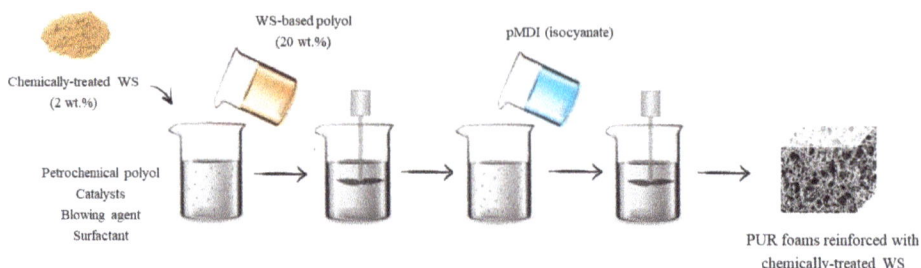

Figure 1. Schematic procedure of PUR foam synthesis.

2.3. Test Methods

The dynamiclLight scattering (DLS) method was used to evaluate the average size of filler particles using a Zetasizer NanoS90 instrument (Malvern Instruments Ltd, Malvern, UK). WS particles were dispersed in a polyol (0.04 g·L^{-1}) and the average of 5 individual measurements was evaluated.

The viscosity of PUR systems was determined according to ISO 2555 [36]. The measurement was performed using a rotatory viscometer (Viscometer DVII+, Brookfield, Berlin, Germany). The viscosity of PUR systems was determined at different share rates—0.5, 5, 10, and 100 rpm (round per minute). The average of 5 individual measurements was evaluated. The standard deviation was calculated.

The chemical structure of fillers was determined by Fourier-transform infrared spectroscopy (FTIR, Nicolet iS50 spectrometer, Thermo Fisher Scientific, Waltham, MA, USA). An average of 64 individual scans was evaluated.

Scanning electron spectroscopy (JEOL JSM-5500 LV, JEOL Ltd., Peabody, MA, USA) was selected to analyze the structure of the fillers and PUR foams. The morphology features of PUR foams were evaluated using ImageJ software (Java 1.8.0, Media Cybernetics Inc., Rockville, MD, USA).

The PUR foam density was calculated as the ratio of PUR mass to volume, following ISO 845 [37]. The average of 5 individual measurements was evaluated. The standard deviation was calculated.

Compressive strength ($\sigma_{10\%}$), flexural strength (ε_f), and impact test were processed using Zwick Z100 Testing Machine (Zwick/Roell Group, Ulm, Germany), following ISO 844 [38], ISO 178 [39], and ISO 180 [40], respectively.

Thermogravimetric analysis (TGA) and differential thermogravimetry (DTG) were applied to determine the thermal properties of PUR samples. A thermal analysis test was conducted in an argon atmosphere using STA 449 F1 Jupiter Analyzer (Netzsch Group, Selb, Germany). The samples with an initial weight of 10 mg were examined in the selected range of temperatures (from 0 to 600 °C).

The dynamic-mechanical characteristic (DMA) was performed using a modular compact rheometer (ARES, TA Instruments, New Castle, DE, USA) under the selected parameters (constant deformation of 0.1%, frequency of 1 Hz). The PUR samples were examined in the selected range of temperatures (from 0 to 250 °C). The average of 5 individual measurements was evaluated. The standard deviation was calculated.

The thermal conductivity of the PUR foams was examined using the LaserComp 50 heat flow meter (HFMA, Westchester, IL, USA). The average of 5 individual measurements was evaluated. The standard deviation was calculated.

3. Results and Discussion

3.1. Characterization of WS-Based Polyol

The properties of WS-based polyol have been widely discussed in our previous study [11]. The selected properties of bio-polyol and polyester polyol (STEPANPOL PS-2352) are shown in Table 2.

Table 2. Selected properties of polyester polyol and walnut shell (WS)-based polyol [11].

Component	Viscosity [mPa s]	Molecular Weight (M_w) [Da]	Hydroxyl Number [mg KOH/g]
WS-based polyol	2550	420	340
STEPANPOL PS-2352	2000–4500	468	230–250

3.2. Characterization of WS Filler

The chemical structure of non-treated, acetylated, and silanized WS was examined using FTIR analysis. The obtained spectra are presented in Figure 2. Bands located at 840, 1035, 1455, and 2900 cm^{-1} are characteristic for the C–H vibration of cellulose, hemicellulose, and lignin of WS, respectively [41]. The new band at ~740 refers to Si–CH$_3$ vibration and confirms the formation of chemical linkage between the silane coupling agent and WS filler [42]. Other bands characteristic for silanized WS occur at 1340, 1080, 3350 cm^{-1}, and refer to Si–O–Si, Si–O, and O–H vibration of silanized WS, respectively [42,43]. The acetylation of WS is confirmed by a reduced intensity of band attributed to OH vibration that occurs at 3300 cm^{-1} [20,44]. Moreover, the presence of new peaks at 1230 cm^{-1} and 1730 cm^{-1}, which are characteristic for C=O stretching of the ester carbonyl group and confirms the successful acetylation of WS filler [45–47].

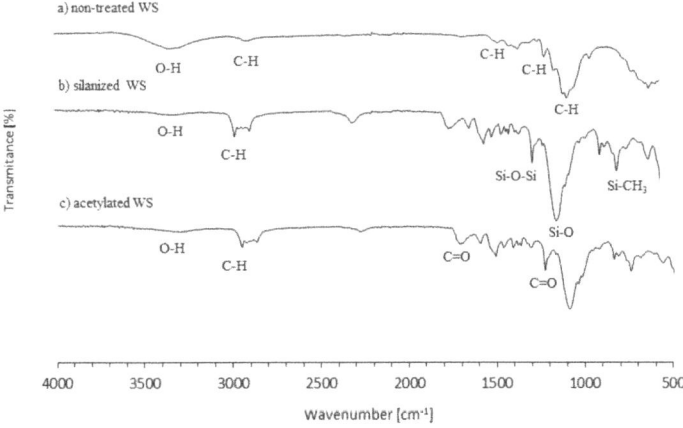

Figure 2. Fourier-transform infrared spectroscopy (FTIR) spectra of (**a**) non-treated WS, (**b**) silanized WS, and (**c**) acetylated WS.

Silanization and acetylation treatments affect the morphology of the WS (Figure 3). The surface of non-treated WS is rough and some cracks are visible on the filler surface. After the silanization and acetylation, the topography of the WS filler becomes rougher, because the chemical treatments, such as alkalization, may remove the waxy substances that smooth the filler surface [48–51]. However, the previous studies have reported that such a rough topography of the filler particles may have a beneficial effect on the further mechanical properties of composites [52,53]. The rough surface of the filler may improve the mechanical interlocking and interphase bonding between the filler surface and polymeric phase, which, in turn, results in better mechanical characteristics of such reinforced composites.

The size of WS's particles was examined in polyol dispersion. The particle size distribution of the non-treated, acetylated, and silanized WS are given in Figure 4. The presented results indicate that the size of non-treated and chemically-treated WS ranges from 300 to 1000 µm in all cases. The highest percentage of particles for non-treated WS, acetylated WS, and silanized WS is observed at ~600, ~650, and ~700 µm, respectively.

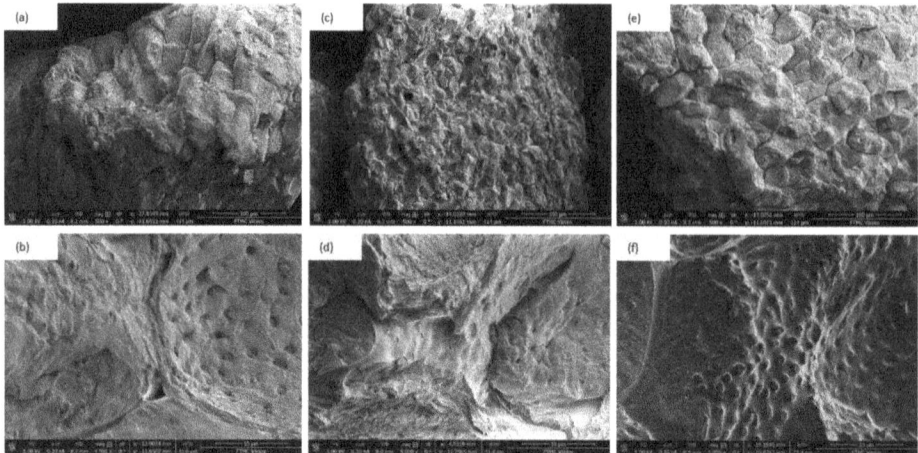

Figure 3. SEM results of (**a,b**) non-treated WS (**c,d**) acetylated WS, and (**e,f**) silanized WS.

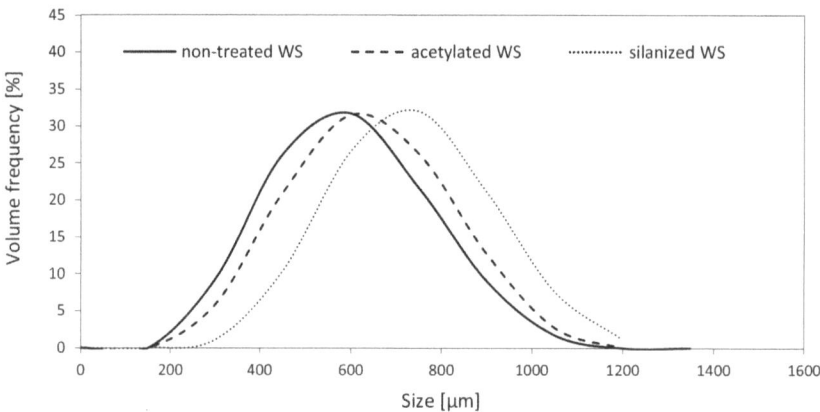

Figure 4. The results of particle size distribution.

As presented in Table 3, comparing to PUR_0, the addition of WS filler results in a greater viscosity. This may be connected with the Van der Waal's forces and hydrogen bonding that occur between the active groups of WS particle and polyester polyol [54]. With the addition of non-treated WS, acetylated WS, and silanized WS, the viscosity increases from 840 mPa·s to 1700, 1900, and 2050 mPa·s, respectively. All PUR mixtures behave as non-Newtonian fluids, and reveal an analog tendency to the PUR mixtures containing various kinds of organic and/or inorganic fillers, which were reported in previous works [36,37].

Table 3. Dynamic viscosity of PUR systems.

Sample	Dynamic Viscosity η [mPa·s]			
	0.5 rpm	5 rpm	10 rpm	100 rpm
PUR_0	840 ± 10	520 ± 8	470 ± 9	230 ± 7
PUR_WS_NT	1700 ± 12	1200 ± 10	750 ± 9	450 ± 8
PUR_WS_A	1900 ± 11	1300 ± 10	1050 ± 10	580 ± 8
PUR_WS_S	2050 ± 10	1450 ± 12	1200 ± 11	650 ± 7

3.3. Foaming Kinetic

The impact of WS filler on the foaming process of PUR foams was monitored by measuring the duration of the cream time, expansion time, and tack-free time. The cream time refers to the rise start of the PUR system, the expansion time refers to the transition of the liquid-state to solid-state, and tack-free time is measured until the foam solidifies completely. The results of characteristic times are presented in Table 4.

Table 4. Processing times of PUR foams.

Sample	Cream Time [s]	Expansion Time [s]	Tack-Free Time [s]
PUR_0	56 ± 3	419 ± 8	340 ± 7
PUR_WS_NT	60 ± 1	495 ± 9	375 ± 8
PUR_WS_A	59 ± 2	510 ± 6	365 ± 9
PUR_WS_S	59 ± 3	515 ± 8	370 ± 8

Due to the incorporation of non-treated, acetylated, and silanized WS, the cream and rise times have increased [55–57]. In the case of PUR foams containing WS filler, the extended expansion time may be connected with a limited expansion of cells. Due to the increased viscosity of the PUR system, the mobility of the molecular chains is reduced, which, in turn, affects the polymerization kinetic of PUR synthesis. As the viscosity of the PUR mixtures increases, the mass transfer of the blowing agent from the solid to the gas phase decreases, and the expansion of PUR foam is slowed down [55–57]. Similar results have been shown by Kurańska et al. [58]. The authors confirmed that the incorporation of the vegetable fillers, such as wood fibers, results in the elongation of processing times due to the reduced reactivity of the modified systems.

3.4. Cellular Structure and Thermal Conductivity

The impact of non-treated WS and chemically-treated WS on the cellular structure of PUR foams was examined using SEM (Figure 5). Mean cell size and closed-cell content are listed in Table 5. When compared with neat PUR_0, the incorporation of each filler leads to a decrease of average cell size, i.e., 380, 360, and 355 µm for PUR_WS_NT, PUR_WS_A, PUR_WS_S against 410 µm for neat PUR_0. Based on this result, it can be concluded that WS particles can promote the nucleation of the air bubbles, and prevent their coalescence during the expansion process. Paciorek-Sadowska et al. [19] have reported a similar tendency—at a higher loading of rapeseed cake, the cellular structure of PUR composites was more heterogeneous, and a reduced cell diameter was observed—for example, the incorporation of 60 wt.% of rapeseed cake has reduced an average diameter from 316 to 250 µm.

Table 5. Structural parameters and thermal conductivity results of PUR foams.

Sample	Cell Size [µm]	Apparent Density [kg m^{-3}]	Closed-Cell Content [%]	Thermal Conductivity [W m^{-1} K^{-1}]
PUR_0	410 ± 9	38 ± 1	86.4 ± 0.6	0.0251 ± 0.0008
PUR_WS_NT	380 ± 9	42 ± 2	83.1 ± 1.1	0.0302 ± 0.0007
PUR_WS_A	360 ± 8	40 ± 3	85.9 ± 1.1	0.0293 ± 0.0009
PUR_WS_S	355 ± 8	39 ± 2	86.0 ± 0.8	0.0284 ± 0.0009

Comparing the PUR foams containing non-treated and chemically-treated PUR foams. it can be seen that the incorporation of silanized and acetylated WS filler leads to the production of PUR foams with regular structure and more uniform cells. Moreover, the incorporation of non-treated and chemically-treated WS affects the closed-cell content of the foams. When compared with PUR_0, the closed-cell content decreases slightly after the incorporation of WS filler, and this trend is more visible in the case of PUR composites containing non-treated WS filler. Thus, the chemical treatment of WS can improve the interphase adhesion between filler particles and polymeric matrix, which leads to the production of PUR composites with a more stable cellular morphology. A cross-linked structure of PUR containing filler particles can prevent the collapse of the cells during their expansion and form additional edges that are able to capture the emitted CO_2 [59]. Sung et al. [60] stated that an increase in pore size is connected with the hydrophobic character of the filler surface-hydrophobic fillers exhibit greater adhesion to the polymer matrix, while the higher the hydrophilicity, the lower the adhesion and greater the cell size of the foam structure. Such a tendency may be also found in our study. Among the examined PUR foams, PUR_WS_NT is characterized by larger cell diameters, due to the highly hydrophilic nature of WS. Acetylation and silanization treatments increase the hydrophobic character of WS, thus the obtained PUR composites are characterized by more uniform cells with reduced diameters.

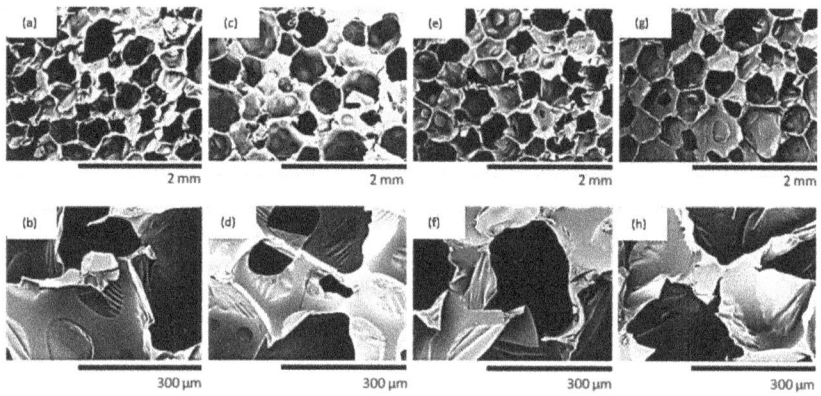

Figure 5. Cellular structure of (**a**,**b**) PUR_0, (**c**,**d**) PUR_WS_NT, (**e**,**f**) PUR_WS_A and (**g**,**h**) PUR_WS_S.

The impact of WS on the density of PUR foams is presented in Figure 6 (for each measurement, the entire experiment was repeated, and the presented value concerns the individual samples). Comparing to PUR_0, the addition of WS fillers increases the density of PUR composites. The average density of the neat PUR_0 is 38 kg m^{-3}, while the density of modified PUR foams oscillates between 39 and 42 kg m^{-3} upon incorporating WS filler. It is clear that non-treated and chemically-treated WS affect the value of apparent density differently, due to their different dispersion in the PUR matrix. The higher level of dispersion, associated with the chemical modification of WS leads to lower nucleation energy and greater adhesion between WS filler and PUR matrix, which determines a finer cell morphology and lower apparent density [61]. As a result of this, PUR_WS_A and PUR_WS_S exhibit lower density than PUR_WS_NT.

Thermal conductivity (λ) is an important parameter that determines the insulation properties of PUR foams [16,62,63]. Generally, the λ value of PUR foams is calculated as a sum of λ of the gas in the cells (λgas), the λ through the solid polymer (λsolid), the radiation heat transfer across the walls of the solid struts (λradiation), and the convection of the gas within the cells (λconvection) [64]. As presented in Table 5, thermal conductivity for PUR_0 is 0.0251 W m^{-1}K^{-1}. The incorporation of non-treated WS increases the value of λ to 0.0302 W m^{-1}K^{-1}, however, it remains almost unchanged for PUR_WS_A and PUR_WS_S—the value of λ increases slightly to 0.0293 and 0.0284 W m^{-1}K^{-1}, and it is still in line with the conditions of insulation materials [1,65].

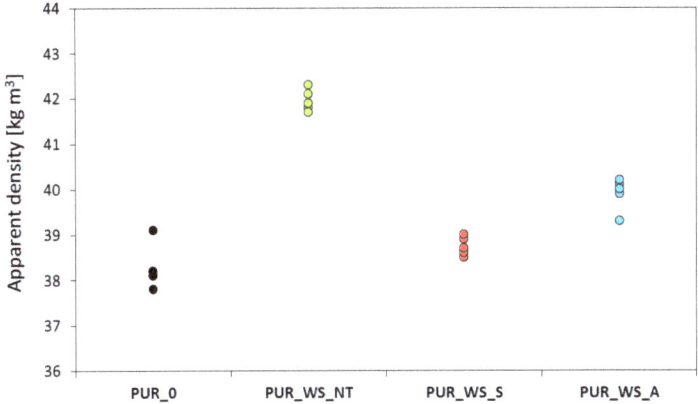

Figure 6. The apparent density of PUR_0, PUR_WS_NT, PUR_WS_S, and PUR_WS_A.

3.5. Mechanical Characteristics

Compressive strength ($\sigma_{10\%}$), flexural strength (σ_f), and impact strength were examined to determine the impact of surface WS fillers on the mechanical characteristics of PUR composites. The results of $\sigma_{10\%}$ are presented in Figure 7. When compared with PUR_0, $\sigma_{10\%}$ increases by ~6% for PUR_WS_NT. The mechanical strength increases after the chemical modifications of WS filler-the value of $\sigma_{10\%}$ increases by ~19% and ~21% for PUR_WS_A and WS_WS_S, respectively. Such a result can be connected with the cellular structure of the resulting composites, which has a great impact on their further mechanical properties. As discussed previously, PUR foams containing acetylated and silanized WS have a more uniform structure, with a greater number of closed-cells (see Figure 5). Moreover, after the silanization and acetylation treatment, the possible chemical reaction may occur between the hydroxyl groups of WS and acetic acid or silane coupling agent, creating chemical covalent linkages that improve the interfacial adhesion between the filler and the polymeric matrix [66,67]. Ciobanu et al. [68] reported that the incorporation of rigid, three-dimensional lignin improved the mechanical characteristics of the resulting products. This was attributed to the fact that incorporated lignin may act as a compatibilizer for PUR segments, successfully enhancing the mechanical properties of the resulting composites.

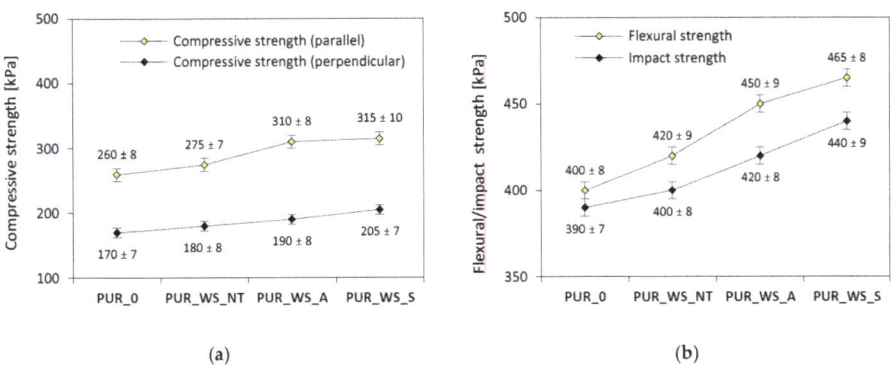

Figure 7. (**a**) Compressive, (**b**) flexural and impact strength of PUR foams.

The results of σ_f are shown in Figure 7. After the incorporation of non-treated, acetylated, and silanized WS the value of σ_f is increased. When WS filler is incorporated without treatment, the σ_f

increases by ~5% when compared with PUR_0. A further improvement is observed for PUR_WS_A and PUR_WS_S—the value of σ_f increases by ~12% and ~16%, respectively. As followed by flexural properties, the impact strength was measured as well (Figure 7). When comparing with PUR_0, the impact strength of PUR foams containing non-treated and chemically-treated WS is increased. The greatest improvement is observed for PUR_WS_S—the impact strength increases by ~13%. Previous studies have shown that the finer cellular structures can improve the mechanical characteristics of porous materials, thus, the PUR foams containing acetylated and silanized WS have a greater flexural strength [69,70]. Moreover, the chemical modification of the filler may result in the rough morphology of that promotes the mechanical interlocking between the filler surface and polymeric matrix, enhancing the mechanical characteristics of the resulting composites. The rigid cellular structure of PUR foams containing solid particles of the filler is able to absorb more energy during impact, resulting in the better mechanical behavior of PUR foams [19].

3.6. DMA Results

DMA results are given in Figure 8. The storage modulus (E') of the PUR foams containing non-treated, acetylated, and silanized WS is higher than the PUR_0. This indicates a reinforcing effect of the WS filler, which results in effective stress transfer from the filler to the PUR matrix. As observed in Figure 8a, PUR_WS_A, and PUR_WS_S have a higher value of E' than that containing non-treated PUR_WS_NT. As reported in previous works, the chemical modification removes waxes and surface impurities from cellulose fillers, thus, increasing its interaction with the polymer matrix [48–51]. Additionally, the chemical treatment helps in the fibrillation of the filler surface. Therefore, it can be concluded that the rough morphology of the WS filler may improve the mechanical interlocking between the filler and polymeric matrix, which, in turn, results in more efficient stress transfer from the matrix to the filler particles and increased storage modulus.

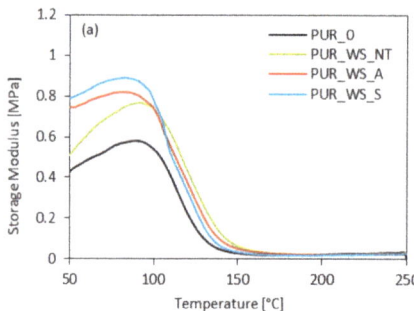 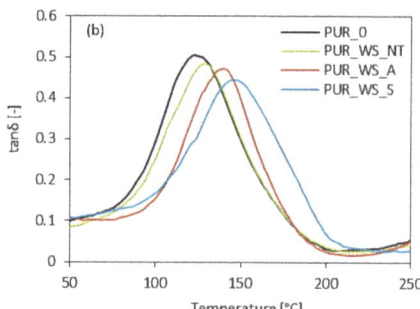

Figure 8. (a) Storage modulus and (b) tanδ of PUR foams.

The glass transition temperature (T_g) of PUR foams refers to the maximum of tanδ in the function of the temperature (Figure 8b). All series of PUR foams show one peak of maximum on the graph, indicating that the resulting composites can be classified as homogeneous blends. Moreover, PUR foams containing WS filler exhibit higher value of T_g, which may be connected with higher viscosity of PUR systems and reduced molecular mobility of polymer chains. As the T_g values of the treated PUR_WS_A and PUR_WS_S are higher than non-treated PUR_WS_NT, it can be concluded that chemical treatment affects the WS surface and improves the interphase contact between the filler and PUR matrix. A lower tanδ peak value for PUR_WS_NT may be connected with weaker linkages between the filler surface and the polymeric matrix, which results in a greater dissipation of energy. Since the tanδ peak value illustrates the filler matrix adhesion, it can be concluded that the higher tanδ peak value observed for PUR_WS_A and PUR_WS_S confirm the better compatibility between filler surface and the polymeric matrix.

3.7. Thermal Stability

TGA results are given in Figure 9 and Table 6. Results showed improved thermal characteristics of PUR composites containing chemically-treated WS, however, there is almost no change in the case of non-treated PUR_WS_NT.

Table 6. Thermogravimetric analysis (TGA) results.

Sample	T_{max} [°C] 1st Stage	T_{max} [°C] 2nd Stage	T_{max} [°C] 3rd Stage	Residue at 450 °C	Residue at 600 °C	DTG [%/min]
PUR_0	218	315	581	48.1	21.9	0.0057
PUR_WS_NT	219	317	587	49.8	23.1	0.0055
PUR_WS_A	224	323	585	50.2	25.5	0.0056
PUR_WS_S	225	325	591	51.5	24.7	0.0056

The first degradation step refers to the temperatures of 5% weight loss ($T_{5\%}$). It was noticed that the value of $T_{5\%}$ increases slightly from 220 °C (for PUR_0) to 224 and 225 °C by the incorporation of acetylated and silanized WS, respectively, while there is almost no change for PUR_WS_NT.

The second stage of degradation refers to the thermal decomposition of soft segments of PUR and occurs in the range of 300–350 °C [71]. For PUR_0, the second stage is observed at 315 °C. A slight improvement in thermal stability exhibits in PUR_WS_A and PUR_WS_S; when compared with PUR_0, the temperature is increased by 8 and 10 °C, respectively. Such an improvement can be connected with a greater crosslink density, due to the addition of acetylated and silanized WS. A more rigid structure of PUR composites may successfully limit the mobility of the polymeric chains, which, in turn, reduces heat transport and improves the thermal stability of PUR composites [72].

The third degradation step of PUR foams in observed in the range of 500–600 °C. The third degradation step is mostly connected with the degradation of organic compounds incorporated into the PUR matrix, such as cellulose, hemicellulose, and lignin [73]. The thermal characteristic of PUR foams containing WS filler is similar to those of the unmodified PUR foam (PUR_0). With the incorporation of acetylated and silanized WS, the mass loss increases due to the presence of cellulose [72]. The char residue mass is another factor that determines the thermal stability of PUR foams. It can be seen that, after the incorporation of each kind of WS filler, the amount of char residue is increased. For PUR_0, the char residue is 48.1% at 450 °C, while, for PUR foams containing non-treated and chemically-treated WS, the amount of char residue increases to 49.8% (for PUR_WS_NT), 50.2% (for PUR_WS_A), and 51.5% (for PUR_WS_S). The rigid structure of lignin can limit the heat transport through the PUR foam structure, which results in better thermal stability of PUR composites [73]. Analog results have been shown in previous studies as well [74,75].

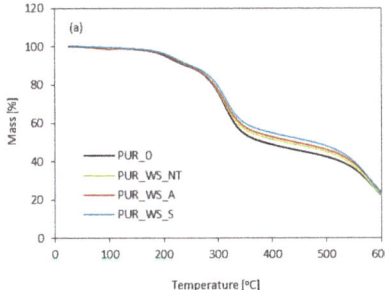 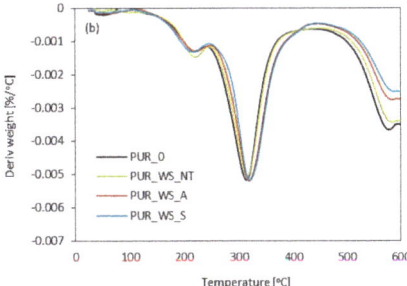

Figure 9. (**a**) TG and (**b**) differential thermogravimetry (DTG) curves obtained for PUR foams.

4. Conclusion

In the present work, the impact of non-treated, acetylated, and silanized WS on morphological, mechanical, thermal, and physical properties of PUR foams has been investigated and discussed. It has been shown that the cellular structure was affected, due to the addition of non-treated and chemically-treated WS. PUR composites with the addition of acetylated and silanized WS exhibited a more uniform structure than PUR foams with the addition of non-treated WS. The mechanical characteristics of PUR foams, such as compressive, flexural, and impact strength, were improved after the incorporation of non-treated and chemically-treated WS. Among all the specimens, the greatest improvement was observed for PUR foams containing silanized WS—the compressive, flexural, and impact strength were improved by 21, 16, and 13%, respectively. From this investigation, it can be concluded that the acetylation and silanization of WS can improve the interfacial bonding between WS and PUR matrix, enhancing the mechanical properties of the PUR foams. Moreover, the addition of non-treated and chemically-treated WS improved the thermal properties of PUR foams—e.g., at 600 °C, the char residue increased from 21.9 to 25.5% for PUR foams, with the addition of acetylated WS. The addition of WS fillers did not affect the thermal conductivity of PUR foams.

Author Contributions: Methodology, Investigation, Data Curation, Writing—Original Draft, Writing—Review & Editing, Visualization, S.C.; Methodology, Investigation, A.S. All authors have read and agreed to the published version of the manuscript.

Funding: This research received no external funding.

Conflicts of Interest: The authors declare no conflict of interest.

References

1. KAIRYTĖ, A.; Kizinievič, O.; Kizinievič, V.; Kremensas, A. Synthesis of biomass-derived bottom waste ash based rigid biopolyurethane composite foams: Rheological behaviour, structure and performance characteristics. *Compos. Part A Appl. Sci. Manuf.* **2019**, *117*, 193–201. [CrossRef]
2. KAIRYTĖ, A.; Vaitkus, S.; Vėjelis, S.; Girskas, G.; Balčiūnas, G. Rapeseed-based polyols and paper production waste sludge in polyurethane foam: Physical properties and their prediction models. *Ind. Crop. Prod.* **2018**, *112*, 119–129. [CrossRef]
3. KAIRYTĖ, A.; Vėjelis, S. Evaluation of forming mixture composition impact on properties of water blown rigid polyurethane (PUR) foam from rapeseed oil polyol. *Ind. Crop. Prod.* **2015**, *66*, 210–215. [CrossRef]
4. Prociak, A.; Szczepkowski, L.; Ryszkowska, J.; Kurańska, M.; Auguścik, M.; Malewska, E.; Gloc, M.; Michałowski, S. Influence of Chemical Structure of Petrochemical Polyol on Properties of Bio-polyurethane Foams. *J. Polym. Environ.* **2019**, *27*, 2360–2368. [CrossRef]
5. Kurańska, M.; Polaczek, K.; Auguścik-Królikowska, M.; Prociak, A.; Ryszkowska, J. Open-cell rigid polyurethane bio-foams based on modified used cooking oil. *Polymer* **2020**, *190*, 122164. [CrossRef]
6. Kurańska, M.; Pinto, J.; Salach, K.; Barreiro, M.F.; Prociak, A. Synthesis of thermal insulating polyurethane foams from lignin and rapeseed based polyols: A comparative study. *Ind. Crop. Prod.* **2020**, *143*, 111882. [CrossRef]
7. Borowicz, M.; Paciorek-Sadowska, J.; Lubczak, J.; Czupryński, B. Biodegradable, Flame-Retardant, and Bio-Based Rigid Polyurethane/Polyisocyanurate Foams for Thermal Insulation Application. *Polymer* **2019**, *11*, 1816. [CrossRef]
8. Soares, B.; Gama, N.V.; Freire, C.S.; Barros-Timmons, A.; Brandão, I.; Silva, R.; Neto, C.P.; Ferreira, A. Spent coffee grounds as a renewable source for ecopolyols production. *J. Chem. Technol. Biotechnol.* **2014**, *90*, 1480–1488. [CrossRef]
9. Kang, J.; Chen, W.; Yao, Y.; Jin, Y.; Cheng, X.; Lü, Q. Optimisation of Bio-polyol Production from Cassava Residue Using Ethylene Glycol as the Liquefaction Reagent. *J. Wuhan Univ. Technol. Sci. Ed.* **2019**, *34*, 945–949. [CrossRef]
10. Huang, G.; Wang, P. Effects of preparation conditions on properties of rigid polyurethane foam composites based on liquefied bagasse and jute fibre. *Polym. Test.* **2017**, *60*, 266–273. [CrossRef]

11. Członka, S.; Strąkowska, A.; KAIRYTĖ, A. Application of Walnut Shells-Derived Biopolyol in the Synthesis of Rigid Polyurethane Foams. *Materials* **2020**, *13*, 2687. [CrossRef] [PubMed]
12. Sarsari, N.A.; Pourmousa, S.; Tajdini, A. Physical and Mechanical Properties of Walnut Shell Flour-Filled Thermoplastic Starch Composites. *Bioresour.* **2016**, *11*, 6968–6983. [CrossRef]
13. Ayrilmis, N.; Kaymakci, A.; Özdemir, F. Physical, mechanical, and thermal properties of polypropylene composites filled with walnut shell flour. *J. Ind. Eng. Chem.* **2013**, *19*, 908–914. [CrossRef]
14. Barczewski, M.; Salasinska, K.; Kloziński, A.; Skórczewska, K.; Szulc, J.; Piasecki, A. Application of the Basalt Powder as a Filler for Polypropylene Composites With Improved Thermo-Mechanical Stability and Reduced Flammability. *Polym. Eng. Sci.* **2018**, *59*, E71–E79. [CrossRef]
15. Barczewski, M.; Mysiukiewicz, O.; Matykiewicz, D.; Kloziński, A.; Andrzejewski, J.; Piasecki, A. Synergistic effect of different basalt fillers and annealing on the structure and properties of polylactide composites. *Polym. Test.* **2020**, *89*, 106628. [CrossRef]
16. Barczewski, M.; Kurańska, M.; Sałasińska, K.; Michałowski, S.; Prociak, A.; Uram, K.; Lewandowski, K. Rigid polyurethane foams modified with thermoset polyester-glass fiber composite waste. *Polym. Test.* **2020**, *81*, 106190. [CrossRef]
17. Zhou, X.; Sethi, J.; Geng, S.; Berglund, L.; Frisk, N.; Aitomäki, Y.; Sain, M.; Oksman, K. Dispersion and reinforcing effect of carrot nanofibers on biopolyurethane foams. *Mater. Des.* **2016**, *110*, 526–531. [CrossRef]
18. KAIRYTĖ, A.; Kirpluks, M.; Ivdre, A.; Cabulis, U.; Vėjelis, S.; Balčiūnas, G. Paper waste sludge enhanced eco-efficient polyurethane foam composites: Physical-mechanical properties and microstructure. *Polym. Compos.* **2016**, *39*, 1852–1860. [CrossRef]
19. Paciorek-Sadowska, J.; Borowicz, M.; Isbrandt, M.; Czupryński, B.; Apiecionek, L. The Use of Waste from the Production of Rapeseed Oil for Obtaining of New Polyurethane Composites. *Polymer.* **2019**, *11*, 1431. [CrossRef]
20. Pappu, A.; Patil, V.; Jain, S.; Mahindrakar, A.; Haque, R.; Thakur, V. Advances in industrial prospective of cellulosic macromolecules enriched banana biofibre resources: A review. *Int. J. Boil. Macromol.* **2015**, *79*, 449–458. [CrossRef]
21. Faruk, O.; Bledzki, A.K.; Fink, H.-P.; Sain, M. Progress Report on Natural Fiber Reinforced Composites. *Macromol. Mater. Eng.* **2013**, *299*, 9–26. [CrossRef]
22. Faruk, O.; Bledzki, A.K.; Fink, H.-P.; Sain, M. Biocomposites reinforced with natural fibers: 2000–2010. *Prog. Polym. Sci.* **2012**, *37*, 1552–1596. [CrossRef]
23. Monteiro, S.N.; Lopes, F.P.D.; Barbosa, A.P.; Bevitori, A.B.; Da Silva, I.L.A.; Da Costa, L.L. Natural Lignocellulosic Fibers as Engineering Materials—An Overview. *Met. Mater. Trans. A* **2011**, *42*, 2963–2974. [CrossRef]
24. Cichosz, S.; Masek, A. Thermal Behavior of Green Cellulose-Filled Thermoplastic Elastomer Polymer Blends. *Molecules* **2020**, *25*, 1279. [CrossRef]
25. Cichosz, S.; Masek, A. Drying of the Natural Fibers as A Solvent-Free Way to Improve the Cellulose-Filled Polymer Composite Performance. *Polymer* **2020**, *12*, 484. [CrossRef]
26. Septevani, A.A.; Evans, D.A.; Annamalai, P.K.; Martin, D.J. The use of cellulose nanocrystals to enhance the thermal insulation properties and sustainability of rigid polyurethane foam. *Ind. Crop. Prod.* **2017**, *107*, 114–121. [CrossRef]
27. Hayati, A.N.; Evans, D.A.C.; Laycock, B.; Martin, D.J.; Annamalai, P.K. A simple methodology for improving the performance and sustainability of rigid polyurethane foam by incorporating industrial lignin. *Ind. Crop. Prod.* **2018**, *117*, 149–158. [CrossRef]
28. Zakaria, S.; Hamzah, H.; Murshidi, J.A.; Deraman, M. Chemical modification on lignocellulosic polymeric oil palm empty fruit bunch for advanced material. *Adv. Polym. Technol.* **2001**, *20*, 289–295. [CrossRef]
29. Wolski, K.; Cichosz, S.; Masek, A. Surface hydrophobisation of lignocellulosic waste for the preparation of biothermoelastoplastic composites. *Eur. Polym. J.* **2019**, *118*, 481–491. [CrossRef]
30. Cichosz, S.; Masek, A. Cellulose Fibers Hydrophobization via a Hybrid Chemical Modification. *Polymer* **2019**, *11*, 1174. [CrossRef]
31. Cichosz, S.; Masek, A. Superiority of Cellulose Non-Solvent Chemical Modification over Solvent-Involving Treatment: Application in Polymer Composite (part II). *Materials* **2020**, *13*, 2901. [CrossRef] [PubMed]

32. Borysiak, S. Fundamental studies on lignocellulose/polypropylene composites: Effects of wood treatment on the transcrystalline morphology and mechanical properties. *J. Appl. Polym. Sci.* **2012**, *127*, 1309–1322. [CrossRef]
33. Neto, J.S.S.; Lima, R.D.A.A.; Cavalcanti, D.; Souza, J.P.B.; Aguiar, R.A.A.; Banea, M.D. Effect of chemical treatment on the thermal properties of hybrid natural fiber-reinforced composites. *J. Appl. Polym. Sci.* **2018**, *136*, 1–13. [CrossRef]
34. Kabir, M.M.; Wang, H.; Lau, K.T.; Cardona, F. Chemical treatments on plant-based natural fibre reinforced polymer composites: An overview. *Compos. Part B Eng.* **2012**, *43*, 2883–2892. [CrossRef]
35. Liu, Y.; Xie, J.; Wu, N.; Wang, L.; Ma, Y.; Tong, J. Influence of silane treatment on the mechanical, tribological and morphological properties of corn stalk fiber reinforced polymer composites. *Tribol. Int.* **2019**, *131*, 398–405. [CrossRef]
36. *ISO 2555-Plastics—Resins in the liquid state or as emulsions or dispersions-Determination of apparent viscosity by the Brookfield Test method*; International Organization for Standardization: Geneva, Switzerland, 1989.
37. *ISO 845-Cellular plastics and rubbers—Determination of apparent density*; International Organization for Standardization: Geneva, Switzerland, 2006.
38. *ISO 844-Preview Rigid cellular plastics—Determination of compression properties*; International Organization for Standardization: Geneva, Switzerland, 2014.
39. *ISO 178-Plastics—Determination of flexural properties*; International Organization for Standardization: Geneva, Switzerland, 2019.
40. *ISO 180-Plastics—Determination of Izod impact strength*; International Organization for Standardization: Geneva, Switzerland, 2019.
41. Kolev, T.M.; Velcheva, E.A.; Stamboliyska, B.; Spiteller, M. DFT and experimental studies of the structure and vibrational spectra of curcumin. *Int. J. Quantum Chem.* **2005**, *102*, 1069–1079. [CrossRef]
42. Al-Oweini, R.; El-Rassy, H. Synthesis and characterization by FTIR spectroscopy of silica aerogels prepared using several $Si(OR)_4$ and $R''Si(OR')_3$ precursors. *J. Mol. Struct.* **2009**, *919*, 140–145. [CrossRef]
43. Wang, X.; Xu, S.; Tan, Y.; Du, J.; Wang, J. Synthesis and characterization of a porous and hydrophobic cellulose-based composite for efficient and fast oil–water separation. *Carbohydr. Polym.* **2016**, *140*, 188–194. [CrossRef]
44. Tang, L.; Huang, B.; Yang, N.; Li, T.; Lu, Q.; Lin, W.; Chen, X. Organic solvent-free and efficient manufacture of functionalized cellulose nanocrystals via one-pot tandem reactions. *Green Chem.* **2013**, *15*, 2369–2373. [CrossRef]
45. Adebajo, M.; Frost, R.L. Infrared and 13C MAS nuclear magnetic resonance spectroscopic study of acetylation of cotton. *Spectrochim. Acta Part A: Mol. Biomol. Spectrosc.* **2004**, *60*, 449–453. [CrossRef]
46. Adebajo, M.; Frost, R.L. Acetylation of raw cotton for oil spill cleanup application: An FTIR and 13C MAS NMR spectroscopic investigation. *Spectrochim. Acta Part A: Mol. Biomol. Spectrosc.* **2004**, *60*, 2315–2321. [CrossRef] [PubMed]
47. Frisoni, G.; Baiardo, M.; Scandola, M.; Lednická, D.; Cnockaert, M.C.; Mergaert, J.; Swings, J. Natural cellulose fibers: Heterogeneous acetylation kinetics and biodegradation behavior. *Biomacromolecules* **2001**, *2*, 476–482. [CrossRef] [PubMed]
48. Rambabu, N.; Panthapulakkal, S.; Sain, M.; Dalai, A.K. Production of nanocellulose fibers from pinecone biomass: Evaluation and optimization of chemical and mechanical treatment conditions on mechanical properties of nanocellulose films. *Ind. Crop. Prod.* **2016**, *83*, 746–754. [CrossRef]
49. Manimaran, P.; Senthamaraikannan, P.; Sanjay, M.; Marichelvam, M.; Jawaid, M. Study on characterization of Furcraea foetida new natural fiber as composite reinforcement for lightweight applications. *Carbohydr. Polym.* **2018**, *181*, 650–658. [CrossRef]
50. Manimaran, P.; Saravanan, S.; Sanjay, M.; Siengchin, S.; Jawaid, M.; Khan, A. Characterization of new cellulosic fiber: Dracaena reflexa as a reinforcement for polymer composite structures. *J. Mater. Res. Technol.* **2019**, *8*, 1952–1963. [CrossRef]
51. Negawo, T.A.; Polat, Y.; Buyuknalcaci, F.N.; Kilic, A.; Saba, N.; Jawaid, M. Mechanical, morphological, structural and dynamic mechanical properties of alkali treated Ensete stem fibers reinforced unsaturated polyester composites. *Compos. Struct.* **2019**, *207*, 589–597. [CrossRef]

52. Yaghoubi, A.; Nikje, M.M.A. Silanization of multi-walled carbon nanotubes and the study of its effects on the properties of polyurethane rigid foam nanocomposites. *Compos. Part A Appl. Sci. Manuf.* **2018**, *109*, 338–344. [CrossRef]
53. Sawpan, M.A.; Pickering, K.; Fernyhough, A. Effect of various chemical treatments on the fibre structure and tensile properties of industrial hemp fibres. *Compos. Part A: Appl. Sci. Manuf.* **2011**, *42*, 888–895. [CrossRef]
54. Amin, K.N.M. Cellulose nanocrystals reinforced thermoplastic polyurethane nanocomposites. Master's Thesis, University of Queensland Library, Brisbane, QL, Australia, 2016.
55. Ma, S.-R.; Shi, L.; Feng, X.; Yu, W.-J.; Lü, B. Graft modification of ZnO nanoparticles with silane coupling agent KH570 in mixed solvent. *J. Shanghai Univ. Engl. Ed.* **2008**, *12*, 278–282. [CrossRef]
56. Marcovich, N.; Kurańska, M.; Prociak, A.; Malewska, E.; Kulpa, K. Open cell semi-rigid polyurethane foams synthesized using palm oil-based bio-polyol. *Ind. Crop. Prod.* **2017**, *102*, 88–96. [CrossRef]
57. Cinelli, P.; Anguillesi, I.; Lazzeri, A. Green synthesis of flexible polyurethane foams from liquefied lignin. *Eur. Polym. J.* **2013**, *49*, 1174–1184. [CrossRef]
58. Kurańska, M.; Aleksander, P.; Mikelis, K.; Ugis, C. Porous polyurethane composites based on bio-components. *Compos. Sci. Technol.* **2013**, *75*, 70–76. [CrossRef]
59. Santiago-Calvo, M.; Tirado-Mediavilla, J.; Ruiz-Herrero, J.L.; Rodriguez-Perez, M.A.; Villafañe, F. The effects of functional nanofillers on the reaction kinetics, microstructure, thermal and mechanical properties of water blown rigid polyurethane foams. *Polymer* **2018**, *150*, 138–149. [CrossRef]
60. Sung, G.; Kim, J.H. Influence of filler surface characteristics on morphological, physical, acoustic properties of polyurethane composite foams filled with inorganic fillers. *Compos. Sci. Technol.* **2017**, *146*, 147–154. [CrossRef]
61. Modesti, M.; Lorenzetti, A.; Besco, S. Influence of nanofillers on thermal insulating properties of polyurethane nanocomposites foams. *Polym. Eng. Sci.* **2007**, *47*, 1351–1358. [CrossRef]
62. Kurańska, M.; Barczewski, M.; Uram, K.; Lewandowski, K.; Prociak, A.; Michałowski, S. Basalt waste management in the production of highly effective porous polyurethane composites for thermal insulating applications. *Polym. Test.* **2019**, *76*, 90–100. [CrossRef]
63. Formela, K.; Hejna, A.; Zedler, Ł.; Przybysz, M.; Ryl, J.; Saeb, M.R.; Przybysz, Ł. Structural, thermal and physico-mechanical properties of polyurethane/brewers' spent grain composite foams modified with ground tire rubber. *Ind. Crop. Prod.* **2017**, *108*, 844–852. [CrossRef]
64. Yang, S.; Wang, J.; Huo, S.; Wang, M.; Wang, J.; Zhang, B. Synergistic flame-retardant effect of expandable graphite and phosphorus-containing compounds for epoxy resin: Strong bonding of different carbon residues. *Polym. Degrad. Stab.* **2016**, *128*, 89–98. [CrossRef]
65. KAIRYTĖ, A.; Vaitkus, S.; Vėjelis, S. Titanate-Based Surface Modification of Paper Waste Particles and its Impact on Rigid Polyurethane Foam Properties. *Key Eng. Mater.* **2016**, *721*, 58–62. [CrossRef]
66. Członka, S.; Strąkowska, A.; Pospiech, P.; Strzelec, K. Effects of Chemically Treated Eucalyptus Fibers on Mechanical, Thermal and Insulating Properties of Polyurethane Composite Foams. *Materials* **2020**, *13*, 1781. [CrossRef]
67. Członka, S.; Strąkowska, A.; KAIRYTĖ, A. Effect of walnut shells and silanized walnut shells on the mechanical and thermal properties of rigid polyurethane foams. *Polym. Test.* **2020**, *87*, 106534. [CrossRef]
68. Ciobanu, C.; Ungureanu, M.; Ignat, L.; Popa, V.; Ungureanu, D. Properties of lignin–polyurethane films prepared by casting method. *Ind. Crop. Prod.* **2004**, *20*, 231–241. [CrossRef]
69. Ciecierska, E.; Jurczyk-Kowalska, M.; Bazarnik, P.; Gloc, M.; Kulesza, M.; Kowalski, M.; Krauze, S.; Lewandowska, M. Flammability, mechanical properties and structure of rigid polyurethane foams with different types of carbon reinforcing materials. *Compos. Struct.* **2016**, *140*, 67–76. [CrossRef]
70. Gu, R.; Konar, S.; Sain, M. Preparation and Characterization of Sustainable Polyurethane Foams from Soybean Oils. *J. Am. Oil Chem. Soc.* **2012**, *89*, 2103–2111. [CrossRef]
71. Xue, B.; Wen, J.-L.; Sun, S.-L. Producing Lignin-Based Polyols through Microwave-Assisted Liquefaction for Rigid Polyurethane Foam Production. *Materials* **2015**, *8*, 586–599. [CrossRef]
72. Gómez-Fernández, S.; Ugarte, L.; Calvo-Correas, T.; Peña-Rodríguez, C.; Corcuera, M.A.; Eceiza, A. Properties of flexible polyurethane foams containing isocyanate functionalized kraft lignin. *Ind. Crop. Prod.* **2017**, *100*, 51–64. [CrossRef]
73. Hu, S.; Li, Y. Two-step sequential liquefaction of lignocellulosic biomass by crude glycerol for the production of polyols and polyurethane foams. *Bioresour. Technol.* **2014**, *161*, 410–415. [CrossRef]

74. Luo, X.; Xiao, Y.; Wu, Q.; Zeng, J. Development of high-performance biodegradable rigid polyurethane foams using all bioresource-based polyols: Lignin and soy oil-derived polyols. *Int. J. Boil. Macromol.* **2018**, *115*, 786–791. [CrossRef]
75. Mahmood, N.; Yuan, Z.; Schmidt, J.; Xu, C. (Charles) Preparation of bio-based rigid polyurethane foam using hydrolytically depolymerized Kraft lignin via direct replacement or oxypropylation. *Eur. Polym. J.* **2015**, *68*, 1–9. [CrossRef]

© 2020 by the authors. Licensee MDPI, Basel, Switzerland. This article is an open access article distributed under the terms and conditions of the Creative Commons Attribution (CC BY) license (http://creativecommons.org/licenses/by/4.0/).

Article

Biocomposites from Rice Straw Nanofibers: Morphology, Thermal and Mechanical Properties

José Carlos Alcántara [1], Israel González [2], M. Mercè Pareta [3] and Fabiola Vilaseca [1,4,5,*]

1. Advanced Biomaterials and Nanotechnology, Department of Chemical Engineering, Polytechnic School, University of Girona, 17003 Girona, Spain; jalcantarac@unitru.edu.pe
2. LEPAMAP Group, Department of Chemical Engineering, University of Girona, 17003 Girona, Spain; israel.gonzalez@udg.edu
3. Department of Architecture and Construction, Polytechnic University of Girona, 17003 Girona, Spain; mm.pareta@udg.edu
4. Department of Fiber and Polymer Technology, KTH Royal Institute of Technology, SE-10044 Stockholm, Sweden
5. Department of Industrial and Materials Science, Engineering Materials, Chalmers University of Technology, SE-412 96 Gothenburg, Sweden
* Correspondence: vilaseca@kth.se or fabvil@chalmers.se

Received: 14 April 2020; Accepted: 1 May 2020; Published: 5 May 2020

Abstract: Agricultural residues are major potential resources for biomass and for material production. In this work, rice straw residues were used to isolate cellulose nanofibers of different degree of oxidation. Firstly, bleached rice fibers were produced from the rice straw residues following chemical extraction and bleaching processes. Oxidation of rice fibers mediated by radical 2,2,6,6-tetramethylpiperidine 1-oxyl (TEMPO) at pH 10 was then applied to extract rice cellulose nanofibers, with diameters of 3–11 nm from morphological analysis. The strengthening capacity of rice nanofibers was tested by casting nanocomposite films with poly(vinyl alcohol) polymer. The same formulations with eucalyptus nanofibers were produced as comparison. Their thermal and mechanical performance was evaluated using thermogravimetry, differential scanning calorimetry, dynamic mechanical analysis and tensile testing. The glass transition of nanocomposites was shifted to higher temperatures with respect to the pure polymer by the addition of rice cellulose nanofibers. Rice nanofibers also acted as a nucleating agent for the polymer matrix. More flexible eucalyptus nanofibers did not show these two phenomena on the matrix. Instead, both types of nanofibers gave similar stiffening (as Young's modulus) to the matrix reinforced up to 5 wt.%. The ultimate tensile strength of nanocomposite films revealed significant enhancing capacity for rice nanofibers, although this effect was somehow higher for eucalyptus nanofibers.

Keywords: rice nanofibers; biocomposites; casting; mechanical properties; thermal properties

1. Introduction

Biomass has become a subject of increasing research and debate over recent times due to its potential for energy and material production [1]. In fact, the current environmental concern is pushing the international community towards policies that aim to displace fossil resources by biobased ones. In this context, agricultural industries are playing a role since they use to produce huge amounts of crop wastes annually. Disposal of these wastes in landfills causes serious problems related to environmental contamination and harmful effects to human and animal health. Depending on the land use and management options, agriculture can be a source or a sink for atmospheric CO_2 [2]. In general, the progressive increase in atmospheric concentration of CO_2 and other greenhouse gases due to agriculture or industrial practices has created a worldwide awareness in identifying new strategies to solve this global problem.

Crop residues are defined as the nonedible plant parts that are left in the field after harvesting. Global production of residues from the six main crops (barley, maize, rice, soybean, sugar cane and wheat) is estimated to be of around 3.7 Pg dry matter yr^{-1}, and this value could increase by 1.3 Pg dry matter yr^{-1} considering the current progress towards highly intensive agriculture [1]. The use of crop residues as a raw material is recommended, especially considering their chemical composition of lignocellulosic fibers, with significant amounts of cellulose in some cases. The mean values of some agricultural residues' chemical composition are presented in Table 1.

Table 1. Chemical composition of agricultural residues.

Agro-Industrial Waste	Chemical Composition (% w/w)				Ref.
	Cellulose	Hemicellulose	Lignin	Ash	
Sugarcane bagasse	30.2	56.7	13.4	1.9	[3]
Rice straw	36.2	23.5	15.6	12.4	[4]
Corn stalks	61.2	19.3	6.9	10.8	[3]
Sawdust	45.1	28.1	24.2	1.2	[3,5]
Sugar beet waste	26.3	18.5	2.5	4.8	[3]
Barley straw	33.8	21.9	13.8	11	[6]
Cotton stalks	58.5	14.4	21.5	10	[6]
Oat straw	39.4	27.1	17.5	8	[5]
Soya stalks	34.5	24.8	19.8	10.4	[7]
Sunflower stalks	42.1	29.7	13.4	11.2	[7]
Wheat straw	32.9	24.0	8.9	6.7	[5,6]

The lignocellulosic character of agricultural residues makes them suitable as source of cellulose fibers, which are the most abundant biopolymers on earth. The hierarchical structure of cellulose fibers is unique, with linear β-1,4-glucan chains forming microfibrils of 3–4 nm diameter (about 30–40 chains), that bond together in bundles resulting in microfibers that intermix with hemicelluloses at the secondary and primary layers of the plant cell wall, with lignin mainly at the primary layer and middle lamella [8]. The extraction of cellulose from nanofibers and its application in composite materials is gaining relevance thanks to their inherent properties such as high strength and stiffness, light weight, biodegradability and renewability of their resources.

Nanocomposites are materials made up of at least two different phases where one component has some dimension in the nano range, typically below 100 nm or even below 50 nm. Structurally, nanocomposites are materials that involve nanosized filler particles (dispersed phase), a matrix (dispersion phase), and an interfacial region. The main advantage of nanocomposites compared to conventional ones is their unique characteristics due to their large surface area to volume ratio [9]. Therefore, nanocomposite materials take advantage of the outstanding properties of the nanosized dispersed phase. The extraordinary mechanical, thermal, and structural properties of cellulose nanofibers give them huge potential to act as a reinforcing agent. Besides, when strong interfacial adhesion with the matrix is attained, proper stress transfer from the matrix to the reinforcing phase is expected. Due to the hydrophilic nature of cellulose polymer, the interfacial region is crucial to allow good dispersion and distribution of nanocelluloses into the polymer matrix, and to promote good adhesion of the two components.

Poly(vinyl alcohol) (PVA) is the largest synthetic water soluble polymer produced in the world [10,11]. It is prepared by the hydrolysis of polyvinyl acetate. The degree of solubility, and the biodegradability as well as other physical properties can be controlled by varying the molecular weight (Mw) and the degree of hydrolysis (saponification) of the original polymer [11]. PVA possesses noticeable features such as water solubility, ease-of-use, film-forming property, and biodegradability [12]. Thanks to all these characteristics, poly(vinyl alcohol) (PVA) has been widely used for more homogeneous distribution of components in the preparation of blends and composites with several natural, renewable polymers like chitosan, nanocellulose, starch, or lignocellulosic fillers [10–17].

In this work, cellulose nanofibers from rice straw residues are obtained and used for the production nanocomposite materials. The performance of rice nanofibers will be compared to other well-known cellulose nanofibers from common sources like wood. Rice straw was first submitted to a chemical bleaching process to obtain rice cellulose fibers, followed by an oxidation mediated by radical 2,2,6,6-Tetramethylpiperidine 1-oxyl (TEMPO) as pretreatment to further extract rice cellulose nanofibers. Nanocellulose fibers from rice straw are tested as reinforcing element of poly(vinyl alcohol) polymer matrix. The hydrophilic character of the matrix is chosen as model for a fully favorable nanofiber–matrix interface in order to analyze the potential of rice cellulose nanofibers. The nanocomposites were prepared by casting to ensure proper dispersion and distribution of the nanofibers into the polymer. The thermal and mechanical properties of the nanocomposites are analyzed and compared to other currently used cellulose nanofibers.

2. Materials and Methods

2.1. Materials

Rice straw from the *Oryza sativa* species and of the appellation "Arroz de Valencia", was used as the raw material to extract cellulose nanofibers for the further preparation of composite materials. Rice straw was air-dried for over 24 h and then cut into lengths of about 3–5 cm. Characterization of the raw material was performed according to the Tappi test methods (Tappi, 2003–2004): hot water extractives (T 207 cm-99), benzene-ethanol extractives (Tappi 204 cm-97), α-cellulose (Tappi 203 om-93), and Klason lignin content (Tappi T222 om-98). The holocellulose was determined by treating the extracted rice straw with $NaClO_2$ solution. For comparison, cellulose nanofibers from eucalyptus were also used as polymer reinforcement.

Poly(vinyl alcohol) (PVA) from Sigma Aldrich (Madrid, Spain) was used as the polymer matrix. PVA, supplied as white powder, had a molecular weight of around 70000 g/mol and was 87–90% hydrolyzed.

For solvent extraction, ethanol, benzene and toluene were used. Bleaching of rice straw was performed with sodium chlorite ($NaClO_2$) and acetic acid (CH_3COOH). Further oxidation reaction of rice straw fibers was carried out using the radical 2,2,6,6-Tetramethylpiperidine 1-oxyl (TEMPO), sodium bromide (NaBr), sodium hypochlorite solution (NaClO) 15% w/v, sodium hydroxide (NaOH) and hydrogen chloride (HCl). For the evaluation of carboxyl groups, methylene blue in powder was used with a buffer solution made from boric acid (H_3BO_3) and potassium chloride (KCl). Finally, for the viscosimetric analysis, cupriethylenediamine ($C_2H_6CuN_2$) was used as cellulose solvent. All chemical reagents were supplied by Sigma Aldrich and used as received.

2.2. Preparation of Rice Fibers

Rice straw was chopped using knives, milled, and meshed with a 40-mesh (400 μm) screen. The retained matter was submitted to solvent extraction with ethanol/toluene (40/60) for 24 h to remove pectin, waxes, and fats. After drying at atmosphere conditions, rice straw fibers were bleached as follows: 50 g of rice fibers were immersed in a solution containing 5 g of sodium chlorite ($NaClO_2$) and 5 g of acetic acid (CH_3COOH) in 1 L of distillate water. Stirring was maintained for 4 h at 80 °C. Rice fibers were afterwards thoroughly rinsed with distillate water. A second bleaching step was performed under the same chemical conditions for 2 h at room temperature. After proper rinsing, the rice fibers were kept at 4 °C in the fridge. The mean fiber length and diameter of the obtained rice straw fibers were analyzed by means of a MorFi Compact fiber analyzer from TECHPAP (Grenoble, France). For this purpose, a dilute suspension of fibers (25 mg/L) was analyzed using an optics and flow cell measurement. Data of more than 3000 fibers were analyzed and the fiber length distribution, diameter distribution, as well as the mean values, were obtained.

2.3. Preparation of Rice Nanofibers

TEMPO-mediated oxidation at basic conditions was used as pretreatment for the extraction of cellulose nanofibers. Rice straw fibers (1 g) were suspended in water (100 mL) in the presence of TEMPO radical (0.016 g, 0.1 mmol) and sodium bromide (0.1 g, 1 mmol). Sodium hypochlorite solution (NaClO) of 15% w/v was added in the suspension and the pH was adjusted to 10 by adding hydrogen chloride 0.1 M. Magnetic stirring was applied to the suspension in order to assure good dispersion of all the substances. The oxidation reaction started when the desired amount of NaClO was added into the system. In this case, different oxidation degrees were tested by adding 3, 5, 8, or 12 mmol of NaClO per gram of cellulose fiber. The addition was dropwise at room temperature with constant stirring of about 500 rpm. The pH was maintained constant at 10 by continuous addition of sodium chloride. The reaction finished when the pH was constant [18,19].

After oxidation, the cellulose fibers were centrifuged at 10,000 rpm for 10 min, to remove all the chemical reagents. The supernatant was discarded and the solids were resuspended in distillate water and centrifuged again. The process was repeated five times. The final suspension was 1% by weight. This suspension was disintegrated using and Ultraturrax IKA T24 digital working at 20.000 rpm during 5 min. The final appearance was a transparent gel-like suspension that was kept at 4 °C in the fridge for further use. This rice cellulose nanofibers were named r-CNF.

The TEMPO-mediated oxidation of cellulose fibers from eucalyptus fibers was carried out at neutral pH conditions according to methodology reported elsewhere [20]. The eucalyptus (hardwood) cellulose nanofibers were named h-CNF.

2.4. Characterization of Rice Nanofibers

Attenuated total reflectance Fourier transform infrared (ATR-FTIR) spectroscopy was performed on a Mattson Satellite spectrometer (Mettler Toledo, L'Hospitalet de Llobregat, Spain) equipped with a MKII Golden Gate Reflection ATR System. Spectra of the different oxidized rice cellulose nanofibers were recorded by co-adding 64 scans at 4 cm^{-1} optical resolution within the range 600–4000 cm^{-1}. The samples were cut in about 5 × 5 cm and were immersed for 5 s in an HCl 0.1 M solution with the aim of acidifying the carboxylate groups.

The carboxyl content of oxidized fibers was measured by UV-visible spectroscopy with the methylene blue method [21,22]. Here, a weighted oxidized cellulose sample (approx. 10–15 mg) was suspended in 25 mL of aqueous methylene-blue chloride solution (300 mg/L) and 25 mL of borate buffer of pH 8.5. The suspension is stirred for 1 h at 20 °C in a 100 mL Erlenmeyer flask. After this time, the suspension is centrifuged at 10,000 rpm for 20 min to isolate the fibers. The supernatant contains the nonadsorbed methylene blue that is determined photometrically, employing a calibration plot. After centrifuging, 1 mL of supernatant is introduced in a 10 mL volumetric flask together with 1 mL of acid chloride HCl 0.1 M. The total amount of free, i.e., nonadsorbed, methylene blue was calculated from experimental results (A) using the UV-VIS Spectrophotometer Shimadzu UV 160 (Thermo Fischer, Bilbao, Spain), working at 664 nm wave length. The final amount of carboxyl groups in mmol per gram of cellulose is given by the equation:

$$COOH\ (mmol/g) = \frac{(7.5 - MB_{na}) \cdot 0.00313}{w} \quad (1)$$

where MB_{na} is the amount of nonadsorbed methylene blue (mg) and w is the dry weight of the sample (g).

The water retention value (WRV) was measured by separating a determined volume of cellulose nanofibers gel into two equal portions, which were then centrifuged in a Sigma Laborzentrifugen model 6 K15 at 2400 rpm for 30 min to eliminate nonbonded water. In order to retain the NFC, a nitrocellulose membrane with a pore diameter of 0.65 µm was used at the bottom of the centrifuge bottles. Once centrifuged, only the NFC in contact with the membrane was removed, weighed, and then dried at 105 ± 2 °C for 24 h in containers of previously measured weight. This methodology is based on Tappi um 256. The average WRV value was then calculated according to the next equation:

$$WRV\ (\%) = \frac{(W_w - W_d)}{W_d} \cdot 100 \qquad (2)$$

where W_w is the wet weight (g) and W_d the dry weight (g).

The degree of polymerization (DP) was determined from intrinsic viscosity measurements, according to UNE 57-039-92 (which agrees with ISO 5351-1:1981) using cupriethylenediamine as solvent. The correlation between the intrinsic viscosity $[\eta]$ and the degree of polymerization (DP) was calculated from the next equation:

$$[\eta] = K' \cdot DP^a \qquad (3)$$

with $K' = 0.42$ and $a = 1$ for DP < 950, and $K' = 2.28$ and $a = 0.76$ for DP > 950 [23,24].

Original fibers were observed by scanning electron microscopy. For this, the used microscope was ZEISS DSM 960 (Zeiss, Jena, Germany) operating at 25 kV. Specimens were previously coated by sputtering 10 Nm gold at the surface. On the other hand, cellulose nanofibers were observed using transmission electron microscopy (TEM) by means of a ZEISS EM 910 Transmission-Electron Microscope (Zeiss, Jena, Germany). Samples were prepared by diluting the gel suspensions 10 times in distilled water. Later, 8 µL of the diluted nanocellulose suspensions were deposited on the membrane grid and, after drying, it was dyed with a 1% solution of uranyl acetate for 3 min. The surplus was removed with absorbent paper for the observation.

2.5. Preparation of PVA-Rice Nanocomposites

For the preparation of nanocomposites, 12.5 g of poly(vinyl alcohol) (PVA) were dissolved in 250 mL of water at 10 °C in high magnetic stirring. The gel of cellulose nanofibers was diluted until 0.1 wt.%. Different amounts of cellulose nanofibers, from rice straw (r-CNF) or from eucalyptus (h-CNF), were mixed with the polymer solution in order to obtain nanocomposites comprising 0.5, 1, 2.5, and 5 wt.% of nanofiber reinforcement. Nanocomposite suspensions of 2 g dry weight were prepared. The PVA-nanocellulose suspensions were homogenized by stirring for 1 h, and cast in petri dishes afterwards. The suspensions were dried in an oven at 37 °C for four to five days. A transparent film was obtained from where the specimens were cut for material characterization. A scheme illustrating the production process of the PVA-CNF nanocomposites is presented in Figure 1.

2.6. Characterization of PVA-Rice Nanocomposites

Thermal characterization of the nanocomposites was conducted from differential scanning calorimetry (DSC) tests using Mettler Toledo DSC822 equipment (Mettler Toledo, L'Hospitalet de Llobregat, Spain). Around 7–9 mg of sample was placed in the 40 µL aluminum capsule. The temperature ramp was from 30 to 240 °C at a 10°/min heating rate under nitrogen flow of 40 µL/min. Values of glass-transition temperature (T_g), melting temperature (T_m), melting enthalpy (ΔH_m), and degree of crystallinity (X_c) were deduced from DSC analysis. Two heating/cooling procedures were applied.

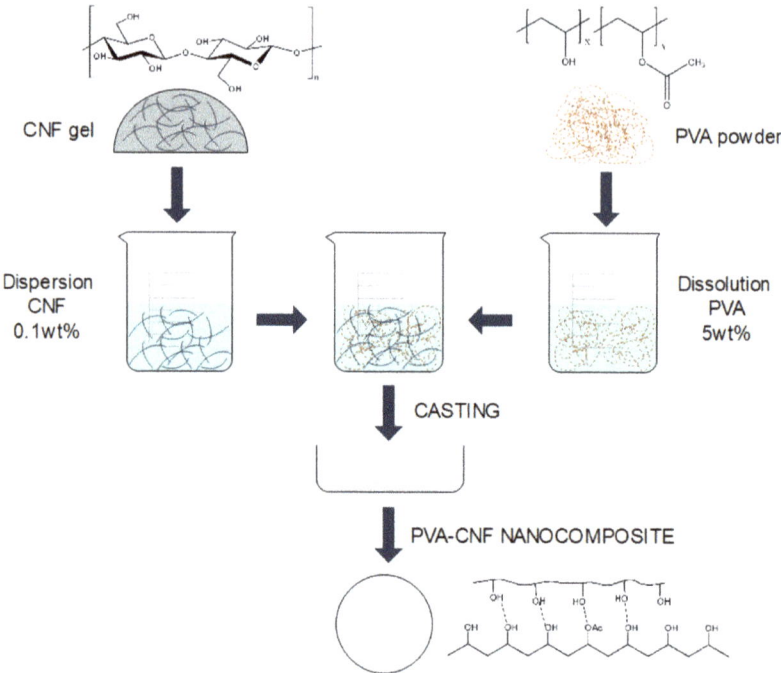

Figure 1. Illustrative scheme on the production mechanism of poly(vinyl alcohol)–cellulose nanofiber (PVA–CNF) nanocomposites.

Thermogravimetric analysis (TGA) was also performed using Mettler Toledo TGA851 equipment (Mettler Toledo, L'Hospitalet de Llobregat, Spain). For this, samples of around 10–15 mg were disposed in a 70 µL aluminum capsule. The assay was carried out from 30 to 650 °C at a 10°/min heating rate and under nitrogen flow of 40 µL/min. The degradation temperatures were obtained from the first derivative of the mass loss curve (DTG).

Prior to mechanical characterization, the specimens were conditioned in a Dycometal (Barcelona, Spain) climatic chamber at 23 °C and 50% relative humidity for 48 h before the tensile test and the dynamic mechanical analysis, according to ASTM D618 standard. Nanocomposite samples were mechanically tested in a universal testing machine Hounsfield model 42 (Tinius Olsen, Salfords, England) equipped with a 2.5 kN load cell. Sample dimensions were 5 mm wide and 45 mm long, and had a thickness of 0.2–0.3 mm. Preload was 0.1 N and cross-head speed was 5 mm/min. The results obtained were the average of at least five tested samples and data scattering within the range of 5–9%.

Dynamic mechanical analyses (DMA) of conditioned samples were performed using a Mettler Toledo DMA/SDTA861 instrument (Mettler Toledo, L'Hospitalet de Llobregat, Spain). The DMA measurements were done operating at tensile mode, at constant frequency of 1 Hz, amplitude of 20 µm and temperature range from −50 to 130 °C at a heating rate of 5°/min. The specimen dimensions were 5 mm width and 22.5 mm length, with thickness of 0.2–0.3 mm.

3. Results and Discussion

3.1. Rice Straw Fibers and Nanofibers

The valorization of agricultural wastes from rice crops is proposed in this study. In particular, we investigate the use of rice straw as potential raw material for cellulose nanofiber production. The chemical composition of the rice straw used in this study is shown in Table 2.

Table 2. Composition of rice straw % (w/w dry weight).

α-Cellulose	Pentosan	Klason Lignin	Benzene-Ethanol Extractives	Hot-Water Extractives	Ashes
41.2	19.5	21.9	0.56	7.3	9.2

These residues can be considered an interesting resource of cellulose fibers complementary to the other types of sources, mainly wood. The use of industrial residues or low-value byproducts as raw materials for the production of materials with high value, such as cellulose nanofibers, has been described in the literature by other authors [25–27]. It is worth noting that the holocellulose content (α-cellulose and pentosans) of rice straw residue is 60.7%. In general, the chemical composition is comparable to literature values, including other straw residues [4,5,27], with lignin values close to those for hardwood fibers [28]. Given the lower lignin content in the crop residues, along with their more porous structure compared to those of hardwoods and softwoods, milder pulping conditions with no sulfur processes can be applied for the lignin extraction [29]. In the current case, two bleaching processes with sodium chloride in acetic acid were employed to remove the lignin, once pectins and waxes were extracted. As a result, rice fibers were obtained with the length and diameter distributions shown in Figure 2. The mean weighted fiber length and fiber diameter were respectively 640 and 22,1 µm, with an aspect ratio of around 29. With this morphology, rice straw fibers can be considered as potential fibers for composite reinforcement.

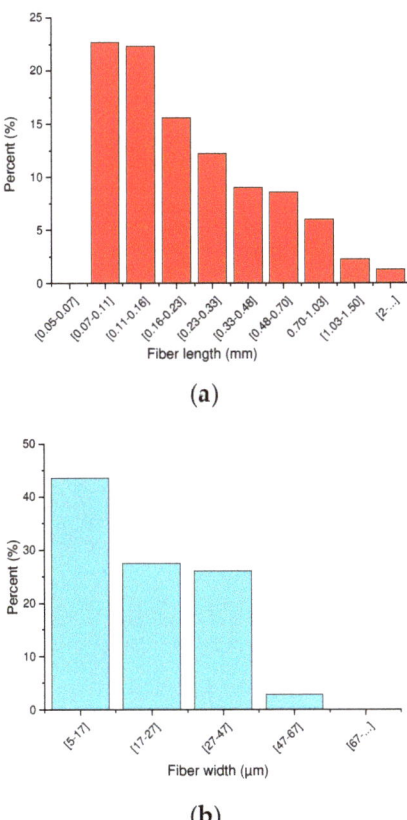

Figure 2. Arithmetic fiber length distribution (**a**) and fiber width distribution (**b**) of rice straw fibers.

Cellulose fibers and nanofibers from wood are often used in comparison with the behavior of other natural fibers [30]. For the current study, micrographs from fibers and nanofibers from rice and eucalyptus are presented in Figure 3. An image of the obtained rice fibers observed by scanning electron microscopy is given in Figure 3a. As comparison, Figure 3b presents microphotography at the same amplification for bleached commercial eucalyptus pulp. Both the rice fibers and eucalyptus fibers were submitted to TEMPO-oxidation as pretreatment to isolate respective cellulose nanofibers, for which micrographs of transmission electron observation are presented in Figure 3c,d respectively. Mean nanofiber diameters were found to be 3–11 nm for rice nanofibers and around 4–6 nm for eucalyptus nanofibers, evidencing that both methodologies were suitable for extracting cellulose nanofibers. The fact that wood delivers somewhat thinner nanofibers is in agreement with results from other authors in the literature, also comparing wood and rice straw nanofibers [31].

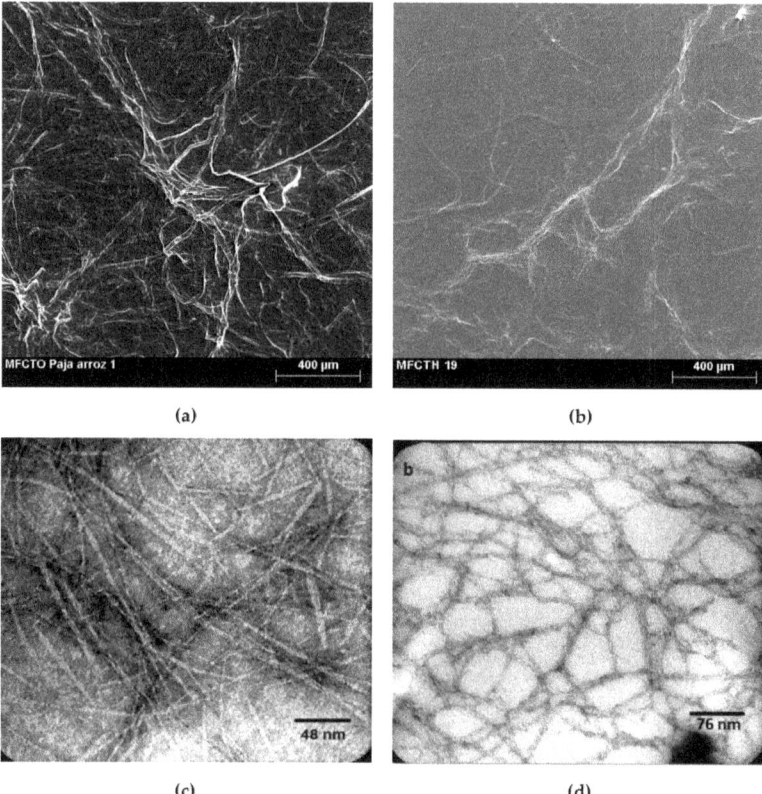

Figure 3. Scanning electron microphotographs of rice straw fibers (**a**) and of eucalyptus fibers (**b**). Transmission electron microscopy of rice straw nanofibers (**c**) and of eucalyptus nanofibers (**d**).

Different oxidation degrees were used on the preparation of cellulose nanofibers from rice straw. Table 3 shows the main characteristics of TEMPO-oxidized rice nanofibers. The oxidation reaction was applied on four different extensions. Different amounts of sodium hypochlorite solution required different reaction times to complete the oxidation step. The resulting rice nanofibers showed different nanofiber lengths (polymerization degree) and different surface charges (carboxylic groups). In these terms, higher degrees of oxidation gave higher amounts of carboxylic groups, but lower cellulose chain lengths. This is expected considering the mechanism associated to TEMPO-oxidation reaction.

Table 3. Characteristics of TEMPO-oxidized rice nanofibers.

Amount of NaClO (mmol/g)	Oxidation Time (min)	Water Retention Value (%)	Carboxylic Groups (mmol/g)	Degree of Polymerization DP	Viscous Molecular Weight (g/mol)
3	110	220	0.23	356	57,600
5	140	290	0.49	330	48,600
8	190	421	0.59	248	40,300
12	220	540	0.99	180	29,200

FTIR is a rapid and nondestructive technique for the qualitative and quantitative determination of biomass components [32]. Moreover, FTIR with ATR (attenuated total reflectance) allows attenuation of the incident radiation and provides IR spectra without the water background absorbance. From ATR-FTIR analysis (Figure 4) it is confirmed the appearance of the absorption peak at 1720 cm^{-1} wavelength, associated to carbonyl groups. From the magnification of the ATR-FTIR spectra between 1680 and 1800 cm^{-1}, rice fibers did not present any absorption band, while a peak appears after the oxidation reaction, more intense for higher extension reaction. In all the cases, the rest of the bands correspond to characteristic cellulose absorption peaks, as follows: –OH broad band between 3600 and 3200 cm^{-1} as well as the peaks at 1335 and 1205 cm^{-1}; C–O–C bond at 1160 cm^{-1}; CH$_2$ stretching and bending vibrations at 2918, 3851, 1427 and 1315 cm^{-1}; and finally bending vibration for CH at 1360 and 1280 cm^{-1}.

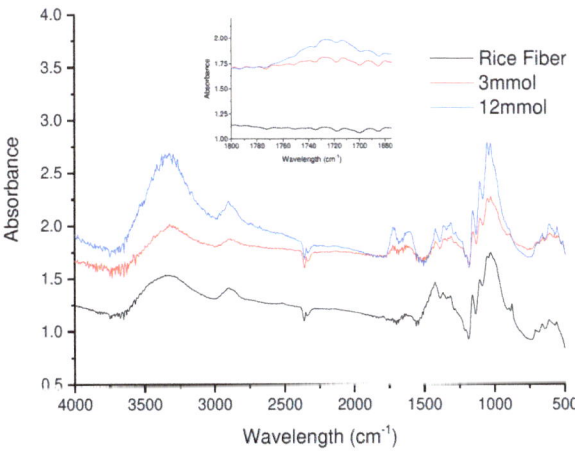

Figure 4. Attenuated total reflectance Fourier transform infrared spectroscopy (ATR-FTIR) spectra of rice fibers and for of nanofibers at two different degrees of oxidation.

The extent of the oxidation reaction impacted on the number of carboxylic groups on the cellulose surface (Table 3). In the current work, carboxylic groups were between 0,23 and 0,99 mmol per gram of rice cellulose fiber. These numbers agree with other values found in the literature for other cellulosic fibers, specifically for softwood [22]. It can be speculated that higher charge means more individualized nanofibers in suspension, although this has not been confirmed. It is clear though that the extent of the oxidation reaction, and so the number of carboxylic groups, influenced the water retention value of the resulting nanofibers. Water retention values relates to the amount of water adsorbed into the fibers. From the processing point of view, higher water retention values mean longer filtration times or larger casting times. Water retention values will combine with larger carboxylic content and also with larger fiber specific surface. Higher fiber individualization will favor the global specific surface and so the water retention values. The higher the oxidation extent, the lower the degree of polymerization was.

The graphical relation between water retention values and polymerization degrees with the carboxylic content is illustrated in Figure 5. Then, fiber shortening was observed when the carboxylic content was higher. The polymerization degree was deduced from the viscous molecular weight analysis of cellulose fibers, using k′ = 0,42 and a = 1 as constant parameters for the current system [23,24]. There was a substantial decrease in the molecular weight of rice cellulose nanofibers with the extent of the oxidation reaction. As described previously, the viscous molecular weight was determined from dissolving cellulose nanofibers with cupriethylenediamine (CED). Some authors [18,33] have shown that the C6 aldehydes, formed as intermediate structures during the TEMPO-mediated oxidation, cause a remarkable depolymerization during dissolution in CED by β-elimination. The values found in our study agree with those found in the literature in similar experimental conditions [33].

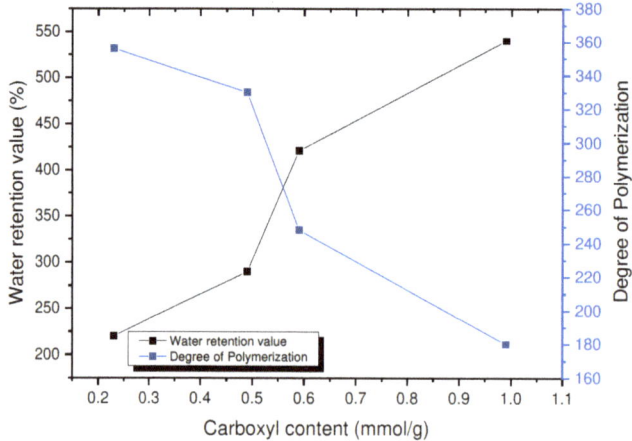

Figure 5. Water retention value and degree of polymerization with the carboxyl content.

3.2. PVA–Rice Nanocomposites

The combination of rice cellulose nanofibers with poly(vinyl alcohol) (PVA) allows PVA–rice nanocomposites to be obtained. Since both components are either water-soluble (polymer) or water-dispersible (rice nanofibers), casting in water was the chosen methodology to ensure proper distribution and dispersion of the reinforcing nanofibers into the polymer matrix and to avoid nanofiber aggregation. One important advantage of using PVA is the polar affinity of the polymer chains with cellulose fibers and nanofibers. It is therefore used as model system in order to analyze the reinforcing effect without side effects due to interface compatibility. Moreover, water-based procedures will avoid nanofiber aggregation during nanocomposite formation, which is one of the main issues in this field. In order to be able to compare the performance of rice nanofibers with others commonly used, PVA nanocomposites reinforced with cellulose nanofibers from eucalyptus (hardwood) were also analyzed.

The thermal characteristics as well as the crystallinity level of the polymer in nanocomposites are summarized in Table 4.

Glass transition temperature was determined from the second heating process applied to the samples. In the first heating process, the peak collided with the presence of water in the sample. Water integration gave 1.48% of water in the PVA sample, and between 0.31 and 0.35% of water in the biocomposites. Only the dry weight of the samples was used in further DSC calculations. In all cases, the addition of either nanofiber into the matrix moved the glass transition towards slightly higher temperature values. The same occurred with maximum melting temperature: biocomposites showed higher melting temperature than unreinforced matrix. This was more evident from the second heating step. In regard to the degree of crystallinity, the nanofibers did not act as nucleating agent, at least from the sample preparation (first heating, x_c^a). Instead, from the second heating (x_c^b), rice nanofibers did promote polymer crystallinity, although this was not observed for the eucalyptus nanofibers. The different nanofiber diameter and morphology of rice and eucalyptus nanofibers can explain this difference. Rice nanofibers were more rigid that eucalyptus nanofibers, better acting as nucleating agents.

Table 4. Thermal characteristics from DSC analysis. Glass transition temperature (T_g), Melting temperature (T_m), Melting enthalpy (ΔH_m), and degree of crystallinity (χ_c) of the polymer. Superscript **a** is for the first heating process, superscript **b** is for the second heating process. (χ_c(%) = (ΔH_m)/((ΔH_m^0) × ω))*100), with ΔH_m^0 is the heat of fusion for the 100% crystalline polymer, which is estimated to be ΔH_m^0 = 139 J/g for PVA-88 hydrolyzed; ω is the weight fraction of polymeric material in the respective composites).

Sample	T_g^b (°C)	T_m^a (°C)	T_m^b (°C)	ΔH_m^a (J·g^{-1})	ΔH_m^b (J·g^{-1})	χ_c^a (%)	χ_c^b (%)
PVA	67.9	193.2	167.8	54.69	30.47	39.3	21.9
PVA/r-CNF2.5	69.0	194.8	176.9	27.73	30.84	19.9	22.2
PVA/r-CNF5	69.3	194.8	178.7	28.36	35.85	20.4	25.8
PVA/h-CNF2.5	69.7	194.9	171.1	27.22	25.66	19.6	18.5
PVA/h-CNF5	70.2	194.3	175.4	26.23	25.02	18.9	18.0

Thermal stability of biomaterials is important for their applicability in biocomposite fields. Figure 6 displays the TG and the differential thermogravimetric curves of the net polymer (Figure 6a), both nanofibers (Figure 6b), and of the respective PVA nanocomposite at 5 wt.% nanofiber content (Figure 6c). PVA polymer showed two main degradation peaks at 339 and 458 °C. Our polymer was 90% hydrolyzed. The literature shows that fully hydrolyzed PVA also shows two main degradation peaks (at 375 and 440 °C), corresponding to the chain-stripping produced by the removal of water molecules (dehydration of the PVA polymer) followed by chain scission and decomposition [34]. For partially acetylated PVA like in the current case, the first peak shifted towards lower temperatures.

The mass loss around 150 °C accounts for volatiles and additives present in the matrix. Two main degradation bands are also found in both types of nanofibers, one around 260–270 °C and another at 325–330 °C (Figure 6b). The first corresponds to hemicelluloses and the second to cellulose itself. The deviation from linear in the curves below 200 °C stand for the slow loss of water molecules kept inside the rice or eucalyptus nanofibers' structure. Once inside the biocomposite, the hemicellulose band was not visible, and only the main peaks from cellulose and the polymer were present (Figure 6c).

Mechanical properties of biocomposites were investigated from DMA and tensile tests. The variation in storage modulus with the temperature of rice and eucalyptus bionanocomposites is presented in Figure 7. Reinforcing of PVA with nanofibers favored a higher storage modulus above room temperature and especially beyond glass transition (rubbery region). Reinforcing with these nanofibers is beneficial when working at higher temperatures. In the glassy region, only rice nanofibers produced some benefit compared to the polymer. Quan et al. also proved the increase in storage modulus for PVA–cellulose nanofiber composites above 30 °C as compared to the plain matrix [35]. It was reported that the difference between the elastic tensile modulus of cellulose nanofibers and that of the matrix is not high enough to benefit from a reinforcement effect in this temperature range. From our study, however, rice nanofibers provided a better reinforcement effect than eucalyptus nanofibers at both glassy and rubbery regions. We do not have a clear explanation for this, but it could be related to the different morphology of both types of nanofibers. Hence, while eucalyptus nanofibers are more flexible, rice nanofibers have a more rigid shape, as seen in the TEM images from Figure 3.

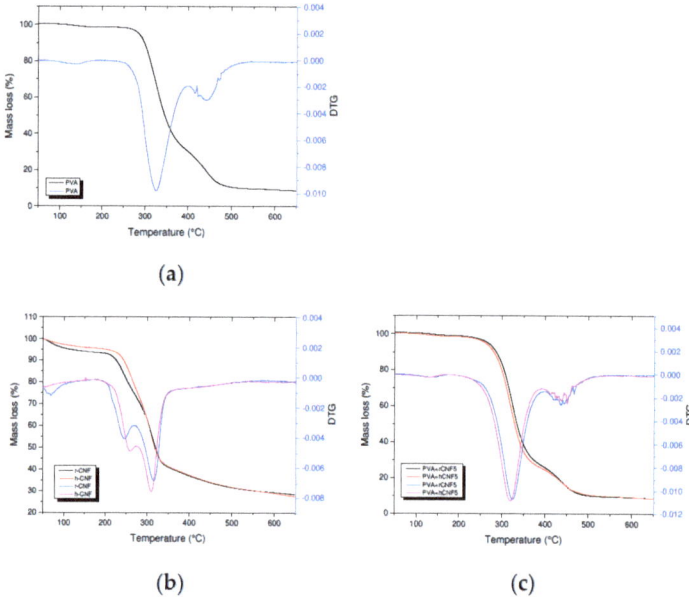

Figure 6. Mass loss (%) and first derivative (DTG) of polyvinyl alcohol (**a**), rice nanofibers (r-CNF) and hardwood nanofibers (h-CNF) (**b**), and of the respective nanocomposites containing 5 wt.% of nanofibers (**c**).

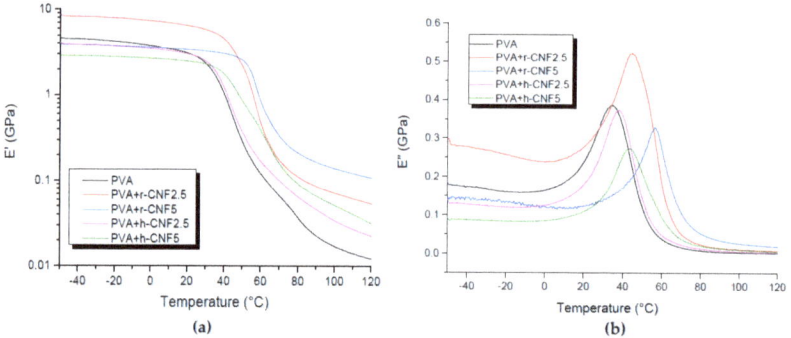

Figure 7. (**a**) Change in storage modulus (E') and (**b**) loss modulus (E") as function of temperature for PVA and biocomposites at 2.5 and 5 wt.% content of rice nanofibers (r-CNF) or eucalyptus nanofibers (h-CNF).

Other authors also found little influence on the storage modulus in eucalyptus CNF nanocomposites, although the methodology for composite preparation was different, and they claimed aggregation of their nanofibers [36]. Values of Tg deduced from the peak of the loss modulus (also in Figure 7) proved the different influence of both type of nanofibers. The dynamic loss modulus is often associated with internal friction and is sensitive to different kinds of molecular motions, relaxation processes, transitions, morphology, and other structural heterogeneities. The determination of the glass transition temperature (Tg), from the loss modulus, gave 35 °C for the net PVA, and 43 and 58 °C for rice nanocomposites at 2.5 and 5 wt.%. Instead, the same formulations with eucalyptus nanofibers gave tan δ of 40 and 42 °C, respectively, so higher than the matrix but below that for rice nanofibers.

The properties from tensile tests of PVA and all PVA biocomposites up to 5 wt.% of rice nanofibers or eucalyptus nanofibers are presented in Table 5. Graphical representation is in Figure 8. The same formulations of rice cellulose nanofibers (r-CNF) and eucalyptus nanofibers (h-CNF) were prepared for comparison. The net PVA matrix was more soft and weaker than the biocomposites, but with tensile strength and Young's modulus comparable to high-density polyethylene or polypropylene. This is good since any improvement will make competitive materials, in terms of mechanical behavior. For biocomposites with rice CNF, mechanical properties, namely tensile strength and Young's modulus, showed a linear increase up to 1 wt.% nanofiber content, and then the increments were less pronounced. Thus, incorporation of 1 wt.% r-CNF gave 1.6 times ultimate tensile strength, whereas incorporation of up to 5 wt.% r-CNF produced only 1.8 times greater ultimate tensile strength, both compared to the plain PVA matrix. Cellulose nanofibers from eucalyptus pulp always produced major improvements, especially for the tensile strength, which were 1.7 and 2.3 times higher for the same respective formulations. The higher increments on the tensile strength for the eucalyptus nanofibers can be explained due to the nanofiber morphology and their lower diameter compared to rice nanofibers. However, both types of nanofibers performed similarly for the Young's modulus with similar increments. It was noticeable that there was a more than 3.5 times increase in rigidity with only 5 wt.% CNF. As expected, elongation at break decreased with the amount of CNF. However, deformations for biocomposites from eucalyptus nanofibers were larger compared to those from rice nanofibers. It seems, therefore, that the different morphology of eucalyptus nanofiber was beneficial for higher elongations at break.

Table 5. Main tensile properties of PVA and bionanocomposites (E_{Young} : Young's modulus, σ : ultimate tensile strength and ε_{break} : elongation at break).

Sample	E_{Young} (GPa)	σ (MPa)	ε_{break} (%)
PVA	1.27 ± 0.1	35.9 ± 1.5	136.5 ± 10.5
PVA/r-CNF0.5	2.37 ± 0.3	43,1 ± 2.0	105.1 ± 5.2
PVA/r-CNF1	3.99 ± 0.1	57.8 ± 1.3	2.54 ± 1.7
PVA/r-CNF2.5	4.05 ± 0.1	62.1 ± 2.0	1.90 ± 1.2
PVA/r-CNF5	4.43 ± 0.2	65.1 ± 1.1	1.76 ± 0.5
PVA/h-CNF0.5	2.67 ± 0.2	44.6 ± 2.9	128.8 ± 7.7
PVA/h-CNF1	3.91 ± 0.1	62.1 ± 2.0	2.99 ± 2.5
PVA/h-CNF2.5	4.10 ± 0.1	70.6 ± 2.1	3.42 ± 1.3
PVA/h-CNF5	4.72 ± 0.2	82.2 ± 1.4	3.66 ± 0.8

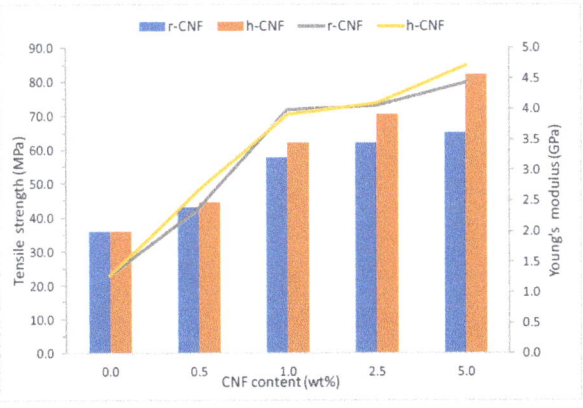

Figure 8. Graphical representation of tensile strength and Young's modulus for all bionanocomposites.

4. Conclusions

In this work, cellulose nanofibers were extracted from rice straw, showing that high-added-value products can be obtained from agricultural residue. This residue contains more than 60% holocellulose, of which about 41% is α-cellulose. Rice fibers were produced following bleaching processes and the obtained fibers had an aspect ratio of 29. From the following procedure, rice nanofibers with diameters ranging between 3–11 nm were extracted. Rice nanofibers were used to reinforce poly(vinyl alcohol) matrix by casting. For comparison, eucalyptus nanofibers were also used to produce nanocomposites. The reinforcing capacity of rice nanofibers was proved, increasing by 3.5 times the Young's modulus of the polymer. Similar stiffening was found for eucalyptus nanofibers. Instead, eucalyptus nanofibers were more favorable in strengthening, with ultimate tensile strengths 2.3 times higher than that from rice nanofibers, with ultimate strength 1.8 times that of the matrix, all values for 5 wt.% nanofiber content. The different morphology of both nanofibers behaved differently with polymer deformation or polymer crystallinity. Higher flexibility for eucalyptus nanofibers resulted in higher biocomposite deformation, but less ability to act as nucleating agent for crystal growing. Conversely, rice nanofibers acted distinctly as a nucleating agent for polymer crystal growing. Similarly, at DMA tests, rice nanofibers were more favorable to keep the storage modulus at a rubbery state, and clearly increased the Tg much more, as determined from the loss modulus peak.

All theses findings should encourage the use of agricultural residues as biomass for added-value materials and help in recommending better solutions for crop waste.

Author Contributions: J.C.A. designed and performed all the experimental work, and wrote the draft of the manuscript. I.G. conducted some characterization of the cellulose nanofibers. M.M.P. performed the thermal analysis. F.V. supervised the research and reviewed the manuscript. All authors have read and agreed to the published version of the manuscript.

Funding: AGAU-Agència de Gestió d'Ajuts Universitaris i de Recerca (AGAUR). Generalitat de Catalunya (2017SGR0516). EMCI-Ministerio de Ciencia e Innovación (MICINN) (CTM2011-28506-C02-01) (CTM2009-06864-E). AGAU-Agència de Gestió d'Ajuts Universitaris i de Recerca (AGAUR). Generalitat de Catalunya (2011 BE1 00860).

Acknowledgments: The authors are thankful to Mas Clarà de Domeny for kindly providing the rice straw.

Conflicts of Interest: The authors declare no conflict of interest.

References

1. Bentsen, N.S.; Felby, C.; Thorsen, B.J. Agricultural residue production and potentials for energy and materials services. *Prog. Energy Combust. Sci.* **2014**, *40*, 59–73. [CrossRef]
2. Lal, R. World crop residues production and implications of its use as a biofuel. *Environ. Int.* **2005**, *31*, 575–584. [CrossRef] [PubMed]
3. El-Tayeb, T.S.; Abdelhafez, A.A.; Ali, S.H.; Ramadan, E.M. Effect of acid hydrolysis and fungal biotreatment on agro-industrial wastes for obtainment of free sugars for bioethanol production. *Braz. J. Microbiol.* **2012**, *43*, 1523–1535. [CrossRef] [PubMed]
4. Bakker, R.; Elbersen, W.; Poppens, R.; Lesschen, J. *Rice Straw and Wheat Straw. Potential Feedstocks for the Biobased Economy*; NL Agency Ministry of Economic Affairs (Netherlands): The Hague, The Netherlands, 2013; pp. 6–30. Available online: http://english.rvo.nl/sites/default/files/2013/12/StrawreportAgNLJune2013.pdf (accessed on 13 April 2020).
5. Guilherme, J.; Martin, P.; Porto, E.; Corrêa, C.B.; De Aquino, L.M. Antimicrobial potential and chemical composition of agro-industrial wastes. *J. Nat. Prod.* **2012**, *5*, 27–36.
6. Nigam, P.S.; Gupta, N.; Anthwal, A. Pre-treatment of Agro-Industrial Residues. In *Biotechnology for Agro-Industrial Residues Utilisation*; Springer: Dordrecht, The Netherlands, 2009; p. 17.
7. Motte, J.C.; Trably, E.; Escudié, R.; Hamelin, J.; Steyer, J.P.; Bernet, N.; Delgenes, J.P.; Dumas, C. Total solids content: A key parameter of metabolic pathways in dry anaerobic digestion. *Biotechnol. Biofuels* **2013**, *6*, 164. [CrossRef]
8. Moser, C.; Henriksson, G.; Lindström, M.E. Structural Aspects on the Manufacturing of Cellulose. *BioResources* **2019**, *14*, 2269–2276.

9. Khattab, M.M.; Abdel-Hady, N.A.; Dahman, Y. *Cellulose Nanocomposites: Opportunities, Challenges, and Applications*; Elsevier Ltd.: Amsterdam, The Netherlands, 2017. [CrossRef]
10. Ramaraj, B. Crosslinked Poly(vinyl alcohol) and Starch Composite Films. II. Physicomechanical, Thermal Properties and Swelling Studiesitle. *J. Appl. Polym. Sci.* **2007**, *103*, 909–916. [CrossRef]
11. Roohani, M.; Habibi, Y.; Belgacem, N.M.; Ebrahim, G.; Karimi, A.N.; Dufresne, A. Cellulose whiskers reinforced polyvinyl alcohol copolymers nanocomposites. *Eur. Polym. J.* **2008**, *44*, 2489–2498. [CrossRef]
12. Tan, B.K.; Ching, Y.C.; Poh, S.C.; Abdullah, L.C.; Gan, S.N. A Review of Natural Fiber Reinforced Poly(Vinyl Alcohol) Based Composites: Application and Opportunity. *Polymers (Basel)* **2015**, *7*, 2205–2222. [CrossRef]
13. Tan, B.K.; Ching, Y.C.; Gan, S.N.; Ramesh, S.; Rahman, M.R. Water absorption properties of kenaf fibre–poly(vinyl alcohol) composites. *Mater. Res. Innov.* **2014**, *18*, S6-144–S6-146. [CrossRef]
14. Zhang, L.; Zhang, G.; Lu, J.; Liang, H. Preparation and Characterization of Carboxymethyl Cellulose/Polyvinyl Alcohol Blend Film as a Potential Coating Material. *Polym.-Plast. Technol. Eng.* **2013**, *52*, 163–167. [CrossRef]
15. Kvien, I.; Oksman, K. Orientation of cellulose nanowhiskers in polyvinyl alcohol. *Appl. Phys. A Mater. Sci. Process.* **2007**, *87*, 641–643. [CrossRef]
16. Zheng, Q.; Cai, Z.; Gong, S. Green synthesis of polyvinyl alcohol (PVA)–cellulose nanofibril (CNF) hybrid aerogels and their use as superabsorbents. *J. Mater. Chem. A* **2014**, *2*, 3110–3118. [CrossRef]
17. Lu, J.; Wang, T.; Drzal, L.T. Preparation and properties of microfibrillated cellulose polyvinyl alcohol composite materials. *Compos. Part A Appl. Sci. Manuf.* **2008**, *39*, 738–746. [CrossRef]
18. Isogai, A.; Saito, T.; Fukuzumi, H. TEMPO-oxidized cellulose nanofibers. *Nanoscale* **2011**, *3*, 71–85. [CrossRef]
19. Saito, T.; Isogai, A. TEMPO-Mediated Oxidation of Native Cellulose. The Effect of Oxidation Conditions on Chemical and Crystal Structures of the Water-Insoluble Fractions. *Biomacromolecules* **2004**, *5*, 1983–1989. [CrossRef]
20. Besbes, I.; Alila, S.; Boufi, S. Nanofibrillated cellulose from TEMPO-oxidized eucalyptus fibres: Effect of the carboxyl content. *Carbohydr. Polym.* **2011**, *84*, 975–983. [CrossRef]
21. Fras, L.; Stana-Kleinschek, K. Quantitative determination of carboxyl groups in cellulose by complexometric titration. *Lenzing. Ber.* **2002**, *81*, 80–88.
22. Vilaseca, F.; Serra, A.; Kochumalayil, J.J. Xyloglucan coating for enhanced strength and toughness in wood fibre networks. *Carbohydr. Polym.* **2020**, *229*, 115540. [CrossRef]
23. Henriksson, M.; Berglund, L.; Isaksson, P.; Lindström, T.; Nishino, T. Cellulose Nanopaper Structures of High Toughness Cellulose Nanopaper Structures of High Toughness. *Biomacromolecules* **2008**, *9*, 1579–1585. [CrossRef]
24. Marx-Figini, M. The control of molecular weight and molecular-weight distribusion in the biogenesis of cellulose. In *Cellulose and Other Natural Polymer Systems*; Springer: Boston, MA, USA, 1982; pp. 243–271.
25. Berglund, L.; Noël, M.; Aitomäki, Y.; Öman, T.; Oksman, K. Production potential of cellulose nanofibers from industrial residues: Efficiency and nanofiber characteristics. *Ind. Crop. Prod.* **2016**, *92*, 84–92. [CrossRef]
26. Feng, Y.H.; Cheng, T.Y.; Yang, W.G.; Ma, P.T.; He, H.Z.; Yin, X.C.; Yu, X.X. Characteristics and environmentally friendly extraction of cellulose nanofibrils from sugarcane bagasse. *Ind. Crop. Prod.* **2018**, *111*, 285–291. [CrossRef]
27. Hassan, M.; Berglund, L.; Hassan, E.; Abou-Zeid, R.; Oksman, K. Effect of xylanase pretreatment of rice straw unbleached soda and neutral sulfite pulps on isolation of nanofibers and their properties. *Cellulose* **2018**, *25*, 2939–2953. [CrossRef]
28. Rodriguez, A.; Moral, A.; Serrano, L.; Labidi, J.; Jiménez, L. Rice straw pulp obtained by using various methods. *Bioresour. Technol.* **2008**, *99*, 2881–2886. [CrossRef] [PubMed]
29. Boufi, S.; Gandini, A. Triticale crop residue: A cheap material for high performance nanofibrillated cellulose. *RSC Adv.* **2015**, *5*, 3141–3151. [CrossRef]
30. Claro, P.I.C.; Corrêa, A.C.; de Campos, A.; Rodrigues, V.B.; Luchesi, B.R.; Silva, L.E.; Mattoso, L.H.C.; Marconcini, J.M. Curaua and eucalyptus nanofibers films by continuous casting: Mechanical and thermal properties. *Carbohydr. Polym.* **2018**, *181*, 1093–1101. [CrossRef]
31. Abe, K.; Yano, H. Comparison of the characteristics of cellulose microfibril aggregates of wood, rice straw and potato tuber. *Cellul.* **2009**, *16*, 1017–1023. [CrossRef]
32. Hospodarova, V.; Singovszka, E.; Števulová, N. Characterization of Cellulosic Fibers by FTIR Spectroscopy for Their Further Implementation to Building Materials. *Am. J. Anal. Chem.* **2018**, *9*, 303–310. [CrossRef]

33. Shinoda, R.; Saito, T.; Okita, Y.; Isogai, A. Relationship between Length and Degree of Polymerization of TEMPO-Oxidized Cellulose Nanofibrils. *Biomacromolecules* **2012**, *13*, 842–849. [CrossRef]
34. Peresin, M.S.; Habibi, Y.; Zoppe, J.O.; Pawlak, J.J.; Rojas, O.J. Nanofiber Composites of Polyvinyl Alcohol and Cellulose Nanocrystals: Manufacture and Characterization. *Biomacromolecules* **2010**, *11*, 674–681. [CrossRef]
35. Qua, E.H.; Hornsby, P.R.; Sharma, H.S.S.; Lyons, G.; McCall, R.D. Preparation and characterization of poly(vinyl alcohol) nanocomposites made from cellulose nanofibers. *J. Appl. Polym. Sci.* **2009**, *113*, 2238–2247. [CrossRef]
36. Lavoratti, A.; Scienza, L.C.; Zattera, A.J. Dynamic Mechanical Analysis of Cellulose Nanofiber/Polyester Resin Composites. In Proceedings of the ICCM International Conference on Composite Materials, Copenhagen, Denmark, 19–24 July 2015.

© 2020 by the authors. Licensee MDPI, Basel, Switzerland. This article is an open access article distributed under the terms and conditions of the Creative Commons Attribution (CC BY) license (http://creativecommons.org/licenses/by/4.0/).

Article

Nanocomposite Polymeric Materials Based on Eucalyptus Lignoboost® Kraft Lignin for Liquid Sensing Applications

Sónia S. Leça Gonçalves [1], Alisa Rudnitskaya [2,*], António J.M. Sales [3], Luís M. Cadillon Costa [3] and Dmitry V. Evtuguin [1,*]

1. CICECO and Department of Chemistry, University of Aveiro, 3810-193 Aveiro, Portugal; leca.sofia@ua.pt
2. CESAM and Department of Chemistry, University of Aveiro, 3810-193 Aveiro, Portugal
3. I3N and Department of Physics, University of Aveiro, 3810-193 Aveiro, Portugal; jsales@ua.pt (A.J.M.S.); kady@ua.pt (L.M.C.C.)
* Correspondence: alisa@ua.pt (A.R.); dmitrye@ua.pt (D.V.E.)

Received: 16 February 2020; Accepted: 30 March 2020; Published: 2 April 2020

Abstract: This study reports the synthesis of polyurethane–lignin copolymer blended with carbon multilayer nanotubes to be used in all-solid-state potentiometric chemical sensors. Known applicability of lignin-based polyurethanes doped with carbon nanotubes for chemical sensing was extended to eucalyptus LignoBoost® kraft lignin containing increased amounts of polyphenolic groups from concomitant tannins that were expected to impart specificity and sensitivity to the sensing material. Synthesized polymers were characterized using FT-MIR spectroscopy, electrical impedance spectroscopy, scanning electron microscopy, thermogravimetric analysis, and differential scanning calorimetry and are used for manufacturing of all solid-state potentiometric sensors. Potentiometric sensor with LignoBoost® kraft lignin-based polyurethane membrane displayed theoretical response and high selectivity to Cu (II) ions, as well as long-term stability.

Keywords: LignoBoost® kraft lignin; potentiometric sensors; carbon nanotubes; impedance spectroscopy; transition metals

1. Introduction

Conducting polymers comprise various types of polymeric materials with electronic and/or ionic conductivity—including doped conjugated polymers, redox polymers, polymer composites, and polymer electrolytes [1,2]. In the field of potentiometric sensors, conducting polymers are primarily used as ion-to-electron transducers for the fabrication of all-solid-state polymeric sensors. Conducting polymers can be employed as a solid inner contact or as a sensitive layer if it is mixed with active substances or contains functional groups capable of ion recognition. The merit of the latter approach is the possibility to diminish leaching of membrane components into aqueous phase, thus, increasing sensor lifetime and reproducibility while decreasing its detection limits [1].

Lignin is one of the main constituents of wood and is available as a waste product of pulp-and-paper industry. Technical lignins are highly branched irregular polymers that consist of phenyl propane units linked by a set of ether and carbon–carbon linkages and contain a variety of functional groups—such as hydroxyl, carbonyl, carboxyl, hydrosulfide, and sulphonate, among others [3]. These groups impart to lignin a capability to complex various compounds, including transition metals, pesticides and polycyclic aromatic hydrocarbons. Lignin also possesses redox activity that is attributed to the quinone structures formed during lignin oxidation serving as a reversible redox couple [4,5]. Both complexing and redox properties of lignin can be exploited in the chemical sensing and numerous applications of lignin as a sensing material in electrochemical sensors have been reported [6]. Impedimetric and amperometric sensors modified by Langmuir–Blodgett lignin films for the detection of copper, lead, cadmium, and humic substances were reported in [7,8]. Electronic tongues comprising lignin thin film sensors were applied to the recognition of taste substances and wines [9], and detection of triazine pesticides [10]. High electrocatalytic activity of oxidized lignin films allowed their use in electrode modification for the amperometric detection of ascorbic acid [4], nicotinamide adenine dinucleotide [5], and ozone [11]. Composites of lignin with nanoparticles and carbon nanotubes were used in electrode modification for the amperometric detection of trinitrotoluene [12] and chlorogenic acid [13].

As a method of sensing material preparation, covalent binding of lignin to the polymer matrix may be preferable to the thin film deposition as it would improve material stability and lifetime by preventing the leaching of low weight lignin fraction into the solution. Though lignin is insulating, its polyunsaturated nature promotes its transformation into conducting material upon appropriate doping. Synthesis of technical lignin-based polyurethanes, doped with carbon nanotubes for the sensing applications, has been previously reported [14–16]. It was demonstrated that doping with carbon nanotubes ensured electrical conductivity increase of kraft lignin-based polyurethane by 5 orders of magnitude [15]. Potentiometric sensors with lignin-based polyurethane membranes displayed selective response to Cr(VI) [16].

The purpose of this work was the synthesis of conducting lignin-based polyurethane and its application as a sensing material using a new type of lignin, LignoBoost® kraft, as raw material. This lignin is isolated from black liquor after kraft cooking of eucalyptus wood by precipitation with CO_2 within an advanced LignoBoost® process. LignoBoost® kraft lignin contains significantly more phenolic hydroxyl (almost double the amount due to the presence of associated tannins) and carbonyl groups compared to the conventional kraft lignin isolated from black liquor by acidification with mineral acid. Therefore, it was expected that the presence of a high amount of functional groups would impart to the sensing material based on Lignooboost® kraft lignin polymer different sensitivity and selectivity properties. As this technical lignin will appear on the market, the evaluation of its applicability in specific areas is of particular importance. This study reports, for the first time, the conducting properties of LignoBoost® lignin-based polyurethane and the potentiometric sensor manufactured therein as a sensitive layer.

2. Materials and Methods

2.1. Reagents

LignoBoost® lignin was produced at the pilot unit from the industrial black liquor after kraft pulping of eucalypt wood (*Eucalyptus globulus*) and supplied by The Navigator Company (Aveiro, Portugal). Poly(propylene glycol), tolylene 2,4-diisocyanate terminated with average Mn ~2300 (narrow MW distribution) with isocyanate content ~3.6 wt % (PPGDI), dibutyltin dilaurate, tris(hydroxymethyl)aminomethane (Tris) and aniline were from Sigma-Aldrich Química S.L. (Lisbon, Portugal). Multi-wall carbon nanotubes (MWCNTs) Nanocyl-3150 (purity > 95%, length 1–5 μm and diameter 5–19 nm) were from Nanocyl, S.A. (Sambreville, Belgium). Hydrochloric acid, sulfuric acid, potassium chromate; nitrates of calcium, cadmium, lead, zinc, copper, and chromium(III); and potassium hexacyanoferrate (II) and (III) were from Panreac Quimica S.A.U., Barcelona (Spain). Mercury(II) nitrate was from Fluka (Riedel-de Haën, Germany). All reagents were p.a. (for analysis) and acids were hyperpur-plus purity. Screen-printed electrodes (SPE) with carbon working, auxiliary electrodes and silver reference electrode were from DropSens (Oviedo, Spain). Ultrapure water (18 MΩ·cm^{-1}) was used for the preparation of all solutions.

2.2. Polymer Synthesis and Sensor Manufacturing

Polycondensation reaction of lignin or ellagic acid with isocyanate was carried out using the procedure adapted from [16] and described in detail elsewhere [14]. Briefly, lignin or ellagic acid powder (250 mg), or their mixture with varying amounts of MWCNTs, was placed in the glass reactor with 1.5 mg of poly(propylene glycol), tolylene 2,4-diisocyanate terminated (PPGDI) and stirred for 40 min at 40 °C in order to obtain an homogenous viscous solution. Amounts of reagents were selected to ensure a ratio of 1.5 between isocyanate (NCO) groups in PPGDI and OH groups in lignin. The reaction was carried under nitrogen atmosphere. Concentration of MWCNTs were 0, 0.2, 0.4, 0.5, 0.8, 1.0, and 1.4% (w/w) of the polymer. After the homogenous mixture was obtained, the temperature was increased to 60 °C and the catalyst (dibutyltin dilaurate, ca 2%) was added. The mixture was stirred for further 10 min until it started to thicken. After that, it was removed from the reactor and used for the preparation of the sensor membranes and polymeric films. Films (about 1 mm thickness) used for the polymer characterization were prepared by pouring still liquid polymer into the flat mold. Polymeric films were cured during 4 h at 60 °C and sensors at room temperature for 24 h. Polymer of each composition was synthetized at least twice.

Potentiometric sensors with and without solid inner contact were fabricated using SPE. Firstly, surface of SPE working electrode was rinsed with ethanol and water and cleaned by cycling potential for 5 cycles between −0.2 and +1.2 V at 5 0 mV·s^{-1} in 50 mmol·L^{-1} sulfuric acid. Solid contact was prepared by electropolymerization of aniline in deaerated aqueous solution of 50 mmol·L^{-1} aniline in 1 mol·L^{-1} hydrochloric acid by cycling potential for 100 cycles between −0.23 and +0.85 V at 50 mV·s^{-1}. Sensors were washed with deionized water, conditioned for 2 h in 1 mmol·L^{-1} hydrochloric acid and dried. All controlled-potential experiments were performed with an EZstat-Pro EIS (NuVant Systems Inc., Crown Point, IN, USA). Platinum wire served as the counter electrode and Ag/AgCl (KCl 3 mol·L^{-1}) served as a reference electrode. Sensors were prepared by placing a drop of liquid polymer on the working electrode of SPE. At least two sensors of the same composition were prepared.

2.3. Polymer Characterization

Lignin-based polyurethanes were characterized by Fourier transform mid-infrared spectroscopy (FT-MIR) using MATTSON 7300 series spectrometer (Mattson Instruments, Madison, WI, USA), equipped with total attenuated reflectance accessory SPECAC Golden Gate-Diamond, in the wavenumber range 4000–600 cm^{-1}, with a resolution of 4 cm^{-1} and 128 scans per sample.

Differential scanning calorimetry (DSC) analysis was carried out using Perkin Elmer Diamond Differential Scanning Calorimeter (Waltham, MA, USA) with power compensation. Measurements were made under nitrogen flow of 40.00 mL·min^{-1} in the temperature range from −100 to 50 °C, with a heating rate of 10 °C·min^{-1} using approx. 10 mg of the sample placed in a platinum cap.

Thermogravimetric analysis (TGA) was carried out using thermogravimetric analyzer SETSYS by Setaram Instrumentation (Caluire, France), equipped with vertical balance. Measurements were made under nitrogen atmosphere, in the temperature range from 20 to 800 °C with a heating rate of 10 °C·min^{-1} using approx. 8 mg of the sample placed in a platinum cap.

SEM images of lignin and lignin-based polyurethanes were recorded using Hitachi S-4100 microscope (Tokyo, Japan) on the carbon coated samples and applying acceleration voltage of 25 kV. SEM images of ellagic acid-based polyurethanes were recorded using bench TM4000Plus Hitachi microscope equipped with backscattered electrons (BSE) detector, using the uncoated samples and applying acceleration voltage of 15 kV.

DC electrical conductivity was measured at the temperatures between −110 and 100 °C using a 617 Keithley electrometer (Cleveland, OH, USA). Electrical contacts were made by painting polymer films on both sides with silver paste, simulating a parallel plate capacitor with a surface area of about 1 cm^2 and 3 mm distance between electrodes.

Dielectric measurements for frequencies between 100 Hz and 1 MHz were carried out using an Agilent 4294A precision impedance analyzer at temperatures between −73 and 127 °C under helium atmosphere. Estimated relative errors on both, real and imaginary parts of the complex permittivity, were below 2%.

2.4. Potentiometric Measurements

Electrochemical measurements were carried out in the following galvanic cell:

Ag|AgCl, KClsat|sample|polymer membrane|carbon

Emf values were measured vs. Ag/AgCl reference electrode with precision of 0.1 mV using a custom-made multichannel voltmeter with high input impedance connected to the PC for data acquisition and processing. Calibration measurements were made in the solutions of zinc nitrate, cadmium, lead, copper, mercury, iron (III), and potassium chromate in the concentration range 10^{-7}–10^{-4} mol·L^{-1}. Tris with concentration of 1 mmol·L^{-1} and pH 7, adjusted by addition of hydrochloric acid, was used as supporting electrolyte. Redox response was studied in the solutions of two redox pairs, Cr (III)/Cr(VI) and Fe(CN)$_6^{3-/4-}$. Total concentration was 1 mM for both pairs, with the ratio of oxidized to reduced form varying from 0.01 to 100. Measurements in the solutions of chromium (III), iron (III), chromate and redox pairs were made at pH 2 on the background of 0.01 mol·L^{-1} HCl. Selectivity was estimated using mixed solution method. At least three replicated calibration measurements were run for each ion. Parameters of Nernst equation, i.e., slope of the electrode function and standard potential were calculated using linear regression and averaged over replicated calibration runs and sensor compositions.

3. Results

3.1. DC Conductivity

Electrical conductivity is one of the requisites for the sensing material to be used as a membrane in potentiometric sensors. According to the previous work, lignin-based polyurethanes are insulating with conductivity of nearly 10^{-8} S·m^{-1}, which can be increased several orders of magnitude by doping with carbon nanotubes [15,16]. Lignin is considered to be an excellent dispersant of carbon nanotubes in the polymeric matrix [14]. As phenolic hydroxyls are much less reactive to isocyanates than aliphatic ones [17], the redox properties of lignin involved in copolymerization may not change significantly. The advantage of this approach is a percolation effect observed in lignin-based polymers doped with MWCNTs, e.g., drastic increase of conductivity in the presence of small amounts of carbon nanotubes. Thus, conducting polymers can be obtained in a cost-effective manner and without changing other properties of the material. A similar approach was adopted in this work to increase the conductivity of the eucalyptus LignoBoost® kraft lignin and ellagic acid-based polyurethanes. Ellagic acid was used in this study as the main eucalyptus kraft lignin contaminant of polyphenolic origin [18]. This tannin contributes substantially to the phenolic functionalities of the isolated lignin.

Firstly, polymers doped with 0.8 and 1.4% (w/w) of MWCNTs were synthesized. DC electrical conductivity (σ_{DC}) of both polymers, as a function of the MWCNTs' concentration at 293 K, is plotted in the Figure 1. The conductivity of both undoped polymers was very low: 1.80×10^{-10} and 7.68×10^{-10} S·m^{-1} for LignoBoost® kraft lignin and ellagic acid-based polymers, respectively (Figure 1). Higher conductivity of polyurethane (PU) based on undoped ellagic acid, when compared to that of based on kraft-lignin, could be explained by the better dispersion of low molecular weight polyphenolic molecules in PPGDI copolymer than the larger lignin oligomers. Incompletely dissolved highly swollen aggregated lignin was previously detected in PU obtained by copolymerization of conventional acid-precipitated eucalyptus kraft lignin and PPGDI [14]. In addition, conjugated aromatic rings of flat ellagic acid structure could positively contribute to the conductivity of the final PU more than kraft lignin, whose structural units are predominantly linked by alkyl–alkyl, alkyl–aryl, and aryl–ether bonds, i.e., having less conjugated aromatic structures [18].

Addition of carbon nanotubes led to a significant increase in the conductivity of the LignoBoost® kraft lignin-based polymer, reaching 5.37×10^{-4} S·m^{-1} after the addition of 1.4% (w/w) of MWCNTs. Despite the fact that conductivity of undoped ellagic acid-based polyurethane was higher compared to lignin-based one, the effect of the MWCNT addition on conductivity was less accentuated. The conductivity of only 1.12×10^{-7} S·m^{-1} was observed at 1.4% (w/w) of MWCNTs (Figure 1).

LignoBoost® kraft lignin-based polymers doped with intermediate concentrations of MWCNTs were synthesized with the aim to confirm that conductivity of this polymer follows percolation scaling law and to determine percolation threshold (Figure 1). Percolation threshold x_c and critical exponent t were calculated by fitting experimental data to the Equation (1)

$$\sigma_{DC} \approx (x - x_c)^t \tag{1}$$

where σ_{DC} is the DC conductivity, x the volume fraction of the conductive phase, x_c the critical concentration and t the critical exponent.

Critical concentration was found to be 0.77% (w/w) of MWCNTs and critical exponent 1.54.

All lignin-polyurethane compositions with MWCNTs' concentration above the percolation threshold have sufficiently high conductivity for electrochemical sensor applications. The percolation threshold for composites with high aspect ratio fillers, such as CNTs, can be calculated using 3D statistical percolation model proposed in a previous study [19]. Nanotubes used in this work had a length of 1–5 μm and a diameter of 5–19 nm, which means that their aspect ratio could vary between 52 and 260 and predicted critical concentration between 0.18% and 1.46%. It is important to point out that statistical model does not account for interactions between filler and polymer matrix, effect of the processing conditions or curved shape of nanotubes. Nevertheless, experimental value of the critical concentration for LignoBoost® kraft lignin-based polyurethane is within the predicted range. Critical concentrations between 0.5 to 4.5% (w/w) of CNTs were reported in the literature for the thermoplastic polymers [20]. Critical concentration obtained for LignoBoost® kraft lignin-based polyurethane is higher compared to the value of 0.18% (w/w) previously reported for the conventional technical kraft lignin doped with MWCNTs with the same characteristics as the ones used in this work [16]. This disparity can be related to the differences in the two lignins' structure and, consequently, the difference in the interaction between nanotubes and polymer matrix.

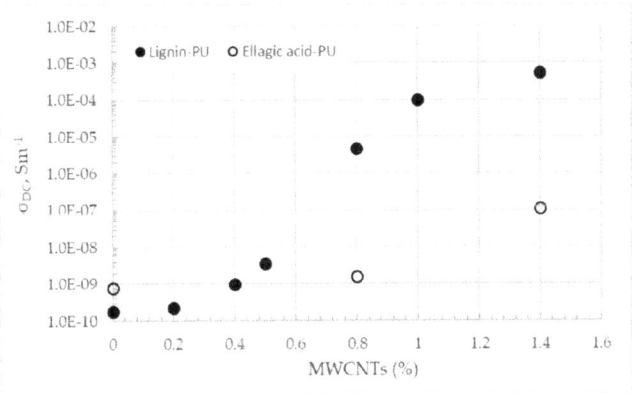

Figure 1. DC electrical conductivity σ_{DC} at constant temperature (293 K) as a function of the MWCNTs concentration of the LignoBoost® kraft lignin and ellagic acid-based polyurethanes.

The LignoBoost® kraft lignin possesses higher molecular weight than the conventional kraft lignin and higher content of phenolic groups [18], two factors favoring the association of lignin molecules in solution with copolymer (PPGDI). This negatively affects the dispersion of lignin in the polymeric matrix and, therefore, the dispersion of associated nanotubes in bulk. As such, the expected threshold occurred at a higher concentration of CNTs in the composite than that observed with conventional lignin kraft. According to previous findings, lignin interacts with CNTs via π-π stacking between aromatic ring and carbon nanotubes side wall, leading to the increase of the electron delocalization and eventual change in lignin chain conformation, thus allowing better π—overlap along the chain of lignin and giving rise to increased electric conductivity [16]. Oriented by lignin, CNTs interact with each other, forming conductive assembles, which are incorporated into the polymeric matrix.

The critical exponent reflects the dimensionality of the system and usually takes values between 1.3 and 1.9, corresponding to two- and three-dimensional percolating systems, respectively [21,22]. Calculated value of the critical exponent for the doped LignoBoost® kraft lignin-based polyurethane was lower than theoretical values, which is often observed for CNTs/polymer composites [20]. Such apparent reduction of system dimensionality was interpreted as a consequence of mutual attraction and realignment of carbon nanotubes during polymer curing. Strong nanotube–nanotube, lignin–nanotube, and nanotube–polymer matrix interactions are a prerequisite for such realignment to take place. Thus, the formation of the conducting network in the polymer is not a true statistical percolation process based on random distribution.

A DC conductivity of both lignin and ellagic acid-based polymers increases exponentially with temperature, as shown in the Figure 2. This behavior, which is typical for polymer composites, indicates that the conductivity is a thermally activated process and can be described by the well-known Arrhenius relation as follows

$$\sigma_{DC} \propto exp[-\frac{E_a}{kT}] \quad (2)$$

where E_a is the activation energy (J mol^{-1} K^{-1}), T is the absolute temperature (K) and k is the Boltzmann constant, 1.380649×10^{-23} J·K^{-1} [23].

Activation energy can be calculated using plots of $\ln\sigma_{DC}$ vs. the inversed temperature (Figure 2). Values of the activation energy for the undoped LignoBoost® kraft lignin and ellagic acid-based polymers were 0.76 eV and 1.05 eV, respectively. After addition of 1.4%(w/w) of MWCNTs, activation energy decreased to 0.08 eV and 0.39 eV for LignoBoost® kraft lignin and ellagic acid-based polymers, respectively. Accordingly, the addition of MWCNTs resulted in a significant decrease of activation energy and increase of DC conductivity in both polymers, being more pronounced in lignin-based polyurethane, indicating active interaction between carbon nanotubes and lignin.

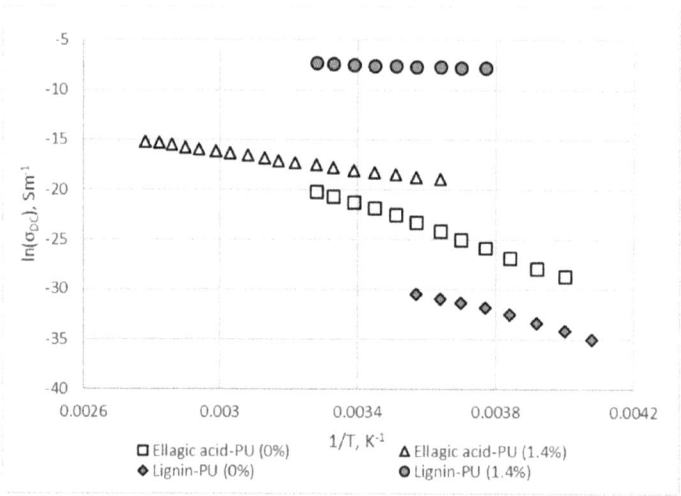

Figure 2. Logarithm of DC conductivity (σ_{DC}) versus the inverse of the temperature for LignoBoost® kraft lignin and ellagic acid based polyurethane polymer composites.

Ellagic acid-based polyurethanes had low conductivity even after doping with 1.4% (w/w) of MWCNTs, indicating that either no percolation occurs in this material or it occurs at relatively high concentrations of carbon nanotubes, making synthesis of such polymer costly and unpractical.

Moreover, higher concentrations of a filler could alter thermoplastic properties of the polymer, which would be undesirable. Though both ellagic acid and lignin have aromatic groups in their structure, that were suggested to be involved in the interaction with carbon nanotubes, in the case of ellagic acid this interaction is certainly insufficient to effectively realign carbon nanotubes. This can be related to the significantly smaller size of the ellagic acid molecule compared to lignin, which prevents effective re-orientation of the carbon nanotubes. Hence, ellagic acid itself does not provide enough performance to be used in conducting blends with MWCNTs. Thus, only LignoBoost® lignin-based polyurethanes were considered a perspective sensing material and studied in more detail.

Interaction between LignoBoost® kraft lignin and MWCNTs was assessed by probing the effectiveness of the nanotube dispersion in lignin-based polyurethane. SEM images of the mixture of LignoBoost® kraft lignin with MWCNTs (1.4% w/w) show an even distribution of disentangled carbon nanotubes, appearing as thin fibers on the surface of the larger lignin particles (Figure 3a,b). SEM images of the transversal cuts of LignoBoost® kraft lignin-based polyurethane and ellagic acid-based polyurethane films, with and without MWCNTs, all differed in their structure. Both undoped polymers are porous, however, while lignin-based polyurethane has several evenly distributed air bubbles of varying sizes (Figure 3c), ellagic acid-based polyurethane is denser with fewer air bubbles in its structure (Figure 3e). Addition of the MWCNTs results in appearance of irregularly shaped pores in lignin-based polyurethane (Figure 3d) and numerous round shaped pores in ellagic acid-based polyurethane (Figure 3f). The structural dissimilarities between undoped and doped polymers confirm interaction between carbon nanotubes and lignin or ellagic acid. Differences in properties of ellagic acid and lignin, the size of the molecule probably being the most important, results in different interaction between these compounds and carbon nanotubes, thus leading to different structure of the respective polyurethanes. Further insights into the interactions between LignoBoost®lignin and carbon nanotubes in the polyurethane matrix were gained using dielectric spectroscopy.

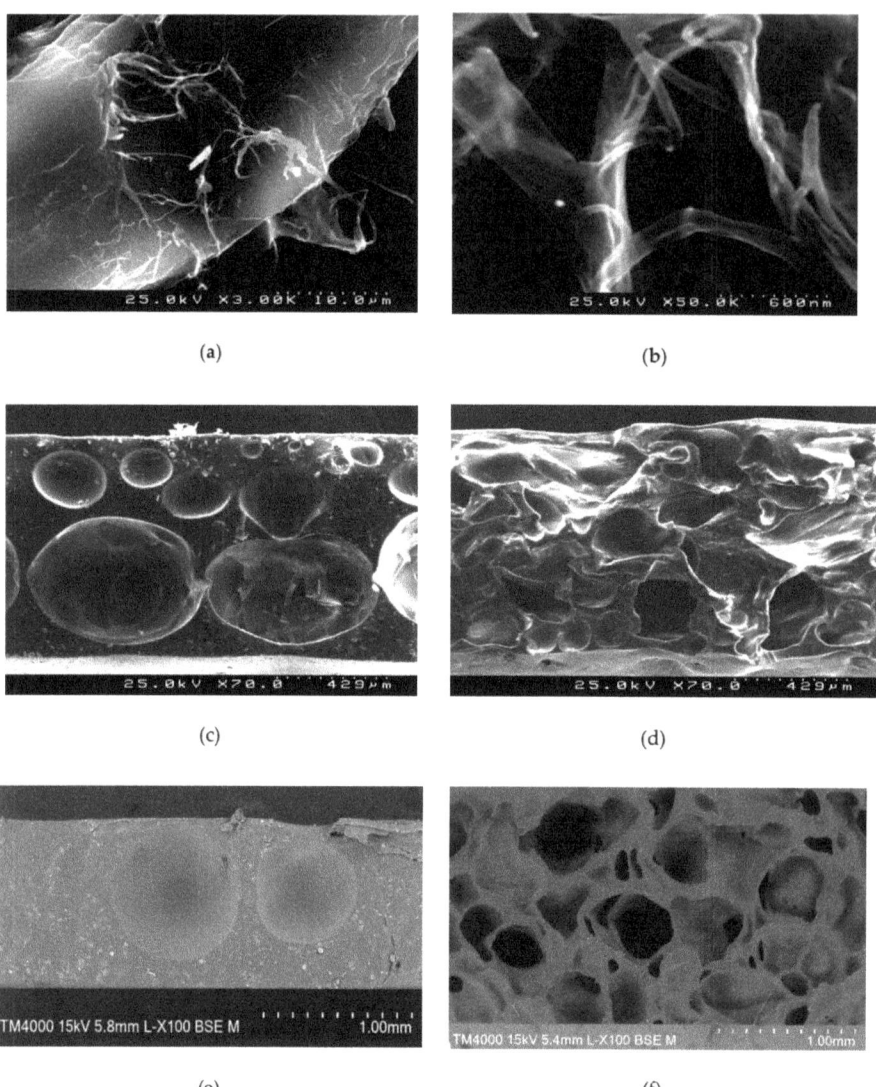

Figure 3. SEM images of: (**a**) and (**b**) LignoBoost® kraft lignin mixture with MWCNTs (1%); (**c**) undoped LignoBoost® kraft lignin-based polyurethane; (**d**) lignin-based polyurethane doped with 1.4% (w/w) of MWCNTs; (**e**) undoped ellagic acid-based polyurethane; (**f**) ellagic acid-based polyurethane doped with 1.4% (w/w) of MWCNTs.

3.2. AC Conductivity

Alternating current (AC) conductivity of LignoBoost® kraft lignin-based copolymers undoped and doped with different amounts of MWCNTs is depicted in the Figure 4a. The effect of the temperature on the AC conductivity variation as a function of frequency for LignoBoost® kraft lignin based polyurethane, doped with 1.4% (w/w) of MWCNTs, is shown in the Figure 4b. Real and imaginary parts of the complex permittivity, $\varepsilon^*(f) = \varepsilon'(f) - i\varepsilon''(f)$, and calculated AC electrical conductivity as a function of frequency for the lignin-based copolymer, with 1.4% (w/w) and without MWCNTs at T = 350 K, are shown in the Figure 5a,b, respectively.

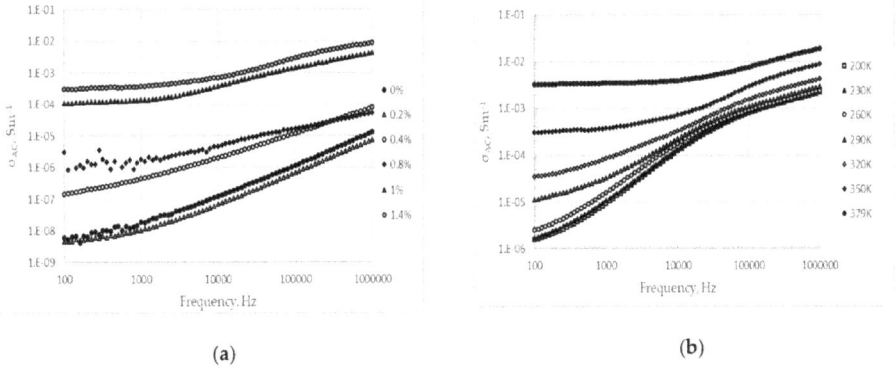

Figure 4. Frequency dependence of AC conductivity (σ_{AC}) for (**a**) lignin-based polyurethane doped with different amounts of MWCNTs at T = 350 K; (**b**) lignin based polyurethane doped with 1.4% (w/w) of MWCNTs at different temperatures.

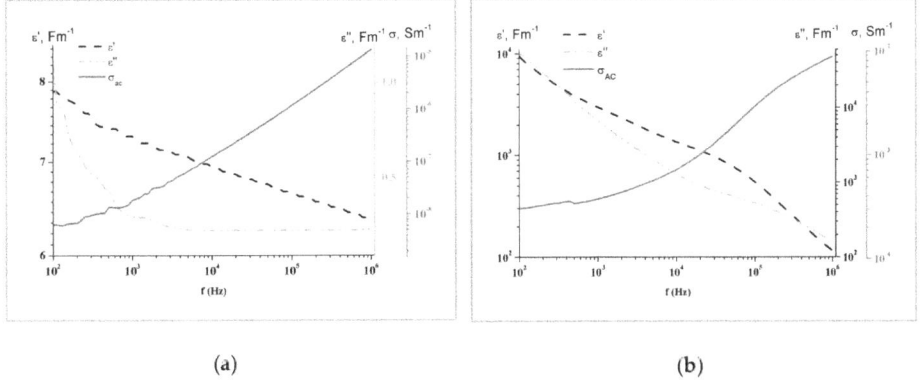

Figure 5. Real and imaginary parts of the complex permittivity, $\varepsilon^*(f) = \varepsilon'(f) - i\varepsilon''(f)$, and the calculated electrical conductivity, σ_{AC}, as function of frequency at T = 350 K for LignoBoost® kraft lignin-based copolymer (**a**) undoped and (**b**) doped with 1.4% (w/w) MWCNTs.

Increase of AC conductivity and decrease of both, real and imaginary parts of permittivity, with an increase of frequency was observed for both polymers (Figure 5a,b). Two distinct domains can be identified in the AC conductivity dependence on frequency [24]. AC conductivity is nearly independent of the frequency at low frequencies (below 1000 Hz), with its value approaching the DC conductivity (Figure 5a,b). This phenomenon, called the anomalous low frequency dispersion, is commonly observed in the disordered low-dimensional materials such as polymer matrices. Anomalous low-frequency dispersion arises from the restricted effective charge transport when charge motions are limited to one- or two-dimensional pathways. At frequencies above the crossover frequency fc, AC conductivity increases to different levels, depending on the material. Both ε' and ε'' decrease in all measured frequency intervals, which also corresponds to anomalous frequency dispersion.

Substantial increase of critical frequency, at which an increase in conductivity occurs, was observed in polymers doped with 0.8% (w/w) and higher concentrations of MWCNTs (Figure 4a). The critical frequency increased concomitantly with MWCNTs' concentration in the polymer, which is common behavior for the polymers doped with conducting particles [25,26]. The effect of the temperature on the AC conductivity dependence of the frequency for lignin based polyurethane doped with 1.4% (w/w) of MWCNTs is shown in the Figure 4b. Concomitantly dependent on applied frequency, with only a small increase observed at high frequencies. At lower temperatures, the increase in conductivity is more pronounced as the frequency increases.

3.3. Polymer Characterization

Apart from electrical conductivity, other LignoBoost® kraft lignin-based polymer properties were not significantly altered by doping with MWCNTs. No differences were observed in FT-MIR spectra of doped and non-doped polymers (Figure S1). Doping with carbon nanotubes also did not affect glass transition temperature of LignoBoost® kraft lignin-based polyurethanes, which was found to be −50 °C and −49 °C, with and without MWCNTs doping, respectively (Figure S2a,b). Low glass transition temperature is associated with the prevalence of flexible segments of the PPGDI chains, which is the major component of the polymer, and the possible disintegration of lignin molecules. It is worth noting that low glass transition temperatures of a polymer are required for its use as self-plasticizing sensing membrane [27].

Incorporation of the Lignoboost® kraft lignin into the polymeric matrix improved its thermal stability as evidenced by the increase of the maximum degradation temperature from 355 °C for lignin to 388 °C for the lignin-based polyurethane (Figure S3a,b, respectively). Increase of the thermal stability of polyurethane is related to the consumption of the hydroxyl groups of lignin in the polymerization reaction with isocyanate groups, and thus, to the decrease of the amount of functional groups susceptible to the thermal degradation [14]. MWCNTs did not significantly change the thermal behavior of lignin-based polyurethane, with maximum degradation temperature of doped polymer and undoped polymers being 387 °C and 388 °C, respectively (Figure S3b). This finding is in agreement with literature data and can be explained by the more heterogeneous structure of doped polyurethane, which was also observed by SEM, resulting from the interaction of MWCNTs with lignin hydroxyl groups, which makes the latter inaccessible for the reaction with isocyanate.

3.4. Sensor Properties

LignoBoost® kraft lignin-based polyurethane doped with 1.4% (w/w) of MWCNTs was used for the preparation of the membrane of potentiometric chemical sensor. Slopes of the electrode function (S) of the sensor in the individual solutions of transition metal salts are depicted in the Figure 6. The sensor displayed no response to sodium, calcium and iron (III), very low response to cadmium, lead, and chromium (VI), and low redox response in the solutions of redox pair $Fe(CN)_6/Fe(CN)_6^{-3/-4}$. Sensor displayed theoretical response of 32 (±1) mV·pM^{-1} to copper (II) with detection limit of 6(±1) × 10^{-6} mol·L^{-1}. Sensor response to mercury (II) decreased with each consecutive calibration from 32 mV·pM^{-1} in the first calibration to 18 mV·pM^{-1} in the third. After exposure to the solutions of mercury (II), the sensor no longer responded to copper (II) ions. This behavior may be associated with irreversible complexation of mercury (II) ions by the polymeric membrane similar to the reported for several copper-sensitive organic ionophores [28]. Thus, mercury must be avoided in the solutions analyzed by the sensor. Sensor displayed high selectivity towards copper in the presence of other transition metals except for mercury (II), to which it was not selective (Table 1). Parameters of the sensor are close to the values reported in the literature for the copper-selective electrodes based on organic ligands (Table S1).

Sensing characteristics of the sensor with LignoBoost® kraft lignin-based membrane differs from the ones with membranes prepared using other technical lignins, such as conventional kraft lignin precipitated by deep acidification of black liquor, lignosulfonate, and organosolv lignin, all of which responded selectively to Cr(VI) in acidic solutions [16]. Mechanism of these sensors' response to Cr(VI) was suggested to be mixed, redox, and ionic, as the sensitivity of sensors was correlated with the content of quinone structures, the dominant redox-active moieties of lignin [29]. Contrary to these findings, LignoBoost® kraft lignin-based sensor did not display redox sensitivity, showing a very low response to Cr(VI) and redox pair, but instead it showed a selective response to Cu(II).

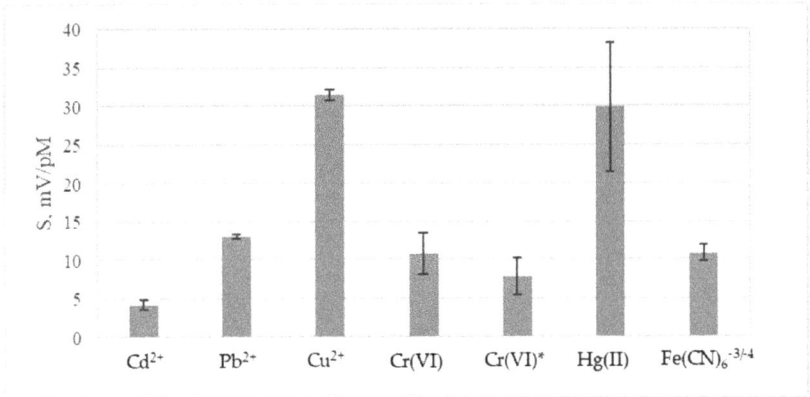

Figure 6. Slopes of the electrode function of LignoBoost® kraft lignin-based sensor. Mean values of three calibrations with standard deviations are shown.

Table 1. Logarithm of the selectivity coefficients, logK_{ij}, of LignoBoost® kraft based sensor to copper(II) determined using mixed solution method. Mean values of four measurements with standard deviations are shown.

Interferent	Cd(II)	Pb(II)	Cr(VI)	Hg(II)
logK_{ij}	−1.68 (±0.1)	−1.49 (±0.05)	−1.6 (±0.2)	0.2 (±0.1)

This behavior can be explained by the differences in the composition of eucalyptus LignoBoost® kraft lignin and other technical lignins [30–32], most noticeably, lower content of redox quinone type moieties and significantly higher content of polyphenolic groups with vicinal hydroxyls originating from concomitant tannins in the former [18]. In particular, the LignoBoost® kraft lignin has a higher total content of total hydroxyl groups and higher relative content of phenolic hydroxyl groups compared to the technical kraft lignin obtained from the cooking of the same wood species, but isolated by the conventional procedure [32]. Considering that sensors based on the polyurethanes, synthesized using kraft lignin isolated using conventional procedure, did not display sensitivity to Cu(II) [16], response of the sensor developed in this work can be attributed to the capability of phenolic hydroxyl groups to complex transition metals with higher specificity towards copper and mercury. This proposition is further corroborated by the reported higher chelating capacity of tannins with vicinal phenolic groups towards Cu(II) when compared to other bivalent transition ions, such as Zn(II) and Fe(II) [33]. However, studies on the exact mechanisms are still necessary for a better understanding of the observed phenomena.

Furthermore, characteristics of sensors with and without solid inner contact, as well as their stability in copper(II) solutions over a 4-week period were evaluated. Potentiometric chemical sensors require inner contact since ion-to-electron transduction between the sensing membrane with ionic conductivity and substrate with electronic conductivity ensures the stability of the sensor potential. In all-solid-state sensors, inner contact is commonly made from conducting polymers that have mixed ion-electronic conductivity [1]. As lignin-based polyurethanes are conducting polymers with mixed ionic and electronic conductivity, it was expected that they could be used for fabrication of the all solid-state sensors without additional inner contact layer. Newly prepared sensors with lignin-based polyurethane membrane, both with and without solid inner contact, displayed responses close to the theoretical ones to copper (II) ions: 33 and 32 mV·pM^{-1}, respectively, being not statistically different for $p = 0.05$ (Figure 7). However, sensor without solid inner contact had higher detection limit than the one with solid inner contact: $2(\pm 1) \times 10^{-5}$ and $6(\pm 1) \times 10^{-6}$ mol·L^{-1}, respectively. After one week, the slope of the sensor with solid inner contact slightly decreased, from 33 to 29 mV·pM^{-1} and remained constant over the next three weeks (slopes were not statistically different for $p = 0.05$). The slope of the sensor without solid inner contact remained close to the theoretical value for two weeks, after which it started to decrease. After the fourth week, this decrease became abrupt, with sensor's response to Cu(II) being only 8 mV·pM^{-1} (statistically different for $p = 0.05$). A slight increase of the detection limit was observed for both types of sensors during this period.

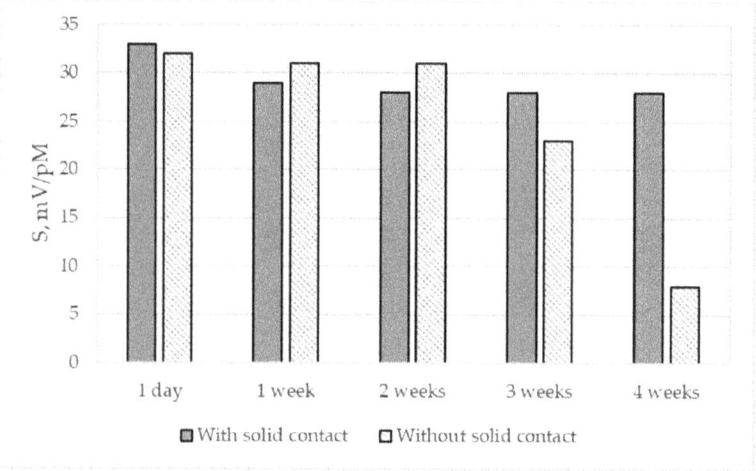

Figure 7. Slopes of the electrode function of the LignoBoost® kraft based sensors with and without solid contact in Cu(II) solutions during four weeks.

Deterioration of properties, such as long-term stability and detection limit, of the sensor without solid inner contact indicates that the interface between electrode and membrane is not well-defined [34]. This may occur due to the poor adhesion of the lignin-based polyurethane to the gold working electrode, leading to the formation of the water layer. Therefore, despite the electronic conductivity of lignin-based polyurethane imparted by MWCNTs, an intermediate conductive polymer layer is required to ensure better temporal stability and a lower detection limit.

4. Conclusions

Conducting composite polyurethane, based on LignoBoost® kraft lignin and doped with MWCNTs, was demonstrated to be a promising material for sensing applications. Electrical conductivity and impedance spectroscopy measurements revealed that the interaction between carbon nanotubes and lignin molecules in the polymer enhances its electrical conductivity. The percolation threshold of 0.77% (w/w) of MWCNTs was observed in LignoBoost® kraft lignin-based polyurethane but not in ellagic acid-based one, highlighting the importance of the presence of conjugated aromatic structure for the percolation effect to occur. LignoBoost® kraft lignin-based polymer doped with MWCNTs was used to manufacture all solid state potentiometric chemical sensors, which displayed high sensitivity and selectivity to Cu(II) and long-term stability. The dissimilarity between sensing properties of developed sensors and potentiometric sensors based on technical lignins was assigned to the higher content of polyphenolic groups originating from tannins, namely the vicinal phenolic hydroxyl groups in

LignoBoost® kraft lignin and lower content of quinone type moieties, which results in enhanced ion-exchanged properties.

Supplementary Materials: The following supplementary data are available online at http://www.mdpi.com/1996-1944/13/7/1637/s1, Figure S1: FT-MIR spectra of LignoBoost® kraft lignin-based polymers undoped and doped with 1.4% (w/w) MWCNTs; Figure S2a,b: DSC curves of LignoBoost® kraft lignin-based polyurethane undoped (a) and doped with 1.4% (w/w) MWCNTs (b); Figure S3a,b: TGA curves of LignoBoost® kraft lignin (a) and LignoBoost® kraft lignin-based polyurethane undoped and doped with 1.4% (w/w) MWCNTs (b); Table S1: Comparison of the performance characteristics of some copper ion sensors based on organic ionophores reported in the literature and developed in this work.

Author Contributions: Conceptualization, D.V.E. and A.R.; Methodology, D.V.E., A.R., and L.M.C.C.; Investigation, S.S.L.G. and A.J.M.S.; Writing—original draft preparation, A.R. and S.S.L.G.; Writing—review and editing, D.V.E., L.M.C.C., and A.J.M.S.; Supervision, D.V.E., A.R., and L.M.C.C.; Funding acquisition, D.V.E. All authors have read and agreed to the published version of the manuscript.

Funding: The funding by FEDER through the COMPETE 2020 Program and National Funds through FCT/MCTES under the projects UID/CTM/50011/2019, UID/AMB/50017/2019, and UID/CTM/50025/2019 is acknowledged. The funding by Portugal 2020 through European Regional Development Fund (ERDF) in the frame of COMPETE 2020 n246/AXIS II/2017, under the Project Inpactus—innovative products and technologies from eucalyptus, project no. 21874 is acknowledged. AR wishes to acknowledge funding by FCT through CEEC-IND program (CEECIND/01873/2017).

Conflicts of Interest: The authors declare no conflict of interest.

Abbreviations

AC conductivity	Alternating current conductivity
DC conductivity	Direct current conductivity
DSC	Differential scanning calorimetry
FT-MIR	Fourier transform–mid-infrared spectroscopy
MWCNT	Multiwall carbon nanotubes
PPGDI	Poly(propylene glycol), tolylene 2,4-diisocyanate terminated
SEM	Scanning electron microscopy
TGA	Thermogravimetric analysis

References

1. Bobacka, J. Conducting Polymer-Based Solid-State Ion-Selective Electrodes. *Electroanalysis* **2006**, *18*, 7–18. [CrossRef]
2. Hatchett, D.W.; Josowicz, M. Composites of Intrinsically Conducting Polymers as Sensing Nanomaterials. *Chem. Rev.* **2008**, *108*, 746–769. [CrossRef]
3. Adler, E. Lignin Chemistry -Past, Present and Future. *Wood Sci. Technol.* **1977**, *11*, 169–218. [CrossRef]
4. Milczarek, G. Preparation and Characterization of a Lignin Modified Electrode. *Electroanalysis* **2007**, *19*, 1411–1414. [CrossRef]
5. Milczarek, G. Lignosulfonate-Modified Electrodes: Electrochemical Properties and Electrocatalysis of NADH Oxidation. *Langmuir* **2009**, *25*, 10345–10353. [CrossRef]
6. Rudnitskaya, A.; Evtuguin, D.V. *Lignin Applications in Chemical Sensing*; Pan Stanford Publishing Inc.: Singapore, 2013. [CrossRef]
7. Martins, G.F.; Pereira, A.A.; Straccalano, B.A.; Antunes, P.A.; Pasquini, D.; Curvelo, A.A.S.; Ferreira, M.; Riul, A.; Constatino, C.J.L. Ultrathin Films of Lignins as a Potential Transducer in Sensing Applications Involving Heavy Metal Ions. *Sens. Actuators B Chem.* **2008**, *129*, 525–530. [CrossRef]
8. Leite, F.L.; Firmino, A.; Borato, C.E.; Mattoso, L.H.C.; da Silva, W.T.L.; Oliveira, O.N. Sensor Arrays to Detect Humic Substances and Cu(II) in Waters. *Synth. Met.* **2009**, *159*, 2333–2337. [CrossRef]
9. Borato, C.E.; Riul, A.; Ferreira, M.; Oliveira, O.N.; Mattoso, L.H.C. Exploiting the Versatility of Taste Sensors Based on Impedance Spectroscopy. *Instrum. Sci. Technol.* **2004**, *32*, 21–30. [CrossRef]
10. Consolin Filho, N.; Medeiros, E.S.; Tanimoto, S.T.; Mattoso, L.H.C. Sensors of Conducting Polymers for Detection of Pesticides in Contaminated Water. *Proc. Int. Symp. Electrets* **2005**, *2005*, 424–427. [CrossRef]

11. Stergiou, D.V.; Veltsistas, P.G.; Prodromidis, M.I. An Electrochemical Study of Lignin Films Degradation: Proof-of-Concept for an Impedimetric Ozone Sensor. *Sens. Actuators B Chem.* **2008**, *129*, 903–908. [CrossRef]
12. Mahmoud, K.A.; Abdel-Wahab, A.; Zourob, M. Selective Electrochemical Detection of 2,4,6-Trinitrotoluene (TNT) in Water Based on Poly(Styreneco-Acrylic Acid) PSA/SiO2/Fe3O4/AuNPs/Lignin-Modified Glassy Carbon Electrode. *Water Sci. Technol.* **2015**, *72*, 1780–1788. [CrossRef] [PubMed]
13. Chokkareddy, R.; Redhi, G.G.; Karthick, T. A Lignin Polymer Nanocomposite Based Electrochemical Sensor for the Sensitive Detection of Chlorogenic Acid in Coffee Samples. *Heliyon* **2019**, *5*, e01457. [CrossRef] [PubMed]
14. Faria, F.A.C.; Evtuguin, D.V.; Rudnitskaya, A.; Gomes, M.T.S.R.; Oliveira, J.A.B.P.; Graça, M.P.F.; Costa, L.C. Lignin-Based Polyurethane Doped with Carbon Nanotubes for Sensor Applications. *Polym. Int.* **2012**, *61*. [CrossRef]
15. Graça, M.P.F.; Rudnitskaya, A.; Faria, F.A.; Evtuguin, D.V.; Gomes, M.T.; Oliveira, J.A.; Costa, L.C. Electrochemical Impedance Study of the Lignin-Derived Conducting Polymer. *Electrochim. Acta* **2012**, *76*, 69–76. [CrossRef]
16. Rudnitskaya, A.; Evtuguin, D.V.; Costa, L.C.; Pedro Graça, M.P.; Fernandes, A.J.S.; Rosario Correia, M.; Teresa Gomes, M.T.; Oliveira, J.A.B.P. Potentiometric Chemical Sensors from Lignin-Poly(Propylene Oxide) Copolymers Doped by Carbon Nanotubes. *Analyst* **2013**, *138*, 501–508. [CrossRef]
17. Evtuguin, D.V.; Andreolety, J.P.; Gandini, A. Polyurethanes Based on Oxygen-Organosolv Lignin. *Eur. Polym. J.* **1998**, *34*, 1163–1169. [CrossRef]
18. Vieira, F.R.; Barros-Timmons, A.; Evtyugin, D.V.; Pinto, P.C.R. Effect of Different Catalysts on the Oxyalkylation of Eucalyptus Lignoboost Kraft Lignin. In Proceedings of the 20th International Symposium on Wood, Fiber and Pulping Chemistry (20th ISWFPC), University of Tokyo, Tokyo, Japan, 9–11 September 2019; C-11.
19. Hu, N.; Masuda, Z.; Yan, C.; Yamamoto, G.; Fukunaga, H.; Hashida, T. The Electrical Properties of Polymer Nanocomposites with Carbon Nanotube Fillers. *Nanotechnology* **2008**, *19*. [CrossRef]
20. Sandler, J.K.W.; Kirk, J.E.; Kinloch, I.A.; Shaffer, M.S.P.; Windle, A.H. Ultra-Low Electrical Percolation Threshold in Carbon-Nanotube-Epoxy Composites. *Polymer (Guildf.)* **2003**, *44*, 5893–5899. [CrossRef]
21. Feng, S.; Halperin, B.I.; Sen, P.N. Transport Properties of Continuum Systems near the Percolation Threshold. *Phys. Rev. B* **1987**, *35*, 197–214. [CrossRef]
22. Grimaldi, C.; Balberg, I. Tunneling and Nonuniversality in Continuum Percolation Systems. *Phys. Rev. Lett.* **2006**, *96*, 15–18. [CrossRef]
23. Macdonald, J.R. *Impedance Spectroscopy Emphasizing Solid Materials and Systems*; Wiley & Sons: Hoboken, NJ, USA, 1987; p. 220.
24. Jonscher, A.K. Review A New Understanding of the Dielectric Relaxation of Solids. *J. Mater. Sci.* **1981**, *16*, 2037–2060. [CrossRef]
25. Barrau, S.; Demont, P.; Peigney, A.; Laurent, C.; Lacabanne, C. DC and AC Conductivity of Carbon Nanotubes–Polyepoxy Composites. *Macromolecules* **2003**, *36*, 5187–5194. [CrossRef]
26. Achour, M.E.; Brosseau, C.; Carmona, F. Dielectric Relaxation in Carbon Black-Epoxy Composite Materials. *J. Appl. Phys.* **2008**, *103*, 094103. [CrossRef]
27. Heng, L.Y.; Hall, E.A.H. Producing "Self-Plasticizing" Ion-Selective Membranes. *Anal. Chem.* **2000**, *72*, 42–51. [CrossRef]
28. Singh, L.P.; Bhatnagar, J.M. Copper(II) Selective Electrochemical Sensor Based on Schiff Base Complexes. *Talanta* **2004**, *64*, 313–319. [CrossRef]
29. Furman, G.S.; Lonsky, W.F.W. *Charge-Transfer Complexes in Kraft Lignin. Part 3: Implications on the Possible Occurrence of Charge-Transfer Complexes in Residual Lignin and In Lignin of Light-Induced Yellowed High-Yield Pulps*; IPC Technical Paper Series; The Institute of Paper Chemistry: Appleton, WI, USA, 1986.
30. Marques, A.P.; Evtuguin, D.V.; Magina, S.; Amado, F.M.L.; Prates, A. Structure of Lignosulphonates from Acidic Magnesium-Based Sulphite Pulping of Eucalyptus Globulus. *J. Wood Chem. Technol.* **2009**, *29*, 337–357. [CrossRef]
31. Pinto, P.C.; Evtuguin, D.V.; Pascoal Neto, C.; Silvestre, A.J.D. Behaviour of *Eucalyptus globulus* lignin during kraft pulping. Part 2. Analysis by NMR, ESI/MS and GPC techniques. *J. Wood Chem. Technol.* **2002**, *22*, 109–125. [CrossRef]
32. Santos, D.A.S.; Rudnitskaya, A.; Evtuguin, D.V. Modified Kraft Lignin for Bioremediation Applications. *J. Environ. Sci. Health Part A* **2012**, *47*, 298–307. [CrossRef]

33. Karamać, M. Chelation of Cu(II), Zn(II), and Fe(II) by Tannin Constituents of Selected Edible Nuts. *Int. J. Mol. Sci.* **2009**, *10*, 5485–5497. [CrossRef]
34. Michalska, A. Optimizing the Analytical Performance and Construction of Ion-Selective Electrodes with Conducting Polymer-Based Ion-to-Electron Transducers. *Anal. Bioanal. Chem.* **2005**, *384*, 391–406. [CrossRef]

© 2020 by the authors. Licensee MDPI, Basel, Switzerland. This article is an open access article distributed under the terms and conditions of the Creative Commons Attribution (CC BY) license (http://creativecommons.org/licenses/by/4.0/).

Article

Composites of Unsaturated Polyester Resins with Microcrystalline Cellulose and Its Derivatives

Artur Chabros *, Barbara Gawdzik, Beata Podkościelna, Marta Goliszek and Przemysław Pączkowski

Department of Polymer Chemistry, Institute of Chemical Sciences, Faculty of Chemistry, Maria Curie-Sklodowska University in Lublin, M. Curie-Sklodowska Sq. 5, 20-031 Lublin, Poland; barbara.gawdzik@poczta.umcs.lublin.pl (B.G.); beatapod@poczta.umcs.lublin.pl (B.P.); marta.goliszek@poczta.umcs.lublin.pl (M.G.); przemyslaw.paczkowski@poczta.umcs.lublin.pl (P.P.)
* Correspondence: artur.chabros@poczta.umcs.lublin.pl

Received: 14 November 2019; Accepted: 19 December 2019; Published: 21 December 2019

Abstract: The paper investigates the properties of unsaturated polyester resins and microcrystalline cellulose (MCC) composites. The influence of MCC modification on mechanical, thermomechanical, and thermal properties of obtained materials was discussed. In order to reduce the hydrophilic character of the MCC surface, it was subjected to esterification with the methacrylic anhydride. This resulted in hydroxyl groups blocking and, additionally, the introduction of unsaturated bonds into its structure, which could participate in copolymerization with the curing resin. Composites of varying amounts of cellulose as a filler were obtained from modified MCC and unmodified (comparative) MCC. The modification of MCC resulted in obtaining composites characterized by greater flexural strength and strain at break compared with the analogous composites based on the unmodified MCC.

Keywords: microcrystalline cellulose (MCC); composites; unsaturated polyester resins; thermogravimetric analysis (TG); mechanical analysis; dynamic mechanical analysis (DMA)

1. Introduction

The pressure exerted by the increase of public awareness to limit human interference in the environment affects also trends in polymer chemistry development. Adaptation of raw materials from renewable resources to the existing technological solutions is one way to achieve this goal [1–4]. The main factor determining the above activities is the renewable nature of the raw material base; easy availability; and, above all, its biodegradability [5,6].

One group are biopolymers, where cellulose is the main representative. Cellulose modifications and preparation of composites are extensively studied in the literature [7–9]. Zadorecki and Flodin [10] investigated the effects of interfacial adhesion of cellulose–polyester composites on their environmental aging behavior. The authors observed that, when the matrix and the cellulose fiber are covalently bonded, no cracks are formed during shrinking. In the other work [11], these authors studied the reinforcement effect of cellulose fibers treated with different coupling agents on unsaturated polyesters. Abdelmouleh et al. [12] tested mechanical properties of unsaturated epoxy and polyester resin matrices filled with silane-treated cellulose fibers. Their reinforcing effect proved to be promising, especially with some silane-treated fibers, and improvement was especially visible for methacrylic silane in contact with an unsaturated polyester resin. Rodrigues et al. [13] studied the sugarcane bagasse fibers modified by estherification to use as reinforcement in a polyester matrix. The authors proved that composites were characterized by better mechanical strength compared with pure polymer. DiLoreto et al. [14] developed a preliminary process for incorporating freeze-dried cellulose nanocrystals powder into polyester resin, where functionalized cellulose particles were used as a potential reinforcement for an unsaturated polyester resin system, a common material for automotive applications. Kargarzadeh et al. [15] prepared new nanocomposites with unsaturated polyester resin (UPR) and silane surface-treated cellulose nanocrystals. Tensile tests showed that both the stiffness and strength of the UPR improved upon the incorporation of cellulose.

Its attractiveness for use is because of the fact that it is an easily available material obtained on a large scale in the paper industry [16]. An important commercial product is also its microcrystalline form obtained by partial depolymerization of cellulose, most often as a result of acid hydrolysis. In this form, it found a number of applications, including the production of medicines, cosmetics, and food, where it mostly acts as a filler [17]. The use of microcrystalline cellulose (MCC) of the same nature was also of significant interest in the preparation of composites based on generally available synthetic polymers [18,19].

However, obtaining homogeneous composites creates problems owing to the hydrophilic nature of the cellulose surface. This is not compatible with the hydrophobic polymer matrix. Uniformity of the filler/matrix surface has a major impact on the load carrying capacity, and thus on the properties of composites. This effect can be achieved using compatibilizers that reduce the tension between phases [20] or modifying the cellulose surface to obtain a less hydrophilic one [21]. Surface modification also allows introducing unsaturated bonds to the filler structure. The development of its surface also has a significant impact on the ability to transfer loads. The combination of this procedure and the use of unsaturated polyester resin as a matrix and nanocellulose modified with unsaturated fatty acids allowed to improve the mechanical properties of composites through the copolymerization of their individual components expressed as increased tensile strength [22]. However, the use of the fillers at nanometer scale in mixtures with resins has limitations. The development of their surface causes the dispersion of the particles, which is accompanied by absorption of the polymer matrix, which causes the viscosity of the systems to increase even at low load, which prevents their practical use [23].

Furthermore, reduction of cellulose particle sizes is associated with additional procedures and costs [24], whereas the microcrystalline form of cellulose is much cheaper and readily available.

The aim of this research was to improve the miscibility of MCC in the unsaturated polyester resin. This effect was obtained by modifying the surface of its microcrystalline form using methacrylic anhydride. The presence of unsaturated bonds additionally allowed the chemical binding of the filler to the structure of the polymer matrix network. In order to develop the filler/matrix interface between the phases and skip the procedure of obtaining cellulose on the nano scale, the decision was made to increase the load on the resin with the filler.

2. Materials and Methods

2.1. Chemicals

The unsaturated polyester resin (UPR) used for the study consisted of 70% of unsaturated polyester and 30% of styrene as a crosslinking monomer. The detailed description of this synthesis and curing procedure was presented in the work of [25].

To modify MCC (Chem Point, Kraków, Poland) pyridine and hydroquinone (Avantor, Gliwice, Poland), 4- (Dimethylamino) pyridine (DMAP) and methacrylic anhydride (Sigma-Aldrich, Saint Louis, MO, USA) were used.

2.1.1. Modification of microcrystalline cellulose MCC

First, 50 g of MCC, 1000 cm^3 of pyridine, 12.5 g of DMAP as catalyst, and 6.9 g of hydroquinone as inhibitor were placed in a 2000 mL flask. The reagents were heated to 80 °C and constantly stirred. Next, 314 cm^3 of methacrylic anhydride was added dropwise over 30 min. Heating at 100 °C was continued for 2 h. The obtained product was filtered under the reduced pressure. The precipitate was washed with water and then acetone. The modified MCC was dried in the air at 40 °C to constant weight.

2.1.2. Preparation of Composites

To the unsaturated polyester resin (UPR), which was pre-accelerated with 0.45 wt% of the cobalt octoate solution and 1.2 wt% of the DMPT solution (N,N-dimethyl-p-toluidine, 10% solution in styrene), 2 wt%, 5 wt%, and 10 wt% of modified MCC (CEL-MOD) was added, respectively. For homogeneity, it was mixed and then placed in an ultrasonic bath for 15 min. This procedure was repeated twice. To this mixture, 3 wt% of Luperox DHD-9 (the methyl ethyl ketone peroxide solution) as an initiator was added. Its amount was calculated for the resin. The prepared mixtures were poured into cuboid-shaped molds with dimensions 14.5 cm × 10 cm × 0.4 cm. Curing was conducted in the molds at room temperature for 24 h. Then, for additional post-curing, the moldings were placed in an oven for 10 h at 90 °C. The same procedure was applied for the composites of unmodified MCC (CEL-UNMOD) and pure resin (UPR).

2.2. Research Methods

Attenuated total reflectance Fourier transform infrared (ATR-FTIR) spectra were obtained with a Bruker TENSOR 27 spectrometer with a diamond crystal (Ettlingen, Germany). The spectra were gathered from 600 to 4000 cm^{-1} with 32 scans per spectrum at a resolution of 4 cm^{-1}.

Mechanical studies were carried out using the mechanical testing machine, ZwickRoell Z010 (Ulm, Germany). With a three-point bending test, the 80 mm × 10 mm × 4 mm sample profiles were used. The support spacing was 64 mm, whereas the bending speed was 5 mm min^{-1}. The final result was the arithmetic averaging of five measurements.

Hardness was determined using a hardness tester HPK (Leipzig, German Democratic Republic) and the Brinell method was applied. The final result was the mean value of 10 measurements.

Thermomechanical properties of materials were determined using the dynamic mechanical analyzer (DMA) Q800 from TA Instruments (New Castle, DE, USA) equipped with a double-cantilever device. The samples of 65 mm × 10 mm × 4 mm dimension were tested. The temperature scanning from 0 to 200 °C with a constant heating rate of 3 °C min^{-1} at a sinusoidal distortion of 10 μm amplitude and 1 Hz frequency was conducted. The glass-transition temperature, damping factor, values of storage modulus, and loss modulus were determined.

Thermal resistance of the obtained materials was measured by the thermogravimetric analysis (TG/DTG) using a NETZSCH STA 449 Jupiter F1 TG analyzer (Selb, Germany), Al_2O_3 crucibles, and the sample mass ca. 10 mg. The analyses were performed in the helium atmosphere with the flow rate of 20 mL min^{-1} in the temperature range of 35–800 °C and the heating rate of 10 °C min^{-1}.

The differential scanning calorimetry (DSC) analyses were performed on a Netzsch DSC 204 calorimeter (Selb, Germany). The scans were carried out in a nitrogen atmosphere (30 cm^3 min^{-1}) at a heating rate of 10 °C min^{-1}. The mass of the sample was approximately ~7 mg. As a reference, an empty aluminum crucible was used.

3. Results and Discussion

3.1. Attenuated total reflectance Fourier transform infrared spectroscopy analysis ATR/FTIR

The MCC used for the preparation of composites was first subjected to the ATR/FTIR analysis. To follow the explanations on the molecular level, the schematic representation of MCC before modification (CEL-UNMOD) and its methacrylic derivative (CEL-MOD) are presented in Figure 1. Figure 2 shows the spectra of methacrylic anhydride (Meth-Anh), CEL-UNMOD, and CEL-MOD. The efficiency of MCC esterification with methacrylic anhydride is generally demonstrated by three ranges: the peak at 1723 cm^{-1} is attributed to the stretching vibrations of C=O carbonyl groups, indicating an ester bond formation. The fuzzy band observed at 1638 cm^{-1} in the spectrum of CEL-UNMOD is connected with the presence of water adsorbed by the crystalline part of cellulose [26]. Its intensity increases after esterification, which may indicate overlapping by the signal of the C=C stretching vibrations from methacrylic groups. A wide peak in the range of 3800–3300 cm^{-1} is associated with the stretching vibrations of hydroxyl groups. The presence of this peak with reduced intensity in the case of CEL-MOD indicates that not all hydroxyl groups were blocked and the esterification process was incomplete. The degree of substitution (DS) = 0.02 was estimated alkacimetrically according to the procedure presented in the work of [27].

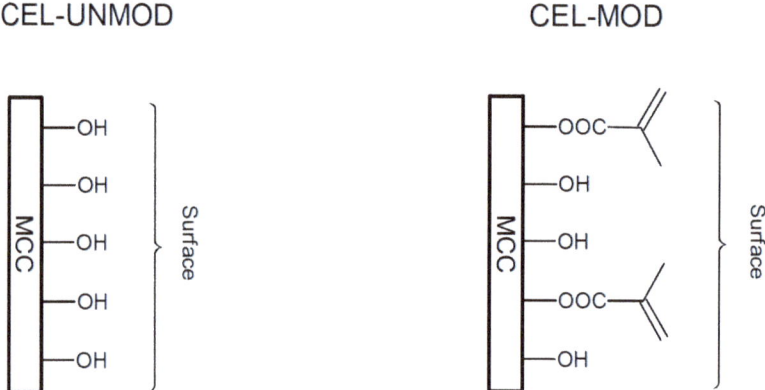

Figure 1. Schematic representation of unmodified microcrystalline cellulose (MCC) (CEL-UNMOD) and modified MCC (CEL-MOD) surface.

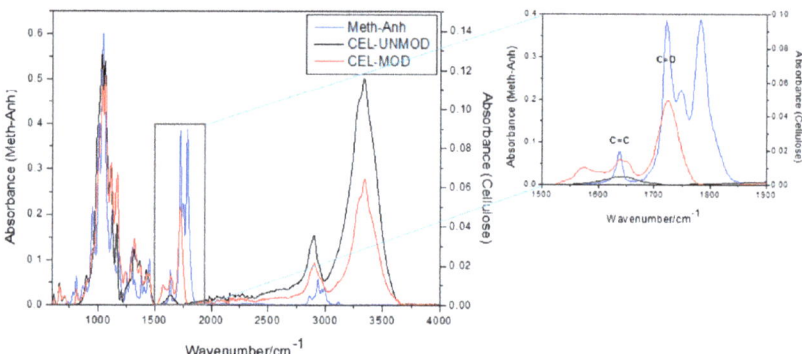

Figure 2. Attenuated total reflectance Fourier transform infrared (ATR/FTIR) spectra of methacrylic anhydride (Meth-Anh) and MCC before and after modification.

UPR and its systems containing 10% filler—both before and after curing—were also subjected to ATR/FTIR analysis. Despite the slight differences in the wide absorption spectrum, discernible differences occur in the range of 1750–1600 cm^{-1} shown in Figure 3. The main point is the disappearance of the peak at 1645 cm^{-1} corresponding to stretching C=C, confirming the effectiveness of the curing process. In addition, the lack of this peak in the case of a composite with 10% CEL-MOD may suggest that, in the heterogeneous curing of the polyester matrix [28], the unsaturated bond of the methacrylic group to CEL-MOD may also be involved. Shifts of signals from the C=O carbonyl group have also been observed. They were related to the intermolecular interaction of hydrogen bonds with hydroxyl groups. The presence of a signal at a higher wavelength for a CEL-UNMOD composite of 1727 cm^{-1} compared with a CEL-MOD composite of 1726 cm^{-1} indicates greater matrix–filler interaction through hydrogen bonds for CEL-UNMOD composites [29,30].

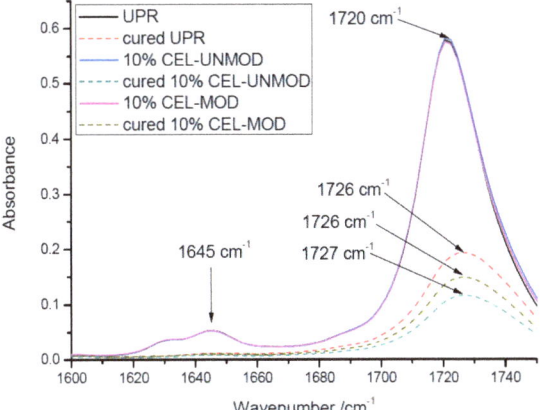

Figure 3. ATR/FTIR spectra of unsaturated polyester resin (UPR) and composites with 10% CEL-MOD/UNMOD before and after curing.

3.2. Mechanical Tests

The mechanical properties of pure resin (UPR) and composites containing unmodified and modified MCC were determined. The results of flexural strength tests are presented graphically in Figure 4 and the numerical data of the flexural modulus, flexural strength, and strain at break are collected in Table 1.

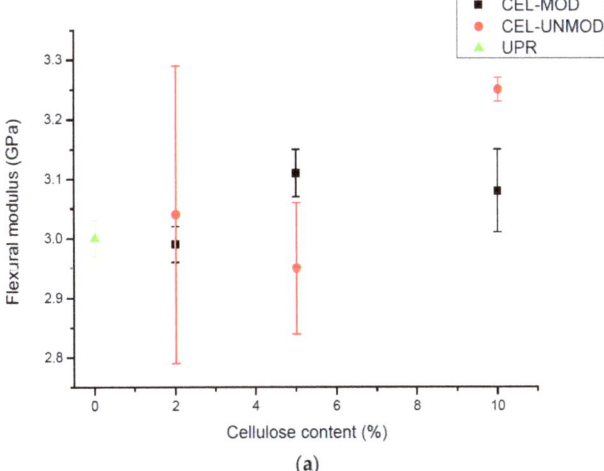

(a)

Figure 4. *Cont.*

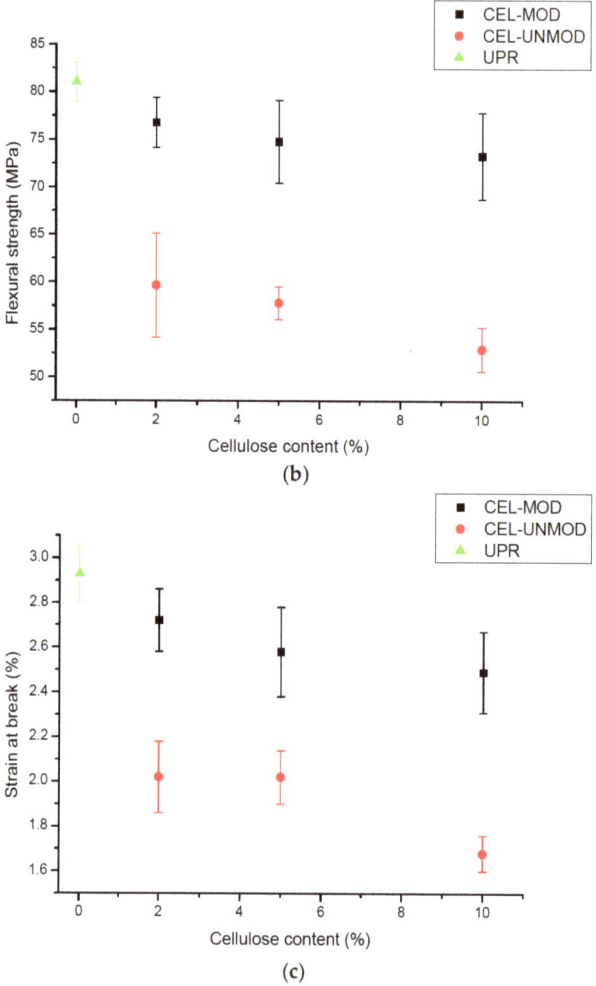

Figure 4. Mechanical data of composites and UPR. (**a**) flexural modulus; (**b**) flexural strength; (**c**) strain at break.

In both the CEL-MOD and CEL-UNMOD systems, an increase in the flexural modulus value is visible as the filler content increases. This is justified by the increased loading of composites by the crystalline phase present in the cellulose structure. However, in the case of systems with CEL-UNMOD, it is higher than for analogous systems with CEL-MOD. This would confirm that the decisive factor affecting its value is the interactions of matrix intermolecular hydrogen bonds (polyester carbonyl groups) with MCC hydroxyl groups. This is particularly visible for systems containing 10% filler. The addition of 5% of CEL-UNMOD reduces the value of the flexural modulus compared with that of pure resin. This can be attributed to the heterogeneous distribution of the filler in the matrix. The low compatibility of the hydrophilic character of the MCC surface with the hydrophobic nature of the resin leads to a decrease in the filler dispersion in the system through the formation of agglomerates [31]. Increasing the content of CEL-UNMOD to 10% makes the clusters' formation more difficult [32]. Comparing the composites containing 5% of the filler, a higher flexural modulus is observed for the system with CEL-MOD. Increasing the hydrophobic character through modification increases the filler compatibility with the matrix. The increase in homogeneity of the CEL-MOD-based composites in comparison with those of CEL-UNMOD indicates that they are characterized by higher flexural strength values as well as higher values of strain at break. The composites with CEL-MOD are less brittle compared with the analogous ones based on CEL-UNMOD. However, the fragility of composites with both CEL-MOD and CEL-UNMOD increases with the increasing amount of filler.

It seems that the hardness of the composites will depend only on the loading of the resin with the filler. The results of the hardness test, presented in Table 1, however, indicate that an upward tendency compared with the cured resin was observed only for composites with CEL-MOD. The hardness of the CEL-UNMOD systems shows some analogy with the values of the flexural modulus. For systems containing 2% and 5%, they are lower than for UPR. In addition, the system with 5% filler is characterized by the lowest hardness value. This relation corresponds well with the results of other studies, thus determining the effect of MCC modification on the quality of the filler dispersion in the matrix.

The results of dynamic-mechanical analysis (DMA) in a wide temperature range allowed for a more detailed assessment of the effect modified MCC (CEL-MOD) on the composites properties. The changes of the storage modulus (E') (Figure 5), loss modulus (E") (Figure 6), as well as the damping factor (tan δ) (Figure 7) are presented. The numerical values of the storage modulus (in the glassy and rubbery regions), glass transition temperature (read with the maximum tg δ), degree of heterogeneity (FWHM, full width at half maximum), and values of the damping factor are presented in Table 1.

In the composites with CEL-UNMOD, the storage modulus in the glassy region E'(25 °C) increases with the increasing amount of filler. This may indicate that the rigidity of the material is mainly owing to the resin loading with the filler. Despite the increase in the filler amount, no modulus changes are observed for 2% and 5% CEL-MOD in comparison with pure resin. This difference may result from the fact that the material stiffness is additionally determined by the filler/matrix interphase interactions. The introduction of groups more compatible with the hydrophobic nature of the resin onto the MCC surface would contribute to the plasticization of the system [33]. However, for the mixtures containing 10% of the filler (CEL-MOD and CEL-UNMOD), the modulus values are the highest and converge. This would suggest that, in this case, the CEL-MOD plasticizing effect is levelled by the dominant influence of the filler amount. The above assumptions are in good agreement with the storage modulus of composites in the rubbery region E'(180 °C). This is related to the material ability to carry loads resulting from the restriction of the movement of matrix polymer chains. The increase in the modulus value as the filler content increases in the systems with CEL-MOD and CEL-UNMOD shows that this movement is more and more difficult. However, a decrease in the modulus value for 5% of CEL-UNMOD would confirm the dominant effect of the heterogeneous dispersion of the filler in the resin. In contrast, lower modulus values for 2% and 5% of CEL-MOD systems in the comparison with pure resin would indicate the overwhelming effect of the plasticizing associated with reduced friction at the filler/matrix interface.

The loss modulus E" (Figure 6) allowed to assess the ability of composites to dissipate energy during deformation in the form of heat, and thus to determine their viscous response. The matrix and filler consistency has a significant impact on it [34]. The maximum modulus values for the composites with CEL-MOD are higher than for the analogous systems with CEL-UNMOD. This may indicate that the unsaturated bonds present in CEL-MOD are embedded in the polymer structure of the matrix. Large values of the loss modulus E" in a wide temperature range indicate the heterogeneity of the structure of the tested materials. However, determination of their degree in full width at the half maximum (FWHM) (Table 1) from the damping factor tg δ (Figure 7) showed insignificant changes. It can thus be assumed that this is associated mainly with the heterogeneous structure of the polymer matrix network visible in the form of two peaks on the tg δ curve. This results from the ability to copolymerize UPR components and the formation of microgels during its curing [35]. However, the addition of filler reduces the damping capacity in the range of 60–110 °C. This may suggest that it limits the formation of microgel clusters and reduces the heterogeneity of the polymer matrix network. In addition, when CEL-MOD is used, a decrease in the damping ability in this temperature range can confirm participation of the double bonds from the filler in the polymer network formation. The consistency of the matrix with the filler affects the glass transition temperature. The highest T_g values were found in 2% of CEL-UNMOD and 5% of CEL-MOD systems.

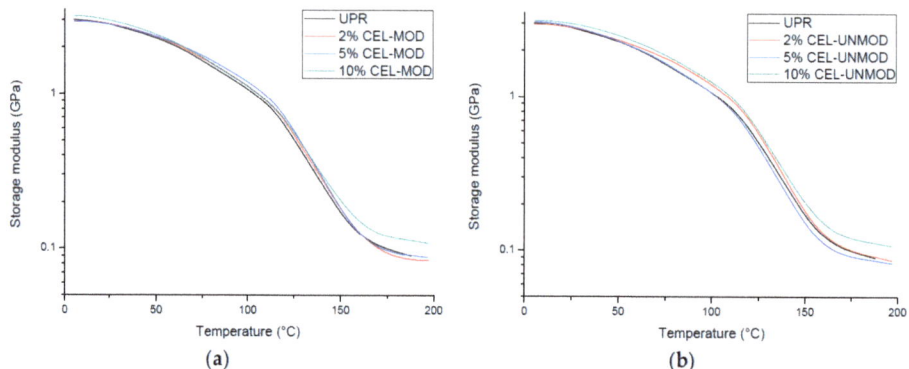

Figure 5. Storage modulus (E′) of UPR and composites with CEL-MOD and CEL-UNMOD. (a) CEL-MOD; (b) CEL-UNMOD.

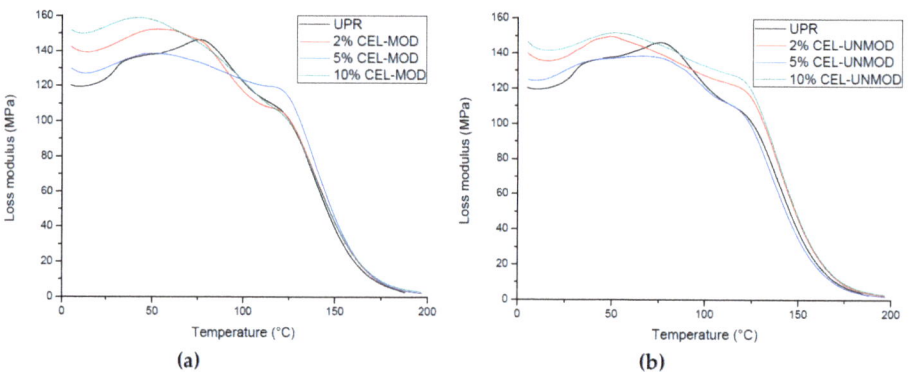

Figure 6. Loss modulus (E″) of UPR and composites with CEL-MOD and CEL-UNMOD. (a) CEL-MOD; (b) CEL-UNMOD.

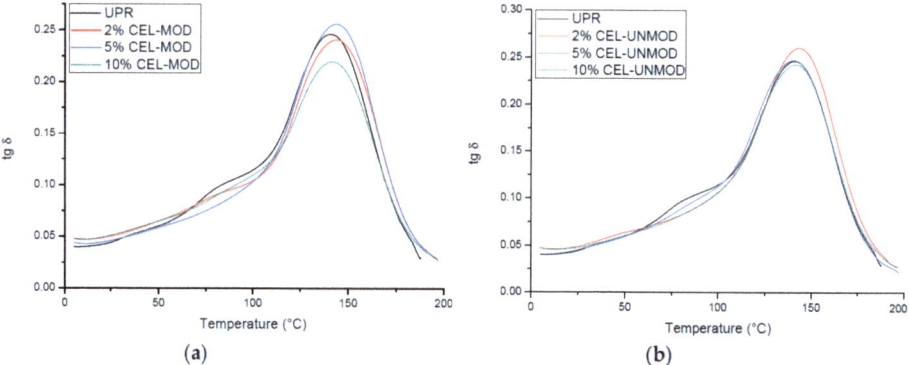

Figure 7. Damping factor (tg δ) of UPR and composites with CEL-MOD and CEL-UNMOD. (a) CEL-MOD; (b) CEL-UNMOD.

Table 1. Mechanical and thermomechanical data of samples.

Properties	UPR	COPOLYMERS					
		CEL-MOD			CEL-UNMOD		
		2%	5%	10%	2%	5%	10%
Flexural modulus/GPa	3.00 ± 003	2.99 ± 003	3.11 ± 0.04	3.08 ± 0.07	3.04 ± 0.25	2.95 ± 0.11	3.25 ± 0.02
Flexural strength/MPa	81 ± 2.1	76.7 ± 2.6	74.8 ± 4.4	73.3 ± 4.6	59.7 ± 5.5	57.8 ± 1.7	53 ± 2.3
Strain at break/%	2.93 ± 0.12	2.72 ± 0.14	2.58 ± 0.2	2.49 ± 0.18	2.02 ± 0.16	2.02 ± 0.12	1.68 ± 0.08
Hardness/MPa [a]	128.1	129.7	130.4	134.4	127.3	126.5	129.7
E'(25 °C)/GPa	2.81	2.82	2.81	2.99	2.82	2.86	2.99
E'(180 °C)/MPa	95.6	89.1	93.6	117.3	96.7	89.7	117
T_g/°C [b]	141.9	143.2	143.3	141.4	143.5	139.9	141
tg δ_{max}	0.206	0.195	0.215	0.174	0.216	0.206	0.198
FWHM/°C [c]	49.4	49.4	49.2	50.2	48	50.8	49

[a] Brinell method; [b] Determined from tg δ_{max}; [c] FWHM, full width at half maximum. CEL-MOD, modified microcrystalline cellulose (MCC); CEL-UNMOD, unmodified MCC.

3.3. Thermal Analysis

Thermogravimetric studies allowed to determine the thermal stability of composites. Characterization of the distribution of individual components of the mixtures (UPR, CEL-MOD, and CEL-UNMOD) (Figure 8) and the systems with CEL-MOD (Figure 9) and CEL-UNMOD (Figure 10) allowed to determine the interactions of the filler with the matrix. The numerical data of thermal analysis are presented in Table 2.

CEL-UNMOD decomposition takes place in one temperature range of ca. 275–400 °C with a maximum decomposition temperature T_{max1} = 340.5 °C. This is attributed to one mechanism by which its degradation proceeds. In the case of CEL-MOD, in addition to the main decomposition, whose maximum occurs at T_{max1} = 356.6 °C, there is an initial step mass loss. The first one in the range of ca. 120–220 °C is associated with the moisture present in the MCC structure, while the second one in the range of ca. 220–300 °C can be attributed to the decomposition of methacrylic groups. The main UPR decomposition associated with the degradation of the spatial structure of the polymer network occurring in the range of about 280–450 °C is preceded by a mass loss in the range of ca. 100–280 °C, associated with the evaporation of unreacted substrates of the synthesis of unsaturated polyester and styrene that were not incorporated into the polymer network during UPR curing and low-boiling and weakly connected components.

The presence of a filler in the resin affects the temperature of the main decomposition of the system. In the case of the composites with CEL-UNMOD, T_{max1} increases for the system with 2% of CEL-UNMOD, whereas 5% and 10% contents of the filler reduce it below the maximum decomposition temperature of pure resin. These changes are in good agreement with the results obtained from the mechanical and thermomechanical tests, and can be related to the degree of CEL-UNMOD dispersion in the matrix. The heterogeneity of the filler distribution can also be reflected in the mass loss for the above systems in the range of about 500–650 °C with T_{max2} = 568.2 °C for 5% of CEL-UNMOD and T_{max2} = 595.2 °C for 10% of CEL-UNMOD. Agglomerates can decompose at higher temperatures as a result of the protective layer formation in the form of carbon derived mainly from the resin pyrolysis. In the CEL-MOD systems, the maximum decomposition temperature T_{max1} increases with the increasing filler content. This can confirm the improved MCC miscibility obtained through modification. In addition, the T_{max1} values, which are higher for composites than for individual components, would indicate a difficult decomposition of the systems resulting from the incorporation of MCC through unsaturated bonds of methacrylic groups in the structure of the polymer matrix network. In addition, the effect of the filler interactions with the resin is also visible in the mass loss of composites compared with its individual components in the range of 220–300 °C. The observed differences in the mass loss can be attributed to the formation of clusters of polymer networks of different compositions in the mixtures. This would correspond to the theory of microgel formation during curing of the resin. For the 2% CEL-UNMOD system, a 0.10% decrease in mass loss could indicate a reduction in its formation.

However, a higher CEL-UNMOD resin load results in an increase in mass loss. The maximum is for the 5% CEL-UNMOD one, being 1.58%. The ability to agglomerate the filler particles would thus facilitate the formation of areas of the polymer network that are more easily degraded. In the case of CEL-MOD, the largest difference in mass loss occurs for the system containing 2% of the filler, amounting to 0.23%. The increase in the CEL-MOD resin loading results in a decrease in the mass loss difference. For the 10% CEL-MOD system, the total mass loss of individual components is the same. Therefore, it can be assumed that the microgels' formation is also affected by the viscosity of the cured mixture.

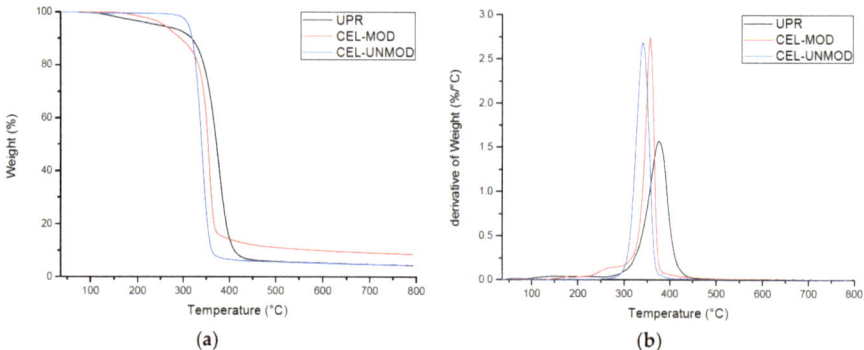

Figure 8. Thermograms (TG/DTG) of composite components. (**a**) TG; (**b**) DTG.

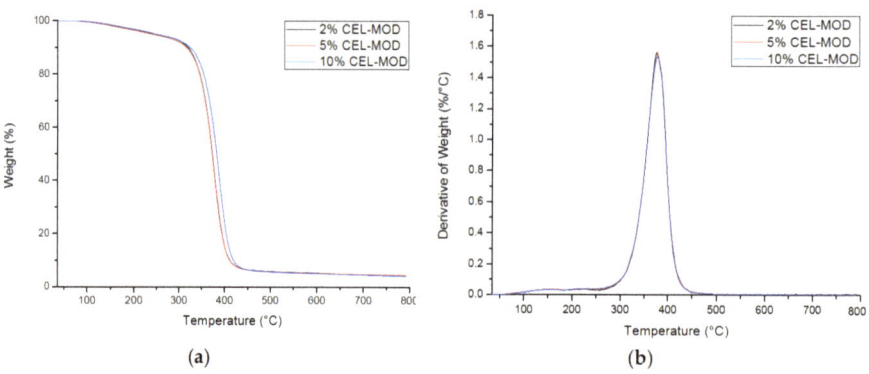

Figure 9. Thermograms (TG/DTG) of composites with CEL-MOD. (**a**) TG; (**b**) DTG.

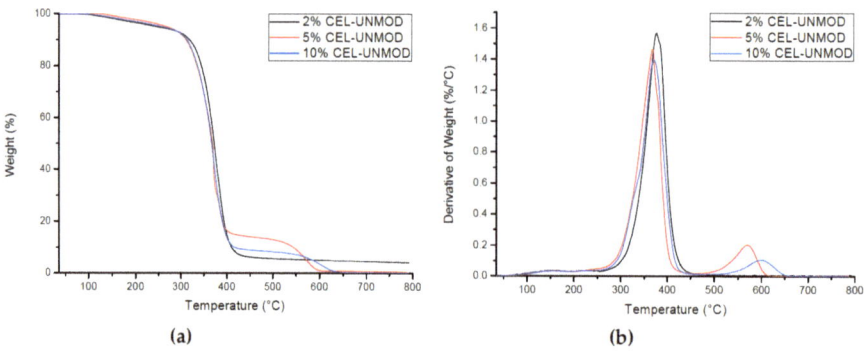

Figure 10. Thermograms (TG/DTG) of composites with CEL-UNMOD. (**a**) TG; (**b**) DTG.

The obtained composites were also investigated by differential scanning calorimetry (DSC). The DSC curves are shown in Figure 11. The values of maximum decomposition (T_d) and enthalpy of decomposition (ΔH_d) are listed in Table 2. The thermal behaviour of the composites is characterized by a well-shaped calorimetric profile containing a single symmetric peak in the range of 300–450 °C. It corresponds to the pyrolytic degradation observed in the TG/DTG study. Information about the caloric effect associated with this process, which is the result of the overlapping of both endo- and exothermic effects, has allowed to confirm the impact of MCC modification on the properties of the obtained materials. The highest value of decomposition enthalpy among composites with CEL-UNMOD was obtained for the system containing 5% filler (Figure 11). The increased share of endothermic effects confirms the largest heterogeneity of the systems with 5% and 10% of CEL-UNMOD and their two-stage distribution noted in the TG/DTG tests. This is because of the formation of a protective layer derived primarily from the pyrolysis of the matrix. T_d may also indicate the quality of the dispersion of the filler in the matrix. The use of CEL-MOD in composites increases T_d, while the use of CEL-UNMOD reduces their value compared with UPR. Additionally, no exothermic effect associated with the post-curing process was observed, which confirms complete cross-linking of the samples.

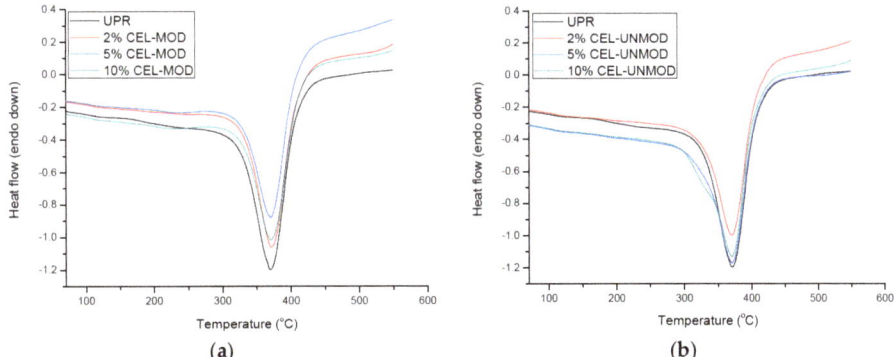

Figure 11. Differential scanning calorimetry (DSC) curves of the UPR and composites with CEL-MOD and CEL-UNMOD. (**a**) CEL-MOD; (**b**) CEL-UNMOD.

Table 2. Thermograms (TG/DTG) and differential scanning calorimetry (DSC) data.

Properties	UPR	CEL-MOD	CEL-UNMOD	COPOLYMERS					
				CEL-MOD			CEL-UNMOD		
				2%	5%	10%	2%	5%	10%
T_{max1} (°C)	375.4	356.6	340.5	375.9	376.7	377.2	375.5	367.7	372.5
T_{max2} (°C)	-	-	-	-	-	-	-	568.2	595.2
Mass change in the range 220–300 °C (%)	3.49	8.60	1.71	3.36	3.66	4.00	3.35	4.98	4.07
Sum of changes in mass of individual components of composites in the range 220–300 °C (%)	-	-	-	3.59	3.75	4.00	3.45	3.40	3.30
The difference in the mass change of the composites and their individual components (%)	-	-	-	−0.23	−0.09	0	−0.10	1.58	0.77
T_d (°C)	369.7	-	-	371.2	369.8	370.6	369.0	369.4	369.3
ΔH_d (J/g)	306.2	-	-	293.5	260.6	276.9	269.8	285.0	280.4

4. Conclusions

Composites of unsaturated polyester resin with the MCC modified with methacrylic anhydride and, comparatively, those with the unmodified MCC were obtained. The presence of methacrylic groups on the MCC surface improved mechanical properties of the composites compared with the systems with the unmodified MCC. The values of flexural strength and strain at break increased. This resulted from the improvement of the filler miscibility in the resin and limiting its ability to form agglomerates. However, MCC modification resulted in a reduction of intermolecular hydrogen bond interactions between matrix (polyester) carbonyl groups and filler hydroxyl groups. This resulted in a smaller increase in flexural modulus, especially visible in systems containing 10% of filler. The improvement of CEL-MOD mixtures homogeneity also affected their thermal stability, which is particularly visible in the case of 5% and 10% of the CEL-UNMOD systems characterized by decomposition in two temperature ranges. At the same time, modification reduced the friction of the matrix with the filler, thus affecting the composites' rigidity. In addition, the presence of unsaturated bonds in the filler structure allowed it to be incorporated into the structure of the polymer matrix formed during curing. As a result, the composite containing 5% of the modified MCC had practically identical damping values and glass transition temperature as the 2% CEL-UNMOD system.

Author Contributions: Conceptualization, A.C. and B.G.; Formal analysis, A.C. and B.G.; Investigation, A.C., B.G., M.G., and P.P.; Methodology, A.C. and B.G.; Project administration, A.C. and B.G.; Resources, A.C., B.G., and B.P.; Supervision, B.G. and B.P.; Visualization, A.C., B.G., B.P., and M.G.; Writing—original draft, A.C.; Writing—review & editing, B.G., B.P., M.G., and P.P. All authors have read and agreed to the published version of the manuscript.

Funding: This research received no external funding.

Conflicts of Interest: The authors declare no conflict of interest.

References

1. Gurunathan, T.; Mohanty, S.; Nayak, S.K. A review of the recent developments in biocomposites based on natural fibres and their application perspectives. *Compos. A Appl. Sci. Manuf.* **2015**, *77*, 1–25. [CrossRef]
2. La Mantia, F.P.; Morreale, M. Green composites: A brief review. *Compos. A Appl. Sci. Manuf.* **2011**, *42*, 579–588. [CrossRef]
3. Gutiérrez, M.C.; De Paoli, M.A.; Felisberti, M.I. Biocomposites based on cellulose acetate and short curauá fibers: Effect of plasticizers and chemical treatments of the fibers. *Compos. A Appl. Sci. Manuf.* **2012**, *43*, 1338–1346. [CrossRef]
4. Bhat, A.H.; Dasan, Y.K.; Khan, I.; Jawaid, M. Cellulosic biocomposites: Potential materials for future. In *Green Biocomposites. Green Energy and Technology*; Jawaid, M., Salit, M., Alothman, O., Eds.; Springer: Cham, Switzerland, 2017; pp. 69–100. [CrossRef]
5. Sayoma, M.; Iji, M. Improving mechanical properties of cardanol-bonded cellulose diacetate composites by adding polyester resins and glass fiber. *Polym. J.* **2017**, *49*, 503–509. [CrossRef]
6. Philp, J.C.; Bartsev, A.; Ritchie, R.J.; Baucher, M.A.; Guy, K. Bioplastics science from a policy vantage point. *New Biotechnol.* **2013**, *30*, 635–646. [CrossRef] [PubMed]
7. Gorade, V.G.; Kotwal, A.; Chandhary, B.U.; Kale, R.D. Surface modification of microcrystalline cellulose using rice bran oil: A bio-based approach to achieve water repellency. *J. Polym. Res.* **2019**, *26*, 217. [CrossRef]
8. Xie, Y.; Cai, S.; Hou, Z.; Li, W.; Wang, Y.; Zhang, X.; Yang, W. Surface hydrophobic modification of microcrystalline cellulose by poly(methylhydro)siloxane using response surface methodology. *Polymers* **2018**, *10*, 1335. [CrossRef]
9. Hill, C.A.S.; Jones, D.; Strickland, G.; Cetin, N.S. Kinetic and mechanistic aspects of the acetylation of wood with acetic anhydride. *Holzforschung* **1998**, *52*, 623–629. [CrossRef]
10. Zadorecki, P.; Flodin, P. Surface modification of cellulose fibers III. Durability of cellulose–polyester composites under environmental aging. *J. Appl. Polym. Sci.* **1986**, *31*, 1699–1707. [CrossRef]
11. Zadorecki, P.; Flodin, P. Surface modification of cellulose fibers. II. The effect of cellulose fiber treatment on the performance of cellulose–polyester composites. *J. Appl. Polym. Sci.* **1985**, *30*, 3971–3983. [CrossRef]
12. Abdelmouleh, M.; Boufi, S.; Belgacem, M.N.; Dufresne, A.; Gandini, A. Modification of cellulose fibers with functionalized silanes: Effect of the fiber treatment on the mechanical performances of cellulose–thermoset composites. *J. Appl. Polym. Sci.* **2005**, *98*, 974–984. [CrossRef]
13. Rodrigues, E.F.; Maia, T.F.; Mulnari, D.R. Tensile strength of polyester resin reinforced sugarcane bagasse fibers modified by estherification. *Procedia Eng.* **2011**, *10*, 2348–2352. [CrossRef]

14. DiLoreto, E.; Haque, E.; Berman, A.; Moon, R.J.; Kalaitzidou, K. Freeze dried cellulose nanocrystal reinforced unsaturated polyester composites: Challenges and potential. *Cellulose* **2019**, *26*, 4391–4403. [CrossRef]
15. Kargarzadeh, H.M.; Sheltami, R.; Ahmad, I.; Abdullah, I.; Dufresne, A. Cellulose nanocrystal: A promising toughening agent for unsaturated polyester nanocomposite. *Polymer* **2015**, *56*, 346–357. [CrossRef]
16. Ummartyotin, S.; Manuspiya, H. A critical review on cellulose: From fundamental to an approach on sensor technology. *Renew. Sustain. Energy Rev.* **2015**, *41*, 402–412. [CrossRef]
17. Ohwoavworhua, F.O.; Adelakun, T.A. Non-wood fibre production of microcrystalline cellulose from Sorghum caudatum: Characterization and tableting properties. *Indian J. Pharm. Sci.* **2010**, *72*, 295–301. [CrossRef]
18. Chakrabarty, A.; Teramoto, Y. Recent advances in nanocellulose composites with polymers: A guide for choosing partners and how to incorporate them. *Polymers* **2018**, *10*, 517. [CrossRef]
19. Abdul Rashid, E.S.; Muhd Julkapli, N.; Yehye, W.A. Nanocellulose reinforced as green agent in polymer matrix composites applications. *Polym. Adv. Technol.* **2018**, *29*, 1531–1546. [CrossRef]
20. Pan, Y.; Pan, Y.; Cheng, Q.; Liu, Y.; Essien, C.; Via, B.; Wang, X.; Sun, R.; Taylor, S. Characterization of epoxy composites reinforced with wax encapsulated microcrystalline cellulose. *Polymers* **2016**, *8*, 415. [CrossRef]
21. Roy, D.; Semsarilar, M.; Guthrie, J.T.; Perrier, S. Cellulose modification by polymer grafting: A review. *Chem. Soc. Rev.* **2009**, *38*, 2046–2064. [CrossRef]
22. Rusmirović, J.D.; Rančić, M.P.; Pavlović, V.B.; Rakić, V.M.; Stevanović, J.; Djonlagić, J.; Marinković, A.D. Cross-linkable modified nanocellulose/polyester resin-based composites: Effect of unsaturated fatty acid nanocellulose modification on material performances. *Macromol. Mater. Eng.* **2018**, *303*, 1700648. [CrossRef]
23. Monti, M.; Terenzi, A.; Natali, M.; Gaztelumendi, I.; Markaide, N.; Kenny, J.M.; Torre, L. Development of unsaturated polyester matrix-carbon nanofibres nanocomposites with improved electrical properties. *J. Appl. Polym. Sci.* **2010**, *117*, 1658–1666. [CrossRef]
24. Yan, Y.; Herzele, S.; Mahendran, A.R.; Edler, M.; Griesser, T.; Saake, B.; Li, J.; Gindl-Altmutter, W. Microfibrillated lignocellulose enables the suspension-polymerization of unsaturated polyester resin for novel composites applications. *Polymers* **2016**, *8*, 255. [CrossRef] [PubMed]
25. Chabros, A.; Gawdzik, B. Methacrylate monomer as an alternative to styrene in typical polyester-styrene copolymers. *J. Appl. Polym. Sci.* **2019**, *136*, 47735. [CrossRef]
26. Li, W.; Cai, G.; Zhang, P. A simple and rapid Fourier transform infrared method for the determination of the degree of acetyl substitution of cellulose nanocrystals. *J. Mater. Sci.* **2019**, *54*, 8047–8056. [CrossRef]
27. Qian, J.; Li, H.; Zou, K.; Feng, X.; Hu, Y.; Zhang, S. Methacrylic acid/butyl acrylate onto feruloylated bagasse xylan: Graft copolymerization and biological activity. *Mater. Sci. Eng. C* **2019**, *98*, 594–601. [CrossRef]
28. Sanchez, E.M.S.; Zavaglia, C.A.C.; Felisberti, M.I. Unsaturated polyester resins: Influence of the styrene concentration on the miscibility and mechanical properties. *Polymer* **2000**, *41*, 765–769. [CrossRef]
29. Rahaman, M.A.; Rana, M.M.; Gafur, M.A.; Mahona, A.A. Preparation and analysis of poly(L-lactic acid) composites with oligo(D-lactic acid)-grafted cellulose. *J. Appl. Polym. Sci.* **2019**, *136*, 47424. [CrossRef]
30. Park, S.J.; Park, W.B.; Lee, J.R. Roles of Unsaturated Polyester in the Epoxy Matrix System. *Polym. J.* **1999**, *31*, 28–31. [CrossRef]
31. Lavoratti, A.; Scienza, L.C.; Zattera, A. Dynamic-mechanical and thermomechanical properties of cellulose nanofiber/polyester resin composites. *Carbohydr. Polym.* **2016**, *136*, 10425. [CrossRef]
32. Zeng, D.; Lv, J.; Wei, C.; Yu, C. Dynamic mechanical properties of sisal fiber cellulose microcrystalline/unsaturated polyester in-situ composites. *Polym. Adv. Technol.* **2015**, *26*, 1351–1355. [CrossRef]
33. Dai, X.; Xiong, Z.; Na, H.; Zhu, J. How does epoxidized soybean oil improve the toughness of microcrystalline cellulose filed polylactide acid composites? *Compos. Sci. Technol.* **2014**, *90*, 9–15. [CrossRef]
34. Mandal, S.; Alam, S. Dynamic mechanical analysis and morphological studies of glass/bamboo fiber reinforced unsaturated polyester resin-based hybrid composites. *J. Appl. Polym. Sci.* **2012**, *125* (Suppl. 1), E328–E387. [CrossRef]
35. Hsu, C.P.; Lee, L.J. Free-radical crosslinking copolymerization of styrene/unsaturated polyester resins: 3. Kinetics-gelation mechanism. *Polymer* **1993**, *34*, 4516–4523. [CrossRef]

© 2019 by the authors. Licensee MDPI, Basel, Switzerland. This article is an open access article distributed under the terms and conditions of the Creative Commons Attribution (CC BY) license (http://creativecommons.org/licenses/by/4.0/).

Article

Biobased Composites from Biobased-Polyethylene and Barley Thermomechanical Fibers: Micromechanics of Composites

Ferran Serra-Parareda [1], Quim Tarrés [1,2], Marc Delgado-Aguilar [1], Francesc X. Espinach [3], Pere Mutjé [1,2] and Fabiola Vilaseca [4,*]

1. LEPAMAP Group, Department of Chemical Engineering, University of Girona, 17003 Girona, Spain; ferran.serra@udg.edu (F.S.-P.); joaquimagusti.tarres@udg.edu (Q.T.); m.delgado@udg.edu (M.D.-A.); pere.mutje@udg.edu (P.M.)
2. Chair on Sustainable Industrial Processes, University of Girona, 17003 Girona, Spain
3. Design, Development and Product Innovation, Dept. of Organization, Business, University of Girona, 17003 Girona, Spain; francisco.espinach@udg.edu
4. Department of Fibre and Polymer Technology, KTH Royal Institute of Technology, SE-10044 Stockholm, Sweden
* Correspondence: vilaseca@kth.se

Received: 21 November 2019; Accepted: 10 December 2019; Published: 12 December 2019

Abstract: The cultivation of cereals like rye, barley, oats, or wheat generates large quantities of agroforestry residues, which reaches values of around 2066 million metric tons/year. Barley straw alone represents 53%. In this work, barley straw is recommended for the production of composite materials in order to add value to this agricultural waste. First of all, thermomechanical (TMP) fibers from barley straw are produced and later used to reinforce bio-polyethylene (BioPE) matrix. TMP barley fibers were chemically and morphologically characterized. Later, composites with optimal amounts of coupling agent and fiber content ranging from 15 to 45 wt % were prepared. The mechanical results showed the strengthening and stiffening capacity of the TMP barley fibers. Finally, a micromechanical analysis is applied to evaluate the quality of the interface and to distinguish how the interface and the fiber morphology contributes to the final properties of these composite materials.

Keywords: bio-polyethylene; barley straw; thermomechanical fibers; interface

1. Introduction

The production of environmentally friendly materials is a key point to more sustainable development. Today's society has become aware of the need to eliminate its dependence on fossil resources. In this sense, the production of materials coming from renewable resources with properties similar to the current ones is becoming a necessity [1–6]. There are two main ways to develop greener plastic materials—bio-based and biodegradable or bio-based and recyclable materials [5,7–10]. Both types must coexist in order to achieve the properties currently provided by materials manufactured from fossil origin resources.

Biopolyethylene is produced from biomass such as sugarcane [11] and is an alternative to polyethylene obtained from oil. However, due to market dimensions and scale economies, the cost and price of bio-based polymers are higher than common plastics. Hopefully, a higher penetration in the market of bio-based materials will reduce its production costs and then its price [12].

In the field of bio-based and recyclable materials, the use of bio-polyethylene reinforced with natural fibers is of great interest. By combining a bio-based matrix such as biopolyethylene with a bio-based fiber reinforcement, it is possible to obtain high performance and recyclable bio-composites [13]. Moreover, if an agroforestry residue is suggested as raw material, a new renewable, cheap, and sustainable reinforcement is used, thus adding value and extending the agroforestry value chain. Barley straw has a world production of 195,000,000 bdmt/year, representing 15.6% of cereal straws world-wide [14]. However, to the best of our knowledge, there are very few publications on the use of this residue as raw material, and none in the field of composite materials [15–18].

The use of natural fibers as reinforcements in plastic matrix composites has a number of challenges. The main ones are the poor interaction at fiber-matrix interface and the low degradation temperatures of natural fibers. The former is related to the incompatibility between highly hydrophilic fibers and the hydrophobic polymer matrices. Under these circumstances, there is not a good interaction between the matrix and the reinforcement, and a strong interface is impossible to attain [19]. This polarity difference can be overcome by adding coupling agents. It is already known in the literature that coupling agents such as maleated polyethylene (MAPE) can improve the strength of the interface between the two phases [20,21]. However, this interaction must be studied in each case because the surface chemical composition of the fibers is also a key factor [22]. On the other hand, the degradation temperature of the cellulosic fibers limits the processing temperatures, which have to be kept below 220 °C [23].

In this work, the production of composite materials of thermo-mechanical barley fibers with bio-polyethylene is studied. Composite materials with MAPE contents ranging from 0 to 8 wt % are produced to elucidate the optimal coupling agent in the formulation that renders the highest tensile strengths. From here, coupled composites from 15–45 wt % of thermo-mechanical barley fibers are obtained, and their mechanical properties are analyzed and discussed. A micromechanics analysis of the tensile strength is carried out in order to assess the strength of the interface and the intrinsic tensile strength of the reinforcements.

2. Materials and Methods

2.1. Materials

Polyethylene based on renewable sources kindly supplied by Braskem (São Paulo, Brazil) was used as polymer matrix. This 100% recyclable biopolyethylene (BioPE) is obtained from sugarcane and therefore is bio-based. The polymer has a molecular weight of 61.9 g/mol, a melt flow index of 20 g/10 min at 190 °C with 2.16 kg, and a density of 0.955 g/cm^3. Maleic anhydride-grafted-polyethylene (MAPE) with a maleic–anhydride substitution of 0.9% (Fusabond MB100D) was provided by DuPont (Wilmington, DE, USA).

Barley straws were kindly provided by Mas Clarà S.A. (Girona, Spain).

2.2. Methods

Initially, the barley straw was chopped by means of a blade mill equipped with a 3 mm mesh. Then, the barley straw was subjected to a thermo-mechanical digestion process. This process was carried out in a pressure reactor at 160 °C and a liquid ratio of 1:6. The fibers extracted from the reactor were washed repeatedly and then passed through a Sprout–Waldron defibrator. The obtained fibers were filtered and dried at 80 °C.

The study of composite materials produced from bio-polyethylene and thermo-mechanical fibers of barley straw was divided in two parts. In the first step, the amount of coupling agent was optimized for composites comprising 30 wt % of fiber content. From these results, a micromechanics analysis allowed us to obtain the orientation factor and the intrinsic resistance of the barley thermo-mechanical fibers. In the second step, the mechanical properties of composite materials adding different percentages of fibers were studied. The analysis was based on the micromechanical values resulting from the optimization of the coupling agent. A flow chart of the composites production and their characterization is shown in Figure 1.

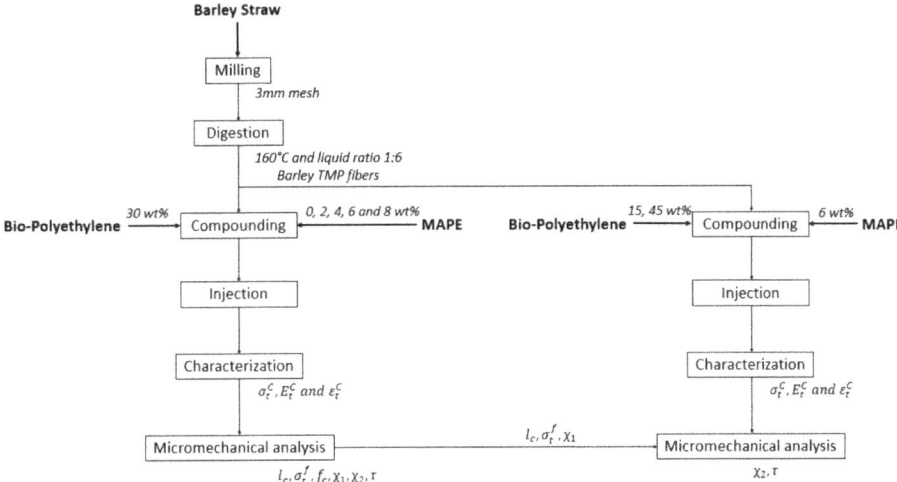

Figure 1. Flow chart of the composites production and characterization.

The compounding process was performed using a Gelimat intensive kinetic mixer (Dusatec, Inc., Ramsey, NJ, USA). The fibers were introduced at a rotor speed of 300 rpm. Subsequently, and maintaining this constant speed, the polymer was incorporated together with the coupling agent. The composites were produced with fiber percentages between 15 and 45 wt %. Once all materials were in the kinetic mixer, the rotor speed was increased up to 3000 rpm. The increase in the rotor speed cause increase in the temperature reaching the melting temperature of the polymer. The material was discharged once the matrix was melted and mixed with the fibers. The blend was then cooled down and subsequently pelletized. The composites were kept in an oven at 80 °C temperature for 24 h before injection molding. The mold injection was carried out using a 220 M 350–90U injection machine (Arburg, Loßburg, Germany). A series of 20 standard test specimens of each composite were mold injected to assess tensile properties.

The morphological study of fibers was carried out using a MORFI equipment (Techpap, Gières, France). The fibers into the composite materials were extracted using a Soxhlet apparatus with Decalin to dissolve the matrix. From here, the mean fiber lengths, fiber diameters as well as fiber length distributions were obtained. The chemical composition of the fibers was determined from the analysis of ash content (ISO 2144:2019 standard [24]), extractives (TAPPI T204 cm-07 [25]), lignin klason (ISO/DIS 21436 [26]), and holocellulose.

Tensile tests were performed according to ASTM D790 standard. All samples were conditioned at 50% of relative humidity and 23 °C in a climatic chamber (Dycometal, Viladecans, Spain) during at least 48 h before testing (ASTM D618 standard [27]). The tensile tests were carried out in an Instron TM 1122 universal testing machine (Instron, Cerdanyola, Spain). This equipment is fitted with a 5 kN load cell. The experimental results are the average of at least of testing five samples.

3. Results and Discussion

3.1. Fibers Assessment

Table 1 shows the results from the chemical analysis of the barley straws, the fibers from submitting barley straws to a thermomechanical process, and spruce fibers subjected to the same process. It is evident from the table that there are significant differences between the chemical composition of virgin raw material (barley straw) and the fibers after a thermomechanical process.

Table 1. Chemical composition of raw material, barley thermomechanical (TMP) fibers, and spruce TMP fibers.

	Holocellulose (%)	Klason Lignin (%)	Extractives (%)	Ash (%)	Length [1] (μm)	Diameter (μm)
Barley straw	70.12 ± 0.54	16.45 ± 0.34	5.90 ± 0.76	7.1 ± 0.2	–	–
Barley TMP fibers	77.67 ± 0.61	15.30 ± 0.46	2.73 ± 0.12	4.3 ± 0.3	745	19.6
Spruce TMP fibers	73.75 ± 0.83	25.80 ± 0.22	0.25 ± 0.34	0.2 ± 0.2	978	24.7

[1] Length weighted in length.

These differences are due to the thermomechanical treatment. More in detail, the thermal treatment acts on the middle lamella and the primary cell wall of the fibers removing part of the ashes, extractives and lignin, mainly present in these areas of the fiber structure. Thereafter the secondary wall is reached (S1, S2, and S3), where higher contents of cellulose and hemicellulose are present. Afterward, when these fibers are subjected to a mechanical defibration process (Sprout–Waldron defibrator), the fiber agglomerates tend to split at the fiber–fiber union. In comparison to a thermomechanical treatment, an exclusively mechanical process would break the structure of the fibers in a disordered way while preserving the same chemical composition of the raw material. On the other hand, as shown in Table 1, the chemical composition of fibers from spruce wood treated by the same thermomechanical process has a higher lignin content. This is due to the different chemical composition of wood fibers with respect to an agroforestry waste. It has been reported, that fibers from non-wood resources present lower lignin contents [28]. This difference in chemical composition can be related to their surface composition [22,29]. Börås and Gatenholm [22] developed a model for the distribution of carbohydrates, lignin, and extractives on the surface of fibers after defibration. This model proposes that lignin is covering in a heterogeneous way the carbohydrates that are in an ordered phase in the form of fibrils. In this model, it is also proposed that the extractives form globular particles spread all over carbohydrates and lignin. These chemical surface properties are essential to understand the difficulty to achieve composites with strong interfaces between natural fibers and polyolefin matrices.

Figure 2 shows the length distributions of barley TMP fibers and spruce TMP fibers. The determination of the morphology of the fibers is important to determine their reinforcement capacity. In the same way as chemical composition, wood fibers tend to show higher aspect ratios (length/diameter) than fibers from annual plants [28]. Nonetheless, the morphology of the fibers is remarkably changed during composite processing. Compounding and mold injection processes cause an important reduction of the mean length of the fibers [29].

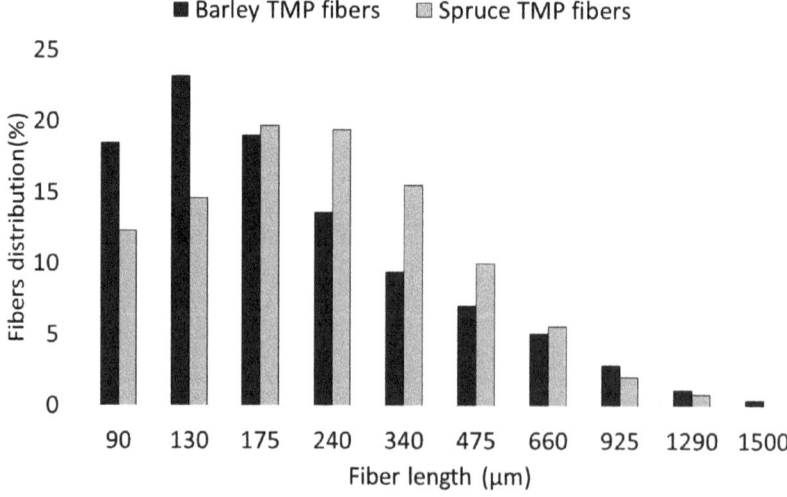

Figure 2. Length distributions of barley TMP and spruce TMP fibers.

3.2. Coupling Agent Optimization

The tensile properties of a composite material are mainly impacted by the nature of the reinforcement and the matrix, the reinforcement content, dispersion and orientation, its aspect ratio, and largely by the quality of the matrix–reinforcement interface [30–32]. The different nature between natural fibers (hydrophilic) and polyethylene (hydrophobic) hinders the achievement of a strong interface [33,34]. Improvement of the interface strength can be achieved by the use of coupling agents such as polyethylene grafted with maleic anhydride (MAPE) [35–37].

Table 2 shows the tensile strength (σ_t^C), Young's modulus (E_t^C), strain at break (ε_{max}), and the contribution of the matrix to the tensile strength of the matrix (σ_t^{m*}) for 30 wt % TMP-reinforced BioPE composites coupled with MAPE percentages ranging from 0 to 8%. The density of the fibers inside the material (1.35 g/cm^3) was determined from the density of the bio-propylene (0.955 g/cm^3) and the density of the resulting composite material. Therefore, the volume fraction (V^F) of fibers in a 30 wt % composite material is 0.233.

Table 2. Tensile properties of barley TMP/ biopolyethylene (BioPE) composite with different coupling agent contents.

MAPE (%)	V^F	σ_t^C (MPa)	E_t^C (GPa)	ε_{max} (%)	σ_t^{m*} (MPa)
0	0	18.05 ± 0.74	1.06 ± 0.08	12.18 ± 0.34	18.05
0		18.82 ± 0.60	1.73 ± 0.10	2.88 ± 0.27	13.29
2		23.51 ± 0.39	1.76 ± 0.05	3.37 ± 0.15	14.19
4	0.233	29.84 ± 0.19	1.85 ± 0.07	5.19 ± 0.22	16.27
6		34.70 ± 0.90	2.14 ± 0.04	5.47 ± 0.31	16.44
8		32.65 ± 0.69	1.93 ± 0.05	5.67 ± 0.17	16.55

The tensile strength of the composites reinforced with 30 wt % of barley TMP composites increased noticeably when MAPE was added in the formulation (Figure 3). However, when the MAPE exceeds 6%, the tensile strength starts to decrease. The increase of the tensile strength of the composites is promoted by the creation of covalent ester bonds. The higher is the MAPE content the higher are the possible bonds. These bonds are formed by the reaction between the anhydride groups of maleic acid and the hydroxyl groups on the surface of the fibers [19,38]. Nonetheless, if a certain percentage of coupling agent is exceeded, grafted polyethylene chains tend to self-entangle and decrease the tensile strength of the composite. In this case, it was resolved that 6% MAPE provides the highest increase in tensile strength of the material (34.70 MPa), which represents an increase over 90% of the tensile strength of the matrix. On the other hand, the Young's modulus of the composite material is not affected by the quality of the interface. In this sense, composite materials with poor interfaces present Young's moduli similar to composites with optimized MAPE contents [39]. For the elongation at the break, the deformation capability increases with the higher coupling agent content as a direct consequence of Hooke's law.

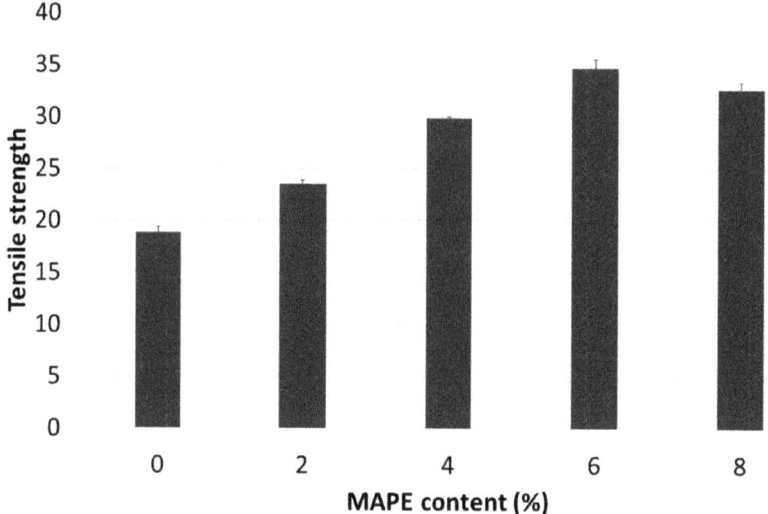

Figure 3. Tensile strength evolution of 30 wt % composite against maleic anhydride (MAPE) content.

Once the elongation at break of the composites are known, it is possible to determine the contribution of the matrix to the strength of these material ($\sigma_t^{m^*}$), as shown in Figure 4. By translating the value of the elongation at rupture of the composites on the stress-strain curve of the matrix, the $\sigma_t^{m^*}$ value can be obtained.

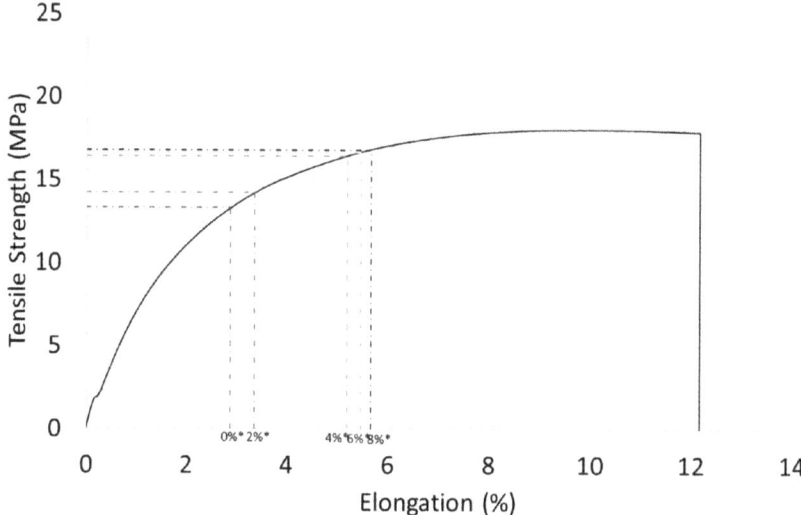

Figure 4. Tensile strength vs elongation diagram of BioPE polymer. Matrix contribution to tensile strength for each MAPE content is also shown.

Mechanical properties of composites (BioPE + 30 wt % barley TMP fibers) related to coupling agent content were modelled by using a modified rule of mixtures for the tensile strength of semi-aligned short fiber reinforced composites (Equation (1)) [40,41].

$$\sigma_t^C = f_c \cdot \sigma_t^f \cdot V^F + \sigma_t^{m^*} \cdot (1 - V^F), \tag{1}$$

where σ_t^C is the tensile strength of the composite, σ_t^f the intrinsic tensile strength of the fibers and f_c a coupling factor. As reported by Sanadi et al. [42], the coupling factor used in the modified rule of mixtures (f_c) is the orientation factor (χ_1) times the length and interface factor (χ_2) of the fibers inside the composite. Nonetheless, the equation cannot be used to evaluate the intrinsic tensile strength of the reinforcements because has another unknown, the value of the coupling factor. Accordingly, the use of the modified Kelly-Tyson equation (Equation (2)) [43].

$$\sigma_t^C = \chi_1 \cdot \left(\sum_{L_c}^{i=0} \left[\frac{\tau \cdot l_i^F \cdot V_i^F}{d^F} \right] + \sum_{\infty}^{j=L_c} \left[\sigma_t^f \cdot V_j^F \cdot \left(1 - \frac{\sigma_t^F \cdot d^F}{4 \cdot \tau \cdot l_j^F} \right) \right] \right) + \sigma_f^{m^*} \cdot (1 - V^F), \qquad (2)$$

where L_c is the critical length, l^f the fiber length and d^f the fiber diameter. The Kelly–Tyson equation divides the contribution of the reinforcements to the tensile strength of the composites between the contributions of the subcritical ($l^f < l_c$) and supercritical ($l^f > l_c$) fibers. The equation introduces an interfacial shear strength (τ). According to the shear-lag model, the matrix transmits the force from the matrix to the reinforcement by shear forces in the interface. This means that the fibers are fully loaded at the center of their length and nil at their ends. According to the length of each fiber inside the composite, the load in its center will be less or equal to its intrinsic tensile strength (Figure 5).

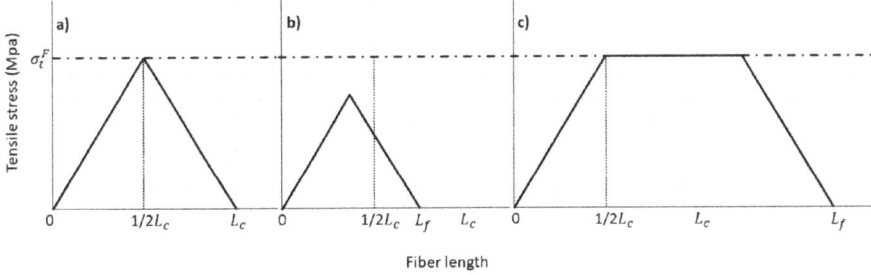

Figure 5. Axial load graphic of (**a**) critical length fibers, (**b**) subcritical length fibers, and (**c**) supercritical length fibers.

Therefore, the length of the reinforcements, its intrinsic tensile strength, and the interfacial shear strength command the value of the critical length. Only those reinforcements with lengths equal or higher than the critical will fully deploy its strengthening capabilities. The critical length is equal to the intrinsic tensile strength of the fibers times the fiber radius, divided by the interfacial shear strength.

The Kelly–Tyson model considers the stress distribution due to a single short fiber embedded in a matrix when the system is subjected to a uniaxial load in the direction of the fiber axis [44]. Nonetheless, Kelly and Tyson's equation cannot be solved as shows three unknowns, the orientation factor, the critical length and the intrinsic tensile strength of the reinforcements. Anyhow, Bowyer and Bader developed a numerical method capable to solve the equation. The Bowyer–Bader solution [45] assumes that the interfacial shear strength is independent of the deformation, and that the effect of the fiber's inclination to the load axis can be explained by a scale factor, and the fiber orientation factor. Then, by using the experimental values of two points of the stress-strain curve χ_1 and the τ can be determined (Figure 6).

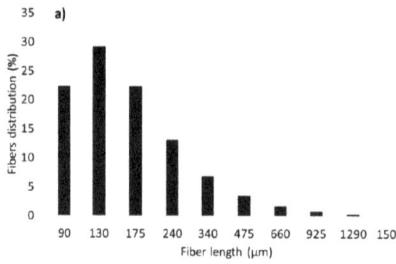 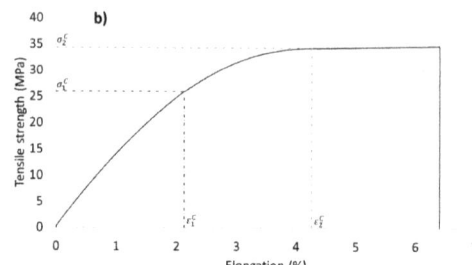

Figure 6. (a) Fibers length distribution from 30% composite and (b) stress-strain curve of a coupled 30% composite.

The results of applying the method to a composite that adds 6 % coupling agent returns an orientation factor of 0.309 and an interfacial shear strength of 10.49. The interfacial shear strength obtained is close to von Misses criteria ($\tau = 10.42$ MPa), where $\tau = \sigma_t^m / \sqrt{3}$ [46]. Von Mises criteria is used to define strong interfaces. Thus, the obtained value can be considered reasonable and corresponding to a strong interface. Then, the critical fiber length was calculated from $l_c = (d^f \cdot \sigma_t^c)/2 \cdot \tau$ [47]. The obtained value of critical fiber length was 409.91 μm. Once obtained the values of χ_1, τ and l_c, it was possible to use of Kelly–Tyson equation (Equation (2)) to determine the value of the intrinsic strength of the fibers. The result for a 30% composite with 6% coupling agent was 521.18 MPa. Then, this intrinsic tensile strength was introduced in the modified rule of mixtures (Equation (1)) to evaluate a coupling factor (f_c) for 30% composite that was found to be 0.18. This value is in the range from 0.18 to 0.20 obtained by composites with optimal interfaces [48].

Equation (1) was used, together with the intrinsic tensile strength of the reinforcement, to evaluate the coupling factors for all the composites reinforced with 30 wt % of fibers and MAPE contents ranging from 0 to 8%. The lower value ($f_c = 0.07$) was obtained for the uncoupled composite.

The following equations show the relation between the coupling factor and the length and interface factor and the orientation factor (Equation (3)). Then, Equations (4) and (5) are used to relate the values of the orientation factor with the interfacial shear strength. Equation (4) is used when the mean length of the reinforcements is higher that the critical length. Otherwise, Equation (5) must be used.

$$f_c = \chi_1 \cdot \chi_2 \qquad (3)$$

$$\tau = \frac{\sigma_t^f \cdot d^f}{4 \cdot l^f \cdot (1 - \chi_2)} \quad for \ l^f \geq l_c \qquad (4)$$

$$\tau = \frac{\sigma_t^f \cdot d^f}{l^f} \chi_2 \quad for \ l^f \leq l_c \qquad (5)$$

The orientation factor is affected by the geometry of the injection mold and the mold injection parameters. As long as all the composites were mold injected in the same mold and under the same parameters, is possible to consider that such composites will share the same orientation factor. Thus, the obtained 0.309 orientation factor was used together with the coupling factors and Equation (4) to obtain the length and interface factors of the composites against MAPE content. Likewise, Equations (4) and (5) were used to evaluate the corresponding interfacial shear strength of the same composites. Figure 7 shows these values against MAPE contents.

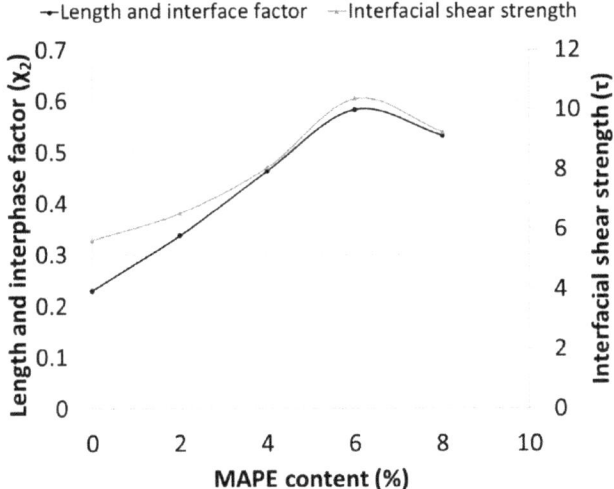

Figure 7. Evolution of the length and interphase factor and the interfacial shear strength as function of coupling agent content.

The figure shows the evolution of both parameters against MAPE content. In some sense, the length and interface factor integrate the strength of the interface along with the influence of the morphology of the reinforcements. The interfacial shear strength accounts only for the strength of the interface. Thus, both curves are expected to be similar in shape. Major dissimilarities will be caused by the effect of the morphology. Figure 7 shows very similar curves, changing only due to scale effects. Thus, in this case, the main parameter ruling the contribution of the fibers against MAPE content will be the strength of the interface. The effect of the morphology of the fibers will be limited and only due to slight decreases of the mean lengths of the fibers against MAPE content. The figure shows the maximum values for the composite with 6 wt % MAPE contents.

3.3. Mechanical Performance of Barley TMP/BioPE Composites

Once the effect of the coupling agent on the tensile strength of the composites was assessed, coupled composites with 15 and 45 wt % reinforcement contents were prepared and tested. Both composites comprise 6% of MAPE, with respect to the fiber content. It can be observed from Table 3 that the increase in the percentage of barley TMP fibers led to a notable increase of the tensile strength, reaching a value of 43.1 MPa for the 45 wt % formulation which represents a 138% increase over the matrix. This increase was slightly higher than those obtained in the bibliography where an addition of 30% natural fibers (maize fibers or corn stalk fibers) does not exceed the 85% increase in tensile strength. This indicates the correct level of interface between fiber and matrix [49,50].

Table 3. Tensile properties of barley TMP/BioPE composites.

Barley TMP (%)	V^F	σ_t^C (MPa)	E_t^C (GPa)	ε_{max} (%)	$\sigma_t^{m^*}$ (MPa)
0	0	18.05 ± 0.74	1.06 ± 0.08	12.18 ± 0.34	18.05
15	0.111	25.2 ± 0.64	1.85 ± 0.06	7.65 ± 0.24	16.37
30	0.233	34.7 ± 0.90	2.59 ± 0.04	6.45 ± 0.31	16.76
45	0.367	43.1 ± 0.57	3.55 ± 0.05	4.69 ± 0.33	15.86

At the same time, the increases in reinforcement contents also leaded to a linear increase of the Young's modulus of the composites, an indicating of good reinforcement dispersion. The addition of natural fibers (stiffer phase) as reinforcement of a polymeric matrix (ductile phase) implies a resulting material with higher stiffness than the plain matrix. The deformation at rupture of the material decreases as the addition of reinforcement increases as a direct consequence of Hooke's law.

The same micromechanics analysis on the length and orientation factor, and on the interfacial shear strength, was made to evaluate the impact of the reinforcement content over the same factors. The orientation factor was assumed to be 0.309 and the intrinsic tensile strength of the fibers 521.18 MPa. Figure 8 shows the evolution of both factors against the reinforcement content.

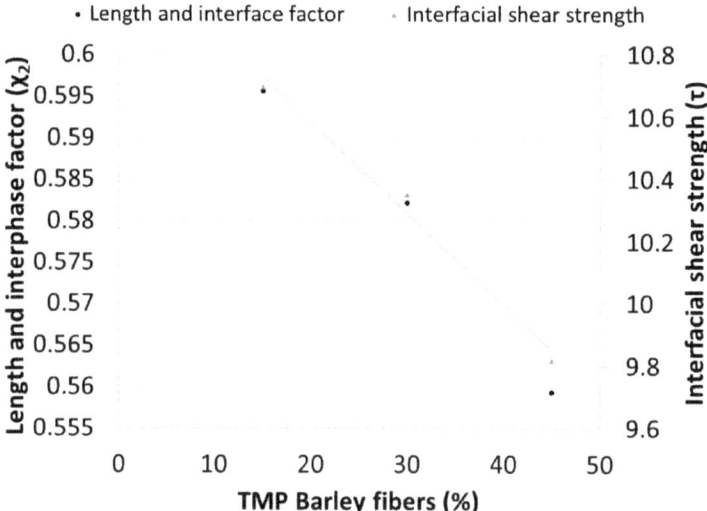

Figure 8. Evolution of the length and interphase factor and the interfacial shear strength as function of TMP barley fiber content.

These two factors decrease almost linearly as the percentage of reinforcement increases. However, the interface strengths obtained between 15–45% reinforcement with the addition of a 6% coupling agent can be considered strong. In this case, both parameters evolved similarly, but a normalization of the scale of both parameters shows how the influence of the length and interface factor changes from positive for the composite adding 15 wt % of reinforcement to negative for the composite adding 45 wt %. This is explained by the decreases of the mean length of the fibers due to attrition phenomena during mixing. The mean length of the fibers decreased with the reinforcement content. Thus, longer fibers have more influence in the tensile strength of a composite than shorter ones. Nonetheless, the deviations of such parameter are slight and the main parameter commanding the tensile strength of the composites is the interfacial shear strength.

4. Conclusions

Thermomechanical barley fibers were used as reinforcement for a BioPE-based composite. A composite adding 30 wt % of reinforcement was used to evaluate the effect of a coupling agent over the tensile strength of the composites. It was found that a 6% MAPE content returned the highest tensile strengths.

A micromechanics analysis allowed us to obtain the intrinsic tensile strength of the reinforcements, with a value of 521.18 MPa. This value is similar to other more commonly used natural reinforcements, ensuring the strengthening capabilities of the barley fibers. The analysis also allowed us to evaluate the strength of the interface. This interface was evaluated by the intrinsic tensile strength, with a value of 10.42 MPa, almost equal to von Mises criteria, and the coupling factor with a value of 0.18 inside the range from 0.18 to 0.20 considered optimum for semi-aligned short fiber–reinforced composites. The analysis also unveiled the higher impact of the strength of the interface over the morphology of the fibers for the contribution of the reinforcements to the tensile strength of a composite, when the percentage of fibers is constant and the percentage of coupling agent changes.

The effect of reinforcement content over tensile strength was also assessed. Composites adding 15 and 45 wt % of reinforcement were prepared and tensile tested. It was found that the tensile strength of the composites evolved linearly with reinforcement contents. It was found that the interfacial shear strength mainly commanded the contribution of the fibers to the tensile strength of the composites. Nonetheless, the length and interface factor also showed a slight effect, as a direct consequence of the shortening of the mean length of the fibers against fiber content.

Author Contributions: M.D.-A. and F.X.E. conceptualized the study. F.S.-P. performed the experimental tests and validated the experimental results. M.D.-A. and Q.T. designed the test case and stablished the analysis conditions. F.X.E. and Q.T. performed the analysis and wrote the paper. P.M. and F.V. supervised the research and corrected the manuscript.

Funding: This research was funded by the Chair on Sustainable Industrial Processes of the University of Girona, ref 01/2019.

Acknowledgments: The authors thank Braskem for kindly providing the bio-polyethylene matrix.

Conflicts of Interest: The authors declare no conflict of interest.

References

1. Peltola, H.; Pääkkönen, E.; Jetsu, P.; Heinemann, S. Wood based PLA and PP composites: Effect of fibre type and matrix polymer on fibre morphology, dispersion and composite properties. *Compos. Part A Appl. Sci. Manuf.* **2014**, *61*, 13–22. [CrossRef]
2. Zierdt, P.; Theumer, T.; Kulkarni, G.; Däumlich, V.; Klehm, J.; Hirsch, U.; Weber, A. Sustainable wood-plastic composites from bio-based polyamide 11 and chemically modified beech fibers. *Sustain. Mater. Technol.* **2015**, *6*, 6–14. [CrossRef]
3. Balart, J.F.; García-Sanoguera, D.; Balart, R.; Boronat, T.; Sánchez-Nacher, L. Manufacturing and properties of biobased thermoplastic composites from poly(lactid acid) and hazelnut shell wastes. *Polym. Compos.* **2018**, *39*, 848–857. [CrossRef]
4. Hoang, D.; Pham, T.; Nguyen, T.; An, H.; Kim, J. Organo-Phosphorus Flame Retardants for Poly(vinyl chloride)/Wood Flour Composite. *Polym. Compos.* **2018**, *39*, 961–970. [CrossRef]
5. Haider, T.P.; Völker, C.; Kramm, J.; Landfester, K.; Wurm, F.R. Plastics of the Future? The Impact of Biodegradable Polymers on the Environment and on Society. *Angew. Chem. Int. Ed.* **2019**, *58*, 50–62. [CrossRef]
6. Dilkes-Hoffman, L.; Ashworth, P.; Laycock, B.; Pratt, S.; Lant, P. Public attitudes towards bioplastics—Knowledge, perception and end-of-life management. *Resour. Conserv. Recycl.* **2019**, *151*, 104479. [CrossRef]
7. Bledzki, A.K.; Franciszczak, P.; Meljon, A. High performance hybrid PP and PLA biocomposites reinforced with short man-made cellulose fibres and softwood flour. *Compos. Part A Appl. Sci. Manuf.* **2015**, *74*, 132–139. [CrossRef]
8. Balart, J.F.; Fombuena, V.; Fenollar, O.; Boronat, T.; Sánchez-Nacher, L. Processing and characterization of high environmental efficiency composites based on PLA and hazelnut shell flour (HSF) with biobased plasticizers derived from epoxidized linseed oil (ELO). *Compos. Part B Eng.* **2016**, *86*, 168–177. [CrossRef]
9. Granda, L.; Tarres, Q.; Espinach, F.X.; Julian, F.; Mendes, A.; Delgado-Aguilar, M.; Mutje, P. Fully biodegradable polylactic composites reinforced with bleached softwood fibers. *Cellul. Chem. Technol.* **2016**, *50*, 417–422.
10. Yusoff, R.B.; Takagi, H.; Nakagaito, A.N. Tensile and flexural properties of polylactic acid-based hybrid green composites reinforced by kenaf, bamboo and coir fibers. *Ind. Crops Prod.* **2016**, *94*, 562–573. [CrossRef]
11. Brodin, M.; Vallejos, M.; Opedal, M.T.; Area, M.C.; Chinga-Carrasco, G. Lignocellulosics as sustainable resources for production of bioplastics—A review. *J. Clean. Prod.* **2017**, *162*, 646–664. [CrossRef]
12. Ferrero, B.; Fombuena, V.; Fenollar, O.; Boronat, T.; Balart, R. Development of Natural Fiber-Reinforced Plastics (NFRP) Based on Biobased Polyethylene and Waste Fibers From Posidonia oceanica Seaweed. *Polym. Polym. Compos.* **2008**, *36*, 1378–1385. [CrossRef]

13. Yang, H.S.; Kim, H.J.; Park, H.J.; Lee, B.J.; Hwang, T.S. Water absorption behavior and mechanical properties of lignocellulosic filler-polyolefin bio-composites. *Compos. Struct.* **2006**, *72*, 429–437. [CrossRef]
14. Hurter, R.W. Nonwood fibres & moulded products. *Pap. Technol.* **2015**, *56*, 14–17.
15. Espinosa, E.; Sánchez, R.; Otero, R.; Domínguez-Robles, J.; Rodríguez, A. A comparative study of the suitability of different cereal straws for lignocellulose nanofibers isolation. *Int. J. Biol. Macromol.* **2017**, *103*, 990–999. [CrossRef] [PubMed]
16. Juárez, M.; Sánchez, R.; Espinosa, E.; Domínguez-Robles, J.; Bascón-Villegas, I.; Rodríguez, A. Environmentally friendly lignocellulose nanofibres from barley straw. *Cellul. Chem. Technol.* **2018**, *52*, 589–595.
17. Serrano, C.; Monedero, E.; Lapuerta, M.; Portero, H. Effect of moisture content, particle size and pine addition on quality parameters of barley straw pellets. *Fuel Process. Technol.* **2011**, *92*, 699–706. [CrossRef]
18. Vargas, F.; González, Z.; Rojas, O.; Garrote, G.; Rodríguez, A. Barley Straw (Hordeum vulgare) as a supplementary raw material for eucalyptus camaldulensis and pinus sylvestris kraft pulp in the paper industry. *BioResource* **2015**, *10*, 3682–3693. [CrossRef]
19. Pickering, K.L.; Efendy, M.G.A.; Le, T.M. A review of recent developments in natural fibre composites and their mechanical performance. *Compos. Part A Appl. Sci. Manuf.* **2016**, *83*, 98–112. [CrossRef]
20. Liu, N.C.; Baker, W.E. Reactive polymers for blend compatibilization. *Adv. Polym. Technol.* **1992**, *11*, 249–262. [CrossRef]
21. Tarrés, Q.; Melbø, J.K.; Delgado-Aguilar, M.; Espinach, F.X.; Mutjé, P.; Chinga-Carrasco, G. Bio-polyethylene reinforced with thermomechanical pulp fibers: Mechanical and micromechanical characterization and its application in 3D-printing by fused deposition modelling. *Compos. Part B Eng.* **2018**, *153*, 70–77. [CrossRef]
22. Böras, L.; Gatenhol, P. Surface composition and morphology of CTMP fibers. *Holzforschung* **1999**, *53*, 188–194. [CrossRef]
23. Yang, H.S.; Wolcott, M.P.; Kim, H.S.; Kim, H.J. Thermal properties of lignocellulosic filler-thermoplastic polymer bio-composites. *J. Therm. Anal. Calorim.* **2005**, *82*, 157–160. [CrossRef]
24. *ISO 2144:2019–Paper, Board, Pulps and Cellulose Nanomaterials—Determination of Residue (Ash Content) on Ignition at 900 °C*; International Organization for Standardization: Geneva, Switzerland, 2019.
25. *TAPPI T204cm-07–Solvent Extractives of Wood and Pulp*; Standard-Specific Interst Group: Atlanta, GA, USA, 2007.
26. *ISO/DIS 21436–Determination of Lignin Content*; International Organization for Standardization: Geneva, Switzerland, 2019.
27. *ASTM D618-13: Standard Practice for Conditioning Plastics for Testing*; ASTM International: West Conshohocken, PA, USA, 2013.
28. Marques, G.; Rencoret, J.; Gutiérrez, A.; del Río, J.C. Evaluation of the Chemical Composition of Different Non-Woody Plant Fibers Used for Pulp and Paper Manufacturing. *Open Agric. J.* **2014**, *4*, 93–101. [CrossRef]
29. López, J.P.; Méndez, J.A.; Espinach, F.X.; Julián, F.; Mutjé, P.; Vilaseca, F. Tensile Strength Characteristics of Polypropylene Composites Reinforced with Stone Groundwood Fibers from Softwood. *BioResources* **2012**, *7*, 3188–3200. [CrossRef]
30. Joffre, T.; Miettinen, A.; Berthold, F.; Gamstedt, E.K. X-ray micro-computed tomography investigation of fibre length degradation during the processing steps of short-fibre composites. *Compos. Sci. Technol.* **2014**, *105*, 127–133. [CrossRef]
31. Huda, M.S.; Drzal, L.T.; Misra, M.; Mohanty, A.K. Wood-fiber-reinforced poly(lactic acid) composites: Evaluation of the physicomechanical and morphological properties. *J. Appl. Polym. Sci.* **2006**, *102*, 4856–4869. [CrossRef]
32. Nygård, P.; Tanem, B.S.; Karlsen, T.; Brachet, P.; Leinsvang, B. Extrusion-based wood fibre-PP composites: Wood powder and pelletized wood fibres—A comparative study. *Compos. Sci. Technol.* **2008**, *68*, 3418–3424. [CrossRef]
33. Vilaseca, F.; Valadez-Gonzalez, A.; Herrera-Franco, P.J.; Pelach, M.; Lopez, J.P.; Mutje, P. Biocomposites from abaca strands and polypropylene. Part I: Evaluation of the tensile properties. *Bioresour. Technol.* **2010**, *101*, 387–395. [CrossRef]
34. Granda, L.A.; Espinach, F.; Méndez, J.A.; Vilaseca, F.; Delgado-Aguilar, M.; Mutjé, P. Semichemical fibres of Leucaena collinsii reinforced polypropylene: Flexural characterisation, impact behaviour and water uptake properties. *Compos. Part B Eng.* **2016**, *97*, 176–182. [CrossRef]

35. Bledzki, A.K.; Gassan, J. Composites reinforced with cellulose based fibers. *Prog. Polym. Sci.* **1999**, *24*, 221–274. [CrossRef]
36. Mohanty, S.; Verma, S.K.; Nayak, S.K. Dynamic mechanical and thermal properties of MAPE treated jute/HDPE composites. *Compos. Sci. Technol.* **2006**, *66*, 538–547. [CrossRef]
37. Oliver-Ortega, H.; Chamorro-Trenado, M.À.; Soler, J.; Mutjé, P.; Vilaseca, F.; Espinach, F.X. Macro and micromechanical preliminary assessment of the tensile strength of particulate rapeseed sawdust reinforced polypropylene copolymer biocomposites for its use as building material. *Constr. Build. Mater.* **2018**, *168*, 422–430. [CrossRef]
38. Mutje, P.; Vallejos, M.E.; Girones, J.; Vilaseca, F.; Lopez, A.; Lopez, J.P.; Mendez, J.A. Effect of maleated polypropylene as coupling agent for polypropylene composites reinforced with hemp strands. *J. Appl. Polym. Sci.* **2006**, *102*, 833–840. [CrossRef]
39. Granda, L.A.; Espinach, F.X.; Tarrés, Q.; Méndez, J.A.; Delgado-Aguilar, M.; Mutjé, P. Towards a good interphase between bleached kraft softwood fibers and poly (lactic) acid. *Compos. Part B Eng.* **2016**, *99*, 514–520. [CrossRef]
40. Korabel'nikov, Y.G.; Rashkovan, I.A. Strength and mechanism of freacture of composites randomly reinforced with short carbon fibres. *Fibre Chem.* **2006**, *38*, 142–146. [CrossRef]
41. Fukuda, H.; Chou, T.-W. A probabilistic theory of the strength of short-fibre composites with variable fibre length and orientation. *J. Mater. Sci.* **1982**, 1003–1011. [CrossRef]
42. Sanadi, A.R.; Young, R.A.; Clemons, C.; Rowell, R.M. Recycled Newspaper Fibers as Reinforcing Fillers in Thermoplastics: Part I-Analysis of Tensile and Impact Properties in Polypropylene. *J. Reinf. Plast. Compos.* **1994**, *13*, 54–67. [CrossRef]
43. Kelly, A.; Tyson, W. Tensile porperties of fibre-reinforced metals-copper/tungsten and copper/molybdenum. *J. Mech. Phys. Solids* **1965**, *13*, 329–338. [CrossRef]
44. Mittal, R.K.; Gupta, V.B.; Sharma, P. The effect of fibre ortientation on the interfacial shear stress in short fibre-reinforced polypropylene. *J. Mater. Sci.* **1987**, *22*, 1949–1955. [CrossRef]
45. Bowyer, W.H.; Bader, H.G. On the reinforcement of thermoplastics by imperfectly aligned discontinuous fibres. *J. Mater. Sci.* **1972**, *7*, 1315–1321. [CrossRef]
46. Pegoretti, A.; Della Volpe, C.; Detassis, M.; Migliaresi, C.; Wagner, H.D. Thermomechanical behaviour of interfacial region in carbon fibre/epoxy composites. *Compos. Part A Appl. Sci. Manuf.* **1996**, *27*, 1067–1074. [CrossRef]
47. Li, Y.; Pickering, K.L.; Farrell, R.L. Determination of interfacial shear strength of white rot fungi treated hemp fibre reinforced polypropylene. *Compos. Sci. Technol.* **2009**, *69*, 1165–1171. [CrossRef]
48. Rodriguez, M.; Rodriguez, A.; Bayer, J.; Vilaseca, F.; Girones, J.; Mutje, P. Determination of corn stalk fibers' strength through modeling of the mechanical properties of its composites. *BioResources* **2010**, *5*, 2535–2546.
49. Trigui, A.; Karkri, M.; Pena, L.; Boudaya, C.; Candau, Y.; Bouffi, S.; Vilaseca, F. Thermal and mechanical properties of maize fibres-high density polyethylene biocomposites. *J. Compos. Mater.* **2013**, *47*, 1387–1397. [CrossRef]
50. Peña, L.; González, I.; Bayer, R.J.; El Mansouri, N.E.; Vilaseca, F. Mechanical behavior of thermo-mechanical corn stalk fibers in high density polyethylene composites. *J. Biobased Mater. Bioenergy* **2012**, *6*, 463–469. [CrossRef]

© 2019 by the authors. Licensee MDPI, Basel, Switzerland. This article is an open access article distributed under the terms and conditions of the Creative Commons Attribution (CC BY) license (http://creativecommons.org/licenses/by/4.0/).

Article

Explorative Study on the Use of Curauá Reinforced Polypropylene Composites for the Automotive Industry

Marc Delgado-Aguilar [1], Quim Tarrés [1], María de Fátima V. Marques [2], Francesc X. Espinach [3], Fernando Julián [3], Pere Mutjé [1] and Fabiola Vilaseca [4,*]

1. LEPAMAP Group, Department of Chemical Engineering, University of Girona, 17003 Girona, Spain; m.delgado@udg.edu (M.D.-A.); joaquimagusti.tarres@udg.edu (Q.T.); pere.mutje@udg.edu (P.M.)
2. Instituto de Macromoléculas, Universidad Federal do Rio de Janeiro, Rio de Janeiro CEP 21941-598, Brasil; fmarques@ima.ufrj.br
3. Design, Development and Product Innovation, Dept. of Organization, Business, University of Girona, 17003 Girona, Spain; francisco.espinach@udg.edu (F.X.E.); fernando.julian@udg.edu (F.J.)
4. Department of Fibre and Polymer Technology, KTH Royal Institute of Technology, SE-10044 Stockholm, Sweden
* Correspondence: vilaseca@kth.se

Received: 21 November 2019; Accepted: 10 December 2019; Published: 12 December 2019

Abstract: The automotive industry is under a growing volume of regulations regarding environmental impact and component recycling. Nowadays, glass fiber-based composites are commodities in the automotive industry, but show limitations when recycled. Thus, attention is being devoted to alternative reinforcements like natural fibers. Curauá (Curacao, *Ananas erectifolius*) is reported in the literature as a promising source of natural fiber prone to be used as composite reinforcement. Nonetheless, one important challenge is to obtain properly dispersed materials, especially when the percentages of reinforcements are higher than 30 wt %. In this work, composite materials with curauá fiber contents ranging from 20 wt % to 50 wt % showed a linear positive evolution of its tensile strength and Young's modulus against reinforcement content. This is an indication of good reinforcement dispersion and of favorable stress transfer at the fiber-matrix interphase. A car door handle was used as a test case to assess the suitability of curauá-based composites to replace glass fiber-reinforced composites. The mechanical analysis and a preliminary lifecycle analysis are performed to prove such ability.

Keywords: automotive industry; natural fiber; composites; polypropylene; stiffness; curauá fibers

1. Introduction

The automotive industry is under a growing volume of regulations, related to safety and security, emissions, recyclability and other aspects [1]. These legislations can change noticeably from one country to the next, but are usually in line with the policies and objectives of the governments and the society. Sometimes, these legislations seem to be contradictory. As an example, increasing the security of the vehicles demands the incorporation of elements that can increase its weight. It is known that around 50% of the fuel consumption of an automobile depends upon its mass [2]. Furthermore, lifecycle analysis in the car industry revealed that the environmental impact of a vehicle mainly occurs in its use phase [2,3]. In order to overcome these apparent conflicts between security and fuel consumption, and in parallel with other solutions, the automotive industry has devoted great efforts in light-weighting its components [2,4].

The light-weighting of a vehicle can be achieved by design changes, modification of the construction processes or the use of light engineering materials [2,5]. In addition to the above-mentioned regulations, the automotive industry has also to fulfill new regulations related to the end of life of the vehicles. Such regulations, like the European Union (EU) End of Life Vehicle directive, restrict the amount of vehicle that can be landfilled at its disposal. This directive enforces recovering and reusing at least 95% of the vehicle [6,7].

The automotive industry has explored alternatives to steel that provide the required mechanical properties while being lighter. These materials include metals like aluminum or magnesium, polymers like polyolefin, or composite materials like glass fiber or natural fiber-reinforced polymers. Moreover, alternative materials must show technical, economic and environmental performances [2].

The use of natural fiber-reinforced polymers has caught the attention of automotive industry, due to certain reasons. On the one hand, natural fibers come from renewable resources, locally available and comparatively cheaper than mineral fibers [8–10]. On the other hand, natural fibers are lighter than glass fibers, and thus, natural fiber-reinforced materials tend to show higher specific properties [11]. Nowadays, the use of natural fiber-reinforced composites has increased [9]. These materials are used for globe boxes, door panels, seat coverings, seat surfaces, trunk panels, trunk floors, spare tire covers, insulation, headliners, or dashboards [9,12]. The use of natural fiber-reinforced composites has been limited to non-structural or semi-structural purposes, and usually as an alternative to glass fiber-reinforced materials [9,12,13]. The most common natural fibers used as reinforcement in the automotive industry are jute, flax, sisal, cotton, wood abaca and kenaf [9,13,14]. The studies on the use of curauá fibers as reinforcement in the automotive industry, to the best knowledge of the authors, are scarce [6].

Curauá is a Bromeliaceae (*Ananás erectifolius*) common to the Amazonas region. Curauá is one of the natural fibers with more potential to be used as composite reinforcement [6,15,16]. Actually, these fibers are being used in the textile and automotive industries [6,17]. Furthermore, curauà fiber is odorless, enabling its use for car interior parts [6].

The use of natural fibers as reinforcement in composites has been extensively reported in the literature [8–10,18–22]. Nonetheless, the use of lignocellulosic reinforcements limits the range of polymers to be used as a matrix, mainly because cellulose starts to degrade fast when exposed to temperatures beyond 200 °C [23].

The mechanical properties of a fibers-reinforced composite are mainly affected by the nature of its phases and its percentages, its compatibility, as the ability to create strong interphases between the reinforcement and the matrix, the morphology of the reinforcements and its dispersion and its main orientation against the loads. While all the above-mentioned aspects have importance, a composite will always fail due to its feeblest phase, usually the interphase [13,24,25]. Obtaining strong interphases in natural fiber-reinforced composites is hindered by the different natures of these phases. Matrices used to be hydrophobic, and natural fibers hydrophilic, hindering a correct wetting of the reinforcements and disabling the creation of chemical bonding [26,27]. Thus, a great effort has been devoted to overcome this problem. The methods that have showed better results were based on fiber treatments or the use of coupling agents [28–31]. These methods enable the correct wetting of the reinforcements and the creation of hydrogen bonds, obtaining strong interphases. It must be mentioned that the use of matrices functionalized with maleic anhydride have reported excellent results in the case of polyolefin [25,32,33].

In the case of curauá-reinforced composites, the matrices that are mainly used are polyolefin, like polypropylene and high-density polyethylene, in some cases biobased. Castro et al. produced biobased composites by reinforcing a biopolyethylene with curauá fibers [15]. These researchers used a liquid hydroxylated polybutadiene as a coupling agent. The produced composites showed decreasing flexural strengths with increasing reinforcement contents. This is usually a sign of a weak interphase. In another research, the authors used castor and canola oils as compatibilizers in curauá-reinforced bio-polyethylene composites [34]. Here the authors observed increases on the flexural strengths of the composites due to the presence of the coupling agents. Nonetheless, the flexural strength tends to decrease with increasing percentages of reinforcement.

The authors blame this on a poor dispersion of the reinforcements in the composite. Nacas et al. produced curauá-reinforced polypropylene composites [35]. These researchers used raw and fibrillated reinforcements and maleic anhydride as their coupling agent. The materials showed noticeable increases of its tensile strength when fibrillated reinforcements and coupling agents were used in the formulation of the composites. The study was limited to 20 wt % reinforcement contents. Mano et al. studied the effect of processing conditions on the mechanical properties of curauá-reinforced polypropylene and polyethylene composites [36]. These authors did not use any coupling agent. The research stated the impact of the morphology of the reinforcements on the mechanical properties of the composites. The reinforcements are exposed to attrition during composite preparation and experience noticeable length shortening. Thus, the processes that prevented the shortening of the reinforcements allowed the obtaining of composites with higher mechanical properties. Another research obtained uncoupled curauá-reinforced polypropylene composites [37]. These materials showed increases of their flexural strengths against reinforcement contents. The maximum content of curauá was limited to 20 wt %, possibly to avoid dispersion problems. The literature shows the importance of using a coupling agent to obtain strong interphases. There is also an unsolved bad dispersion of the fibers for percentages of curauá higher than 20 wt %.

In this work, composites based on a polypropylene matrix reinforced with curauá fibers were formulated and prepared. The composites added 20 wt %, 30 wt %, 40 wt % and 50 wt % of reinforcement to assess the effect of these percentages over the tensile strength, Young's modulus and strain at the break of the composites. In order to obtain strong interphases, the materials added a 6 wt % against the reinforcement content of the coupling agent. The results are compared with glass fibers-reinforced composites. The tensile strength of the composites shows a linear increase up to 50 wt % cuarauá contents.

A door car handle was modeled, and a finite element analysis was used to assess the potential use of the studied composites. The results are presented and compared with other potential materials, in terms of specific properties and component lightweighting.

Finally, a preliminary lifecycle analysis was performed to assess the environmental impact of the components depending on the materials they are made of.

The main objective of the research is showing the potential of curauá fibers as polyolefin reinforcement. An automotive component has been chosen due to the global impact of such industry and its innovative nature. The results show the advantages of this natural reinforcement in front of glass fiber, in terms of environmental impact, mechanical properties and the possibility of obtaining good fiber dispersions and strong interphases at high fiber contents.

2. Materials and Methods

2.1. Materials

A polypropylene (PP) homopolymer with trademark Isplen PP099 G2M by REPSOL YPF (Tarragona, Spain) was used as our matrix. This polymer has a 0.905 g/cm^3 density and a 55 g/10min melt flow index (230 °C, 2.16 kg). This melt flow index makes the polymer especially suitable for mold injection.

A polypropylene functionalized with maleic anhydride (MAPP), with trademark Epolene G3015 by Eastman Chemical Products (San Roque, Spain) was used as the coupling agent. The MAPP has a 15 mg KOH/g acid number as a 24800Da atomic mass.

Cuarauá fibers were obtained from the Instituto de Macromoléculas Professora Eloisa Mano Universidade Federal do Rio de Janeiro (Rio de Janeiro, Brazil). Glass fiber (GF) by Vetrotex (Chamberley, France) was provided by Maben, S.L. (Banyoles, Spain).

Other reactants used during fiber treatment were: Sodium hydroxide (NaOH) by Merck, KGaA (Darmstadt, Germany) and anthraquinone by Basf AG (Tarragona, Spain). The reactants were used as received, without any further purification.

2.2. Fiber Treatment

The fibers were provided as long strands, unable to be used for mold injection. Curauá strands were chopped in a knives mill and then sorted, to obtain fibers with a mean length of 2 mm. Next, the fibers were defibrated in a Sprout-Waldron equipment. The process took place at room temperature and under aqueous conditions, obtaining a curauá mechanical pulp (CF). The process rendered a 99% yield with respect to the raw strands. The CF showed a density of 1.44 g/cm^3.

2.3. Composite Preparation

The composites were formulated with 20 wt %, 30 wt %, 40 wt %, 50 wt % and 60 wt % CF contents. All the composites added a 6 wt % of MAPP, against CF content. The percentage of MAPP was based on previous works and had the objective of maximizing the tensile strength of the composites [26]. MAPP was added with the PP. Composites were prepared in a kinetic internal mixer by Gelimat®(Ramsey, USA). The process took place at 3000 rpm and lasted approximately 2 min, until a 210 °C discharge temperature was reached. Although compounding decreases the mean length of the reinforcements due to attrition, Gelimat®mixers have less impact on the mean length of the reinforcements than other equipment does, and obtains reinforcements with better aspect ratios [38].

GF-based composites were prepared in a Brabender®plastograph mixer (Duisburg, Germany). The equipment was operated at 20 rpm during 10 min and at 180 °C. Coupled (GFe) and uncoupled (GFs) composites were prepared. In the case of the coupled composites, the coupling agent was added with the PP.

Before its use, the materials were pelletized in a knife mill and stored at 80 °C during at least 24 h.

2.4. Specimen Obtention

Standard dog-bone specimens were mold injected. The equipment was a Meteor 40 injection mold machine by Mateu & Solé (Barcelona, Spain). At least 10 valid specimens for all composite formulations were obtained. The temperature profile was 175, 175 and 190 °C for the three heating areas. The last one corresponds with the injection nozzle. First and second pressures were 120 and 37.5 kgf/cm^2, respectively.

2.5. Tensile Testing

The specimens were stored in a Dycometal (Barcelona, Spain) climatic chamber at 23 °C and 50% relative humidity the 48 h before the tensile test, in agreement with ASTM D638 and ASTM D618 [39,40]. The specimens were tensile tested in an Instron 112 universal machine (Norwood, MA, USA). The machine is equipped with a 5 kN load cell and was operated at 2 mm/min. The results were the mean of at least five measurements.

2.6. Car Door Handle Modelling and Analysis

The original component was measured in a Mitutoyo Crysta Apex 544 coordinate measuring machine (Elgoibar, Spain). These measures were used to obtain a digital mockup with SolidWorks® by Dassault Systemes (Vélizy-Villacoublay, France).

The interface conditions were obtained from the literature [13]. Use conditions were based on a normal use of a car door handle. A finite element analysis was performed by using the Simulation 2017 x64 SP3.0 module of Solid Works. The preliminary life cycle analysis (LCA) was performed with the sustainability module of Solid Works.

3. Results and Discussion

3.1. Tensile Properties of Curauá Reinforced Polypropylene Composites

The composites were submitted to tensile test to obtain its tensile strength (σ_t^C), Young's modulus (E_t^C) and strain at break (ε_t^C). Table 1 shows the obtained values. In the table, V^F and ρ^C account for the reinforcement volume fraction and the composite density, respectively.

Table 1. Tensile properties of coupled curauá-reinforced polypropylene composites against reinforcement contents.

Sample	V^F	ρ^C (g/cm³)	σ_t^C (MPa)	E_t^C (GPa)	ε_t^C (%)
PP	0	0.905	27.6 ± 0.5	1.5 ± 0.1	9.3 [1] ± 0.2
PP + 20CF	0.136	0.977	36.2 ± 0.6	3.1 ± 0.1	2.8 ± 0.1
PP + 30CF	0.212	1.019	41.5 ± 0.7	4.1 ± 0.1	2.3 ± 0.1
PP + 40CF	0.295	1.063	47.5 ± 0.8	5.1 ± 0.2	2.1 ± 0.1
PP + 50CF	0.386	1.111	53.8 ± 1.2	6.2 ± 0.2	1.9 ± 0.1

[1] This is the strain at maximum strength.

The results show a sustained increase of the tensile strength of the composites against CF contents. The composites increased the tensile strength of the matrix 31%, 50%, 72%, 95% and 121% by adding 20 wt %, 30 wt %, 40 wt % and 50 wt % of CF, respectively. Moreover, Figure 1a shows that the evolution of the tensile strength is also linear against CF content. The literature indicates that such behavior is possible when a proper dispersion of the reinforcements inside the composite and a strong interphase between the fibers and the matrix is also present [41].

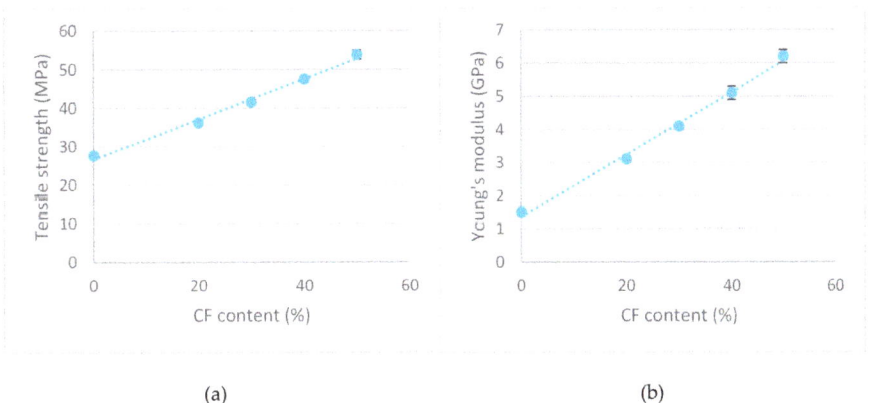

(a)   (b)

Figure 1. Evolution of the tensile properties of the composites against reinforcement contents: (a) Tensile strength; (b) Young's modulus.

In the case of the Young's modulus of the composites, a sustained and linear increase against CF content was also observed (Figure 1b). The Young's moduli of the composites increased noticeably with the CF contents. The composites increased the modulus of the matrix by 106%, 173%, 240%, 313% and 393% for the composites with 20 wt %, 30 wt %, 40 wt % and 50 wt % contents, respectively. It is known that the Young's modulus of a composite is little affected by the strength of the interphase, but strongly affected by the dispersion of the fibers [42]. Thus, the results back the hypothesis of a proper dispersion of the fibers, but the use of a technique that permits a direct measurement of the fiber dispersion is needed to fully back such a hypothesis. Notwithstanding, a micromechanical analysis is needed to assess the strength of the interphase between CF and PP.

The elongation at break of the composites decreased as a direct consequence of the higher percentages of reinforcement, a fragile phase. Thus, although the tensile strength increased, the composites were unable to sustain high elongations.

The tensile properties of the composites are similar to other strand-reinforced composites like hemp- or jute-reinforced PP [26,43]. The results were higher than those obtained by reinforcing the same matrix with wood fibers [38,42,44]. Nonetheless, the composites that are more used in the automotive industry are GF-based. Thus, in order to assess the possibilities of using CF-based composites in the automotive industry, its properties must be compared to GF-based materials. Table 2 shows the tensile properties of GF-reinforced PP composites. The table shows the results obtained for uncoupled (GFs) and coupled (GFe) composites [38,44,45].

Table 2. Tensile properties of uncoupled and coupled glass fiber-reinforced polypropylene composites against reinforcement contents.

Sample	V^F	ρ^C (g/cm^3)	σ_t^C (MPa)	E_t^C (GPa)	ε_t^C (%)
PP + 20GFs	0.084	1.036	50.9 ±4.3	4.6 ± 0.1	3.1 ± 0.1
PP + 30GFs	0.136	1.116	58.5 ± 4.3	5.9 ± 0.2	3.0 ± 0.2
PP + 20GFe	0.084	1.036	67.6 ± 0.9	4.5 ± 0.2	4.7 ± 0.2
PP + 30GFe	0.136	1.116	79.6 ± 1.2	6.0 ± 0.1	4.4 ± 0.2

GF-based composites were prepared with the same matrix and the same equipment than CF-based materials, in order to discard the effect of such parameters on the tensile properties of the materials. The content of GF was limited to a maximum of 30%, because on the one hand, this is the standard in the industry, yet on the other hand, GF is highly affected by attrition during compounding. These phenomena increase with the amount of reinforcement and they reduce the length of GF [46].

The differences between the coupled and uncoupled GF composites were noticeable. Both kinds of composites increased the tensile strength of the matrix, but while uncoupled composites increased the tensile strength of the matrix by an 85% and a 112% for 20 wt % and 30 wt % GF contents, respectively, the coupled composites increased 145% and 188% for the same GF contents. This shows the effect of a strong interphase on the tensile properties of a composite. In the case of the Young's modulus, there were few differences between the moduli of coupled and uncoupled GF composites at the same reinforcement contest, showing the above-mentioned limited effect of the interphase in the Young's modulus. In any case, GF-based composites showed tensile strengths superior to CF-based materials. Only the PP + 50CF was superior to the PP + 20GFs. In the case of the Young's modulus, CF showed a notable stiffening capability, and the composites. Composites with 40 wt % of CF showed a Young's modulus similar to composites with a 20 wt % of GF, and so on.

As commented upon in the introduction, one of the goals of the automotive industry is lightweighting. The densities of CF (Table 1) and GF (Table 2) composites show how at the same reinforcement contents, CF composites are lighter than GF, but only slowly. Figure 2 shows the specific tensile strength and Young's modulus of the composites.

GF-based composites showed higher specific properties at the same reinforcement contents. Figure 2 show how only a composite with a 50 wt % CF content can reach more similar specific tensile strengths than an uncoupled GF composite with 20 wt % contents. In the case of the Young's modulus, the, CF composites with 50 wt % contents showed values superior to any GF material.

Automotive components are usually designed to sustain reasonable deformation under use conditions. Thus, the stiffness of the composites can prove, to a limit, more important than its tensile strength when applied to a test case.

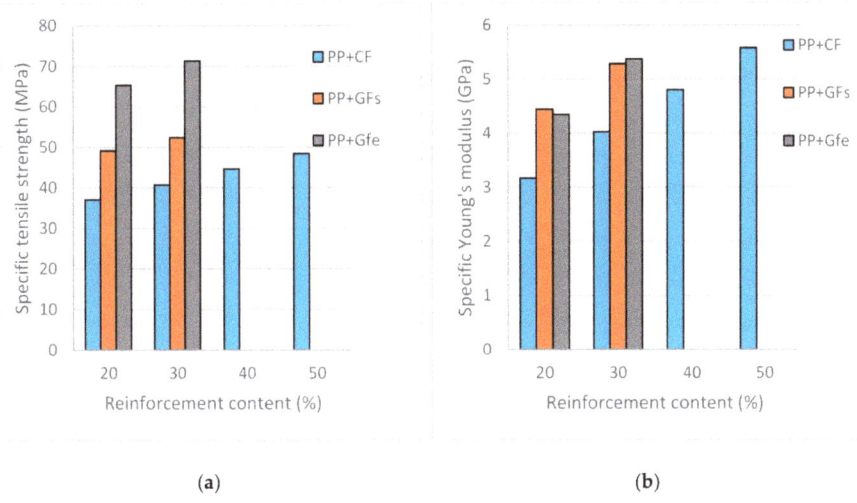

Figure 2. Specific tensile properties of curauá, and uncoupled and coupled glass fiber-reinforced polypropylene composites: (**a**) Specific tensile strength; (**b**) Specific Young's modulus.

3.2. Test Case

3.2.1. Car Interior Door Handle

The test case is a car interior door handle. This component was chosen because is a widely known mechanism, with few variations in its operating principle. Figure 3 shows different views of the modeled component.

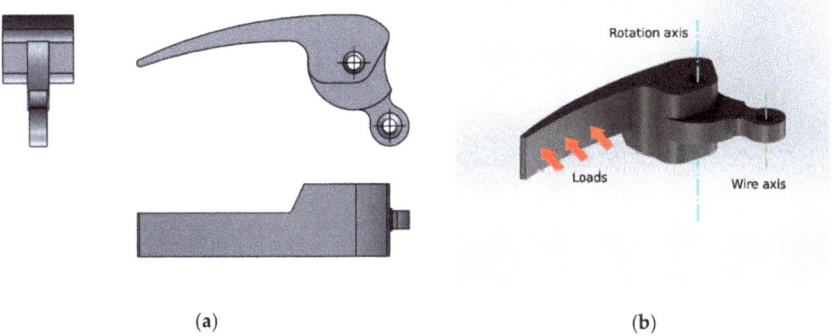

Figure 3. Digital mockup of the car interior door handle used as our test case: (**a**) Normalized views; (**b**) Perspective view.

A similar design was used by some of the authors in a recent study on the application of natural fiber-reinforced polyamide 11 composites [13]. The reference model was made of an uncoupled GF-reinforced PP composite, with a 20 wt % GF content. The mechanical properties of such material are shown in Table 2.

The operating principle of the mechanism is based on a class 1 lever. The fulcrum is the rotational axis, inserted in the handle and in the car door. The load is applied with the fingers in one side (Figure 3b). The resistance is placed in the wire axis, which conveys the loads to the opening mechanism. The wire is under tensile loads, ensuring the return of the mechanism to the designed neutral position.

The load needed to operate the mechanism was measured with a dynamometer on five car doors. The highest value was found to be 15 N. This value was multiplied by a 1.5 safety factor to stablish 20 N as a load, under which almost any car door handle can be operated under normal conditions. These loads are always exerted with one or two fingers and the thumbs. The literature shows that it is possible to exert up to 70 N loads with two fingers [47]. These two loads define the analyzed use conditions. The 20 N hypothesis defines a proper use of the mechanism, the 70 N a misuse. Nonetheless, both situations are possible, and must be taken into account in the following analysis.

3.2.2. Analysis of the Use of CF-Reinforced PP Composites.

The loads and restrictions used to simulate the mechanism were:

- Loads were applied in the interior face of the lever (Figure 3b)
- The hole around the rotation axis was limited to a rotation degree of freedom, hindering all the other movements.
- The wire axis was limited in all of its degrees of freedom to simulate the reaction forces when the lever is loaded.

The analysis was defined as static. Then, the solid was meshed with standard quad point elements. A mesh of 8235 elements, with a mean size of 2.3 ± 0.11 mm, was created before some refining operations. Almost all of the elements (98.4%) showed aspect ratios below 3.

Tables 1 and 2 show the mechanical properties used to perform the analyses. All the materials were applied to test the model under normal and misuse conditions. The results were collected in the shape of von Mises strengths (MPa), percentage displacements (%), net displacement (mm) and safety factors. The safety factor is defined as the ratio between the yield stress and the working stress. CF-based composites are not ductile materials, and yield and ultimate stresses are very similar.

Figure 4 shows the output provided by the analysis software for a component made of a CF-reinforced PP composite with a 50 wt % CF content.

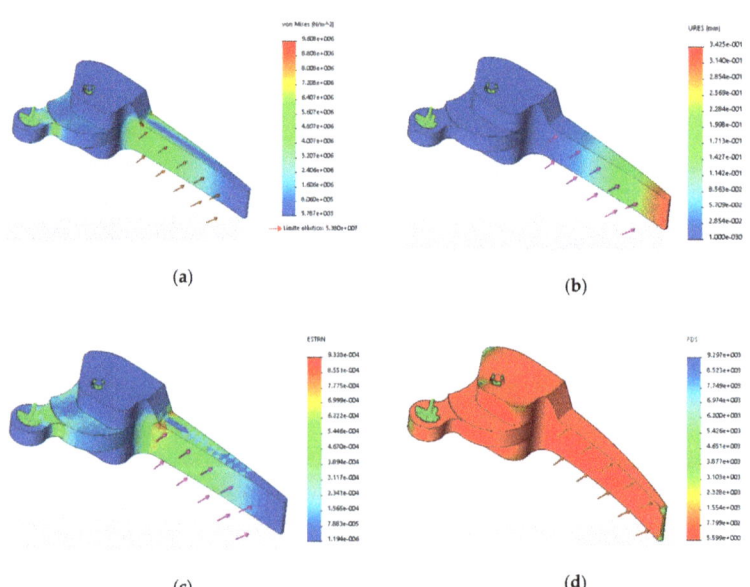

Figure 4. Graphical output obtained after assaying a component made of a 50 wt % CF-reinforced composite under 20 N loads: (**a**) Von Mises; (**b**) Net displacements; (**c**) Strain; (**d**) Safety factor.

The von Mises diagram obtained for all the studies was very similar, with only slight changes, as shows Table 3. The maximum von Mises stress was located at a zone were the area of the handle changes. This area coincides with an edge, a known stress raiser. It must be pointed out that the model submitted to analysis lacked some rounding operations that could mitigate such stress concentration but ease the meshing. Nonetheless, the maximum stresses were located at the expected areas. Thus, the analysis was considered accurate. The net displacements changed noticeably form one composite to the other, as expected due to the different Young's moduli of such materials. The strains evolved similar to the displacements, and were always inferior to the stresses at break of the materials; thus no collapse was previewed. Safety factor diagrams showed a very regular coloring, mostly on the low safety factor area. This shows that the design is balanced, and no waste of material was done [13,22].

Table 3 shows the results obtained for the models under normal use conditions (20 N).

Table 3. Main outputs of the analysis of the test case under 20 N loads.

Sample	Safety Factor	Net Displacement (mm)	Percentage Displacement (%)	Von Mises (MPa)	Mass (g)
PP	2.9	1.4	0.4	9.6	10.2
PP + 20GFs	5.3	0.4	0.1	9.6	11.7
PP + 30GFs	6.1	0.4	0.1	9.6	12.6
PP + 20GFe	7.0	0.5	0.1	9.6	11.7
PP + 30GFe	8.3	0.3	0.1	9.6	12.6
PP + 20CF	3.8	0.7	0.2	9.6	11.0
PP + 30CF	4.3	0.5	0.1	9.6	11.5
PP + 40CF	4.9	0.4	0.1	9.6	12.0
PP + 50CF	5.6	0.3	0.1	9.6	12.5

Under normal circumstances, the component must be far from its collapse or breaking point. This is shown by the obtained safety factors, all noticeably above 2. Nonetheless, normal use conditions must ensure that the deformation of the component does not compete with the developments of its function. The handle is submitted to a load that tends to deform the element. The maximum deformations are shown in Table 4 as net displacements. It was found that all the composite materials ensure deformations below 1mm, and pp matrix deformations around 1.4 mm. Any of these deformations endangers the handle function deployment, but affects the perceived quality of such component. Thus, having in account that the original component was made of PP + 20GFs, the net displacement of such composite was used as a reference. In order to avoid placing a too strict limit, a ±0.1 tolerance was proposed. Thus, all the materials that ensure net displacements lower that 0.5 mm were accepted as suitable as replacement materials.

Thus, all the GF-based materials can be used, and also, the CF-based composites, where CF contents from 30 wt % and above can be also candidates.

Table 4 shows the values obtained after submitting the model to misuse conditions (70 N).

Table 4. Main outputs of the analysis of the test case under 70 N loads.

Sample	Safety Factor	Net Displacement (mm)	Percentage Displacement (%)	Von Mises (MPa)
PP	0.8	5.0	1.3	33.7
PP + 20GFs	1.5	1.6	0.4	33.7
PP + 30GFs	1.7	1.3	0.3	33.7
PP + 20GFe	2.0	1.7	0.5	33.7
PP + 30GFe	2.4	1.2	0.3	33.7
PP + 20CF	1.1	2.4	0.7	33.7
PP + 30CF	1.2	1.8	0.5	33.7
PP + 40CF	1.4	1.4	0.4	33.7
PP + 50CF	1.6	1.2	0.3	33.7

The values increased in consonance with the increased load. All neat deformations were above 1 mm, and the safety factors decreased noticeably. In fact, a car handle made of PP was unable to endure the loads, and the analysis previews a breakage of the element. The rest of these materials returned a safety factor above 1. Alike the normal use conditions case, in the misuse hypothesis, the value obtained for the PP + 20GFs composite was used as a reference. In this case, a component was ruled suitable if it showed safety factors around 1.5. A similar tolerance to the applied for normal use conditions was applied to the safety factor, allowing all the components with safety factors above 1.4. Thus, all the GF-based composites fulfill the condition, and CF-based materials with 40 wt % or 50 wt % reinforcement contest returned favorable values.

Two criteria were defined: On the one hand, the components under normal use conditions must show maximum displacements below 0.5 mm. On the other hand, the handles under misuse conditions must return safety factors above 1.5. Figure 5 shows the materials that fulfill both condition.

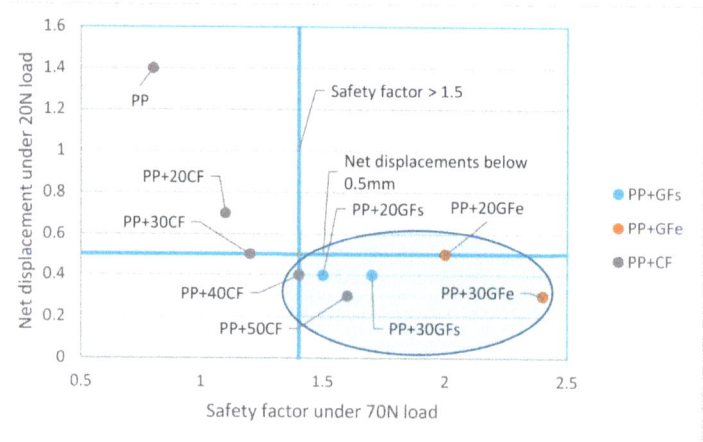

Figure 5. Combined requirements for the normal use and misuse conditions. The ellipse encircles the suitable materials that fulfill both requirements.

The figure shows how all the GF-based composites (uncoupled and coupled) fulfill the conditions necessary to be used for a car door handle. On the other hand, only the composites with 40 wt % and 50 wt % contents were able to substitute the original GF-based material.

The mentioned criteria were based on the mechanical properties of the materials. Nonetheless, other criteria, like lightweighting and environmental impact, must be taken into account.

Table 3 shows the mass of the components. The densities of GF and CF are 2.46 and 1.44 g/cm^3, respectively. These densities are higher than the matrix (0.905 g/cm^3); thus, adding reinforcement contents increased the density of the composites and the mass of the handles. The reference model shows a mass of 11.7 g. The handle with a 30 wt % of GF shows a 7.7% weight increase. The materials adding a 40 and 50 wt % of CF increased the weight of the reference models by 2.6% and 7.2%. Thus, replacing the GF by CF increased the weight of the components and disagreed with the lightweighting criteria. Notwithstanding, the differences are slight, and the environmental impact criteria must be considered.

The environmental impact analysis was performed under the following conditions:

- Manufacturing process, injection molding
- The elements are manufactured in Europe, to be consumed in Europe
- The lifespan is of 15 years
- At the end of life only 5% of the total is dumped

Table 5 shows the results of the preliminary LCA analysis. The analysis was performed only for the composites suitable to substitute the original PP + 20GFs composite. The analysis does not distinguish between coupled and uncoupled GF-based materials because the database lacks information on the environmental impact of the coupling agents. The environmental impacts of a fully PP component were added as control.

Table 5. LCA analysis of a door car handle made with the considered materials.

Sample	Carbon Footprint (kg CO_2)	Energy Consumption (MJ)	Atmospheric Acidification (kg SO_2)	Eutrophication (kg PO_4)
PP	0.048	1.10	1.40×10^{-4}	1.10×10^{-5}
PP + 20GF	0.079	1.34	2.86×10^{-4}	2.34×10^{-5}
PP + 30GF	0.095	1.46	3.59×10^{-4}	2.96×10^{-5}
PP + 40CF	0.032	0.67	8.70×10^{-4}	7.40×10^{-6}
PP + 50CF	0.028	0.56	7.37×10^{-4}	6.50×10^{-6}

The results show how the impact increases noticeably when the percentage of GF is increased, but also shows how the same impact decreases fast when the percentage of CF increases. In order to compare the environmental impacts, Figure 6 shows the percentage increases and decreases against PP.

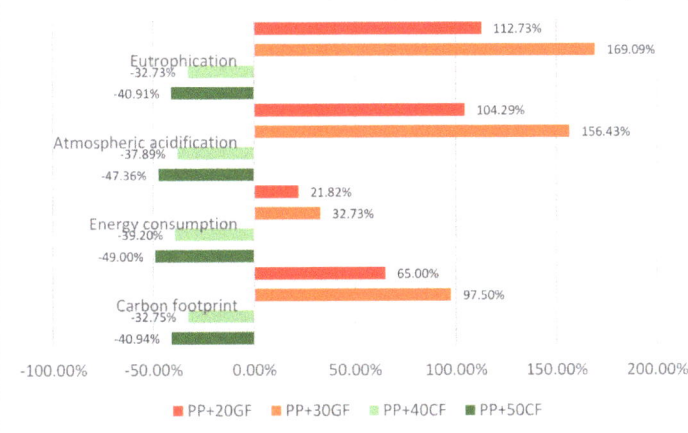

Figure 6. Percentage differences between the environmental impact of a polypropylene (PP) component and curauá mechanical pulp (CF)- and glass fiber (GF)-based composites.

The figure clearly shows the impact of adding glass fibers to a composite. Adding 20 wt % and 30 wt % of GF increases 65% and 97% the carbon footprint of a PP-based component. On the other hand, including curauá fibers decreases the carbon footprint. This behavior is similar for all the other considered impacts. Thus, though CF-based composites do not lightweight the door handle, they do contribute noticeably to decrease its environmental impact. As cited in the introduction, alternative materials must show technical, economic and environmental performances [2]. Curauá-reinforced PP composites showed technical performance equal to GF-reinforced PP composites, and better environmental behavior. The economic performance is out of the scope of the article, and further research is needed to compare the costs between CF- and GF-reinforced materials. Nonetheless, if these natural fibers-based composites are adopted by the automotive industry, surely industrial partners that will produce such composite pellets will provide such materials. Although the mixing equipment used to produce the composites is an industrial scale equipment, other processes like twin extruders can be used, and this can affect the costs, the properties of the materials (due to the morphology of the fibers) and the environmental impact.

4. Conclusions

Coupled Curauá fiber-reinforced polypropylene materials were formulated, mixed and textile tested. The linear evolution of the tensile strength and the Young's modulus of the composites against reinforcement content indicated a good dispersion of the reinforcements and the presence of chemical interactions in the interphase. The composites added 20 wt % to 50 wt % reinforcement contents.

The tensile properties of the curauà-based composites were similar to uncoupled glass fiber-based composites with 20 wt % less reinforcement contents.

A car interior door handle was proposed as the test case to assess the possibility to change form a glass fiber-reinforced polypropylene composite to a curauá fiber-reinforced material. Normal use and misuse conditions were defined, and a finite element analysis of the test case was performed. The results showed that it was possible to replace an uncoupled 20 wt % glass fiber-reinforced polypropylene by a 40 wt % or 50 wt % curauà-reinforced material.

Curauá-based materials able to replace glass fiber showed higher densities. Thus, it was not possible to lightweight the component by changing from GF to CF.

The environmental impact of CF-based composites was noticeably lower than GF-based materials. From an environmental point of view, changing from mineral reinforcements to curauá fibers makes sense. Nonetheless, a more accurate life cycle analysis is needed to explore the sensitivity of the environmental impact to all the possible variables.

A micromechanics analysis of the interphase is needed to assess the strength of such interphases. From this information it is possible to identify whether or not it is also possible to further increase the mechanical properties of curauá fibers-reinforced materials.

More research is needed to evaluate the effects of water absorption on the mechanical properties of the composites.

Author Contributions: M.D.-A. and F.X.E. conceptualized the study. Q.T. and M.d.F.V.M. performed the experimental tests and validated the experimental results. F.J. designed the test case and stablished the analysis conditions. F.X.E. performed the analysis and wrote the paper. P.M. and F.V. supervised the research and corrected the manuscript.

Funding: This research was funded by the Chair on Sustainable Industrial Processes of the University of Girona, ref 01/2019.

Acknowledgments: The authors thank REPSOL YPS for kindly provide the polypropylene matrix.

Conflicts of Interest: The authors declare no conflict of interest.

References

1. Hottle, T.; Caffrey, C.; McDonald, J.; Dodder, R. Critical factors affecting life cycle assessments of material choice for vehicle mass reduction. *Transp. Res. Part. D Transp. Environ.* **2017**, *56*, 241. [CrossRef] [PubMed]
2. Lewis, G.M.; Buchanan, C.A.; Jhaveri, K.D.; Sullivan, J.L.; Kelly, J.C.; Das, S.; Taub, A.I.; Keoleian, G.A. Green Principles for Vehicle Lightweighting. *Environ. Sci. Technol.* **2019**, *53*, 4063–4077. [CrossRef] [PubMed]
3. Kojima, K.; Ryan, L. IEA Enger Papers. *Transp. Energy Effic.* **2010**. [CrossRef]
4. Puri, P.; Compston, P.; Pantano, V. Life cycle assessment of Australian automotive door skins. *Int. J. Life Cycle Assess.* **2009**, *14*, 420–428. [CrossRef]
5. Kollamthodi, S.; Kay, D.; Skinner, I.; Dun, C.; Hausberger, S. The potential for mass reduction of passenger cars and light commercial vehicles in relation to future CO2 regulatory requirements. *Rep. Eur. Comm. -DG Clim. Action (CLIMA. C. 2.2/FRA/2013/0006/SR1)* **2015**.
6. Zah, R.; Hischier, R.; Leao, A.L.; Braun, I. Curaua fibers in the automobile industry–A sustainability assessment. *J. Clean. Prod.* **2007**, *15*, 1032–1040. [CrossRef]
7. Directive, E. 53/EC of the European Parliament and of the Council of 18 September 2000 on end-of life vehicles. *Off. J. Eur. Union L Ser.* **2000**, *21*, 34–42.
8. Fazita, M.R.N.; Jayaraman, K.; Bhattacharyya, D.; Haafiz, M.K.M.; Saurabh, C.K.; Hussin, M.H.; Khalil, A.H.P.S. Green Composites Made of Bamboo Fabric and Poly (Lactic) Acid for Packaging Applications-A Review. *Materials* **2016**, *9*, 435. [CrossRef]

9. Kumar, R.; Ul Haq, M.I.; Raina, A.; Anand, A. Industrial applications of natural fibre-reinforced polymer composites - challenges and opportunities. *Int. J. Sustain. Eng.* **2019**, *12*, 212–220. [CrossRef]
10. Yan, L.; Kasal, B.; Huang, L. A review of recent research on the use of cellulosic fibres, their fibre fabric reinforced cementitious, geo-polymer and polymer composites in civil engineering. *Compos. Part. B Eng.* **2016**, *92*, 94–132. [CrossRef]
11. Reixach, R.; Espinach, F.X.; Arbat, G.; Julián, F.; Delgado-Aguilar, M.; Puig, J.; Mutjé, P. Tensile Properties of Polypropylene Composites Reinforced with Mechanical, Thermomechanical, and Chemi-Thermomechanical Pulps from Orange Pruning. *BioResources* **2015**, *10*, 4544–4556. [CrossRef]
12. Pandey, J.K.; Ahn, S.; Lee, C.S.; Mohanty, A.K.; Misra, M. Recent advances in the application of natural fiber based composites. *Macromol. Mater. Eng.* **2010**, *295*, 975–989. [CrossRef]
13. Oliver-Ortega, H.; Julian, F.; Espinach, F.X.; Tarrés, Q.; Ardanuy, M.; Mutjé, P. Research on the use of lignocellulosic fibers reinforced bio-polyamide 11 with composites for automotive parts: Car door handle case study. *J. Clean. Prod.* **2019**, *226*, 64–73. [CrossRef]
14. Kumar, N.; Das, D. Fibrous biocomposites from nettle (Girardinia diversifolia) and poly(lactic acid) fibers for automotive dashboard panel application. *Compos. Part. B Eng.* **2017**, *130*, 54–63. [CrossRef]
15. Castro, D.; Ruvolo-Filho, A.; Frollini, E. Materials prepared from biopolyethylene and curaua fibers: Composites from biomass. *Polym. Test.* **2012**, *31*, 880–888. [CrossRef]
16. Monteiro, S.N.; Lopes, F.P.D.; Barbosa, A.P.; Bevitori, A.B.; Da Silva, I.L.A.; Da Costa, L.L. Natural Lignocellulosic Fibers as Engineering Materials-An Overview. *Metall. Mater. Trans. A-Phys. Metall. Mater. Sci.* **2011**, *42a*, 2963–2974. [CrossRef]
17. Pimenta, P.; Borges, L.M.; Oliveira, F.R.; Silva, S.; Souto, A.P. Pimenta 2016 New. *New Text. Fibre: Curaua* **2016**, 50–57.
18. Kian, L.K.; Saba, N.; Jawaid, M.; Sultan, M.T.H. A review on processing techniques of bast fibers nanocellulose and its polylactic acid (PLA) nanocomposites. *Int. J. Biol. Macromol.* **2019**, *121*, 1314–1328. [CrossRef]
19. Singh, N.; Hui, D.; Singh, R.; Ahuja, I.P.S.; Feo, L.; Fraternali, F. Recycling of plastic solid waste: A state of art review and future applications. *Compos. Part. B Eng.* **2017**, *115*, 409–422. [CrossRef]
20. Pickering, K.L.; Efendy, M.A.; Le, T.M. A review of recent developments in natural fibre composites and their mechanical performance. *Compos. Part. A Appl. Sci. Manuf.* **2016**, *83*, 98–112. [CrossRef]
21. Dunne, R.; Desai, D.; Sadiku, R.; Jayaramudu, J. A review of natural fibres, their sustainability and automotive applications. *J. Reinf. Plast. Compos.* **2016**, *35*, 1041–1050. [CrossRef]
22. Serrano, A.; Espinach, F.X.; Tresserras, J.; Pellicer, N.; Alcala, M.; Mutje, P. Study on the technical feasibility of replacing glass fibers by old newspaper recycled fibers as polypropylene reinforcement. *J. Clean. Prod.* **2014**, *65*, 489–496. [CrossRef]
23. Sena Neto, A.R.; Araujo, M.A.M.; Souza, F.V.D.; Mattoso, L.H.C.; Marconcini, J.M. Characterization and comparative evaluation of thermal, structural, chemical, mechanical and morphological properties of six pineapple leaf fiber varieties for use in composites. *Ind. Crop. Prod.* **2013**, *43*, 529–537. [CrossRef]
24. Tarrés, Q.; Vilaseca, F.; Herrera-Franco, P.J.; Espinach, F.X.; Delgado-Aguilar, M.; Mutjé, P. Interface and micromechanical characterization of tensile strength of bio-based composites from polypropylene and henequen strands. *Ind. Crop. Prod.* **2019**, *132*, 319–326. [CrossRef]
25. Salem, S.; Oliver-Ortega, H.; Espinach, F.X.; Hamed, K.B.; Nasri, N.; Alcalà, M.; Mutjé, P. Study on the Tensile Strength and Micromechanical Analysis of Alfa Fibers Reinforced High Density Polyethylene Composites. *Fibers Polym.* **2019**, *20*, 602–610. [CrossRef]
26. Vilaseca, F.; Del Rey, R.; Serrat, R.; Alba, J.; Mutje, P.; Espinach, F.X. Macro and micro-mechanics behavior of stifness in alkaline treated hemp core fibres polypropylene-based composites. *Compos. Pt. B-Eng.* **2018**, *144*, 118–125. [CrossRef]
27. Serra, A.; Tarrés, Q.; Llop, M.; Reixach, R.; Mutjé, P.; Espinach, F.X. Recycling dyed cotton textile byproduct fibers as polypropylene reinforcement. *Text. Res. J.* **2019**, *89*, 2113–2125. [CrossRef]
28. Suardana, N.; Piao, Y.; Lim, J. Mechanical Properties of Hemp Fibers and Hemp/PP composites: Effects of chemical surface treatment. *Mater. Phys. Mech.* **2011**, *12*, 113–125.
29. Saha, P.; Manna, S.; Chowdhury, S.R.; Sen, R.; Roy, D.; Adhikari, B. Enhancement of tensile strength of lignocellulosic jute fibers by alkali-steam treatment. *Bioresour. Technol.* **2010**, *101*, 3182–3187. [CrossRef]

30. Franco-Marques, E.; Mendez, J.A.; Pelach, M.A.; Vilaseca, F.; Bayer, J.; Mutje, P. Influence of coupling agents in the preparation of polypropylene composites reinforced with recycled fibers. *Chem. Eng. J.* **2011**, *166*, 1170–1178. [CrossRef]
31. Pickering, K.L.; Beckermann, G.W.; Alam, S.N.; Foreman, N.J. Optimising industrial hemp fibre for composites. *Compos. Part. a-Appl. Sci. Manuf.* **2007**, *38*, 461–468. [CrossRef]
32. Oliver-Ortega, H.; Chamorro-Trenado, M.À.; Soler, J.; Mutjé, P.; Vilaseca, F.; Espinach, F.X. Macro and micromechanical preliminary assessment of the tensile strength of particulate rapeseed sawdust reinforced polypropylene copolymer biocomposites for its use as building material. *Constr. Build. Mater.* **2018**, *168*, 422–430. [CrossRef]
33. Espinach, F.X.; Granda, L.A.; Tarrés, Q.; Duran, J.; Fullana-i-Palmer, P.; Mutjé, P. Mechanical and micromechanical tensile strength of eucalyptus bleached fibers reinforced polyoxymethylene composites. *Compos. Part. B Eng.* **2017**, *116*, 333–339. [CrossRef]
34. Castro, D.; Passador, F.; Ruvolo-Filho, A.; Frollini, E. Use of castor and canola oils in "biopolyethylene" curauá fiber composites. *Compos. Part. A Appl. Sci. Manuf.* **2017**, *95*, 22–30. [CrossRef]
35. Nacas, A.M.; Silva, R.L.; De Paoli, M.A.; Spinacé, M.A. Polypropylene composite reinforced with fibrillated curaua fiber and using maleic anhydride as coupling agent. *J. Appl. Polym. Sci.* **2017**, *134*. [CrossRef]
36. Mano, B.; Araújo, J.; Spinacé, M.; De Paoli, M.-A. Polyolefin composites with curaua fibres: Effect of the processing conditions on mechanical properties, morphology and fibres dimensions. *Compos. Sci. Technol.* **2010**, *70*, 29–35. [CrossRef]
37. Bispo, S.J.L.; Freire Júnior, R.C.S.; Aquino, E.M.F.D. Mechanical properties analysis of polypropylene biocomposites reinforced with curaua fiber. *Mater. Res.* **2015**, *18*, 833–837. [CrossRef]
38. Lopez, J.P.; Mendez, J.A.; Espinach, F.X.; Julian, F.; Mutje, P.; Vilaseca, F. Tensile Strength characteristics of Polypropylene composites reinforced with Stone Groundwood fibers from Softwood. *BioResources* **2012**, *7*, 3188–3200. [CrossRef]
39. International, A. *Standard Test Method for Tensile Properties of Plastics*; ASTM International: West Conshohocken, PA, USA, 2010; p. D638-10.
40. International, A. *Standard Practice for Conditioning Plastics for Testing*; ASTM International: West Conshohocken, PA, USA, 2013; p. D618-13.
41. Granda, L.A.; Espinach, F.X.; Lopez, F.; Garcia, J.C.; Delgado-Aguilar, M.; Mutje, P. Semichemical fibres of Leucaena collinsii reinforced polypropylene: Macromechanical and micromechanical analysis. *Compos. Pt. B-Eng.* **2016**, *91*, 384–391. [CrossRef]
42. Granda, L.A.; Espinach, F.X.; Mendez, J.A.; Tresserras, J.; Delgado-Aguilar, M.; Mutje, P. Semichemical fibres of Leucaena collinsii reinforced polypropylene composites: Young's modulus analysis and fibre diameter effect on the stiffness. *Compos. Pt. B-Eng.* **2016**, *92*, 332–337. [CrossRef]
43. Serra, A.; Tarrés, Q.; Claramunt, J.; Mutjé, P.; Ardanuy, M.; Espinach, F. Behavior of the interphase of dyed cotton residue flocks reinforced polypropylene composites. *Compos. Part. B: Eng.* **2017**, *128*, 200–207. [CrossRef]
44. Lopez, J.P.; Mendez, J.A.; El Mansouri, N.E.; Mutje, P.; Vilaseca, F. Mean intrinsic tensile properties of stone groundwood fibers from softwood. *BioResources* **2011**, *6*, 5037–5049. [CrossRef]
45. Lopez, J.P.; Mutje, P.; Pelach, M.A.; El Mansouri, N.E.; Boufi, S.; Vilaseca, F. Analysis of the tensile modulus of PP composites reinforced with Stone grounwood fibers from softwood. *BioResources* **2012**, *7*, 1310–1323. [CrossRef]
46. Patel, H.K.; Ren, G.; Hogg, P.J.; Peijs, T. Hemp fibre as alternative to glass fibre in sheet moulding compound Part 1-influence of fibre content and surface treatment on mechanical properties. *Plast. Rubber Compos.* **2010**, *39*, 268–276. [CrossRef]
47. DTI. *Strength Data for Design Safety*; D.o.t.a., Ed.; Industry: London, UK, 2002.

© 2019 by the authors. Licensee MDPI, Basel, Switzerland. This article is an open access article distributed under the terms and conditions of the Creative Commons Attribution (CC BY) license (http://creativecommons.org/licenses/by/4.0/).

Article

Radiation Synthesis of Pentaethylene Hexamine Functionalized Cotton Linter for Effective Removal of Phosphate: Batch and Dynamic Flow Mode Studies

Jifu Du [1,†], Zhen Dong [2,†], Zhiyuan Lin [1], Xin Yang [1] and Long Zhao [2,*]

1. School of Nuclear Technology and Chemistry & Biology, Hubei University of Science and Technology, Xianning 437100, China; duzidedu@163.com (J.D.); zhiyuanlin12318@126.com (Z.L.); sophieyangyifan@163.com (X.Y.)
2. State Key Laboratory of Advanced Electromagnetic Engineering and Technology, School of Electrical and Electronic Engineering, Huazhong University of Science and Technology, Wuhan 430074, China; zhendong@hust.edu.cn
* Correspondence: ryuuchou@hotmail.com or zhaolong@hust.edu.cn; Tel./Fax: +86-15021474065
† These authors contributed equally to this work.

Received: 18 September 2019; Accepted: 15 October 2019; Published: 17 October 2019

Abstract: A quaternized cotton linter fiber (QCLF) based adsorbent for removal of phosphate was prepared by grafting glycidyl methacrylate onto cotton linter and subsequent ring-opening reaction of epoxy groups and further quaternization. The adsorption behavior of the QCLF for phosphate was evaluated in a batch and column experiment. The batch experiment demonstrated that the adsorption process followed pseudo-second-order kinetics with an R^2 value of 0.9967, and the Langmuir model with R^2 value of 0.9952. The theoretical maximum adsorption capacity reached 152.44 mg/g. The experimental data of the fixed-bed column were well fitted with the Thomas and Yoon–Nelson models, and the adsorption capacity of phosphate at 100 mg/L and flow rate 1 mL/min reached 141.58 mg/g. The saturated QCLF could be regenerated by eluting with 1 M HCl.

Keywords: radiation grafting; cotton linter; phosphate adsorption; dynamic studies

1. Introduction

Eutrophication means the enrichment of water in nutrients by nitrogen and phosphorus compounds, which cause an accelerated growth of algae and superior forms of vegetable life, thus leading to degradation of the aquatic ecosystem [1]. In most cases, the concentration of phosphorus is the key factor in eutrophication control. To protect eutrophication from phosphorus, regulations and guidelines of many countries have set limitations on phosphorus concentrations in discharging waters. For example, the US has recommended that the average phosphorus concentration should not surpass 0.05 mg/L in streams discharging into lakes or reservoirs [2]. Therefore, it is very important to remove low-concentration phosphorus from waters. Many physical, chemical and biological technologies were investigated for phosphate removal. Adsorption is considered an efficient technique for removing low-concentration contamination from wastewater. The advantages of adsorption are simple in design and operation, cheap to implement and effective at low concentrations [3,4].

Inorganic adsorbents (activated carbon, metal oxides, silicates, Ca and Mg carbonates), organic adsorbents (anion exchange resins) and industrial by-products (red mud, slags, fly ash) were used for phosphate removal [2]. However, these particle-based adsorbents have several disadvantages such as high cost, fragile, post-usage disposal difficulty and lower adsorption capacity [5]. The fiber-shaped adsorbents can overcome the brittleness weakness due to the radial expansion at swollen state, and have larger surface area and huge interspaces than particle adsorbents, which can improve the adsorption performance [6,7]. In recent years, the natural fibers were recognized as an efficient, cost-effective and environmentally friendly adsorbents for contamination removal purposes. Many kinds of native cellulose fiber such as wool fiber [8], cotton linter [9], protein fibers and jute fibers [10] and kapok fiber [11] have been modified by various functional groups to remove oil [12], heavy metal ions [10,13], dyes [14], Au(III) [15], fluoride and arsenic [16–18] and humic acid [9] from aqueous solution. Cotton linter, the relatively short fuzz left on cotton seed after the cotton ginning process, has a high cellulose content which makes it very suitable to be an adsorbent due to its biodegradability, biocompatibility and non-toxicity. However, few reports have been studied on the removal of phosphate by modified cotton linter.

Amine groups were mostly used for anion ions adsorption through electrostatic interaction [15,17]. Pentaethylene hexamine was selected as the monomer for cotton linter modification due to its abundant amino groups and excellent thermal stability. Radiation-induced graft polymerization (RIGP) has been widely used to modify various fibers with the aim to introduce various functional groups onto the cellulose substrate [19–21]. The adsorbents prepared by RIGP had a high adsorption velocity because the functional groups were mainly concentrated on the surface of the substrates. In this paper, a quaternary ammonium group functionalized cotton linter was prepared by RIGP and its adsorption performance to phosphate was investigated. The newly cotton linter based adsorbent prepared by RIGP are expected to have good application prospects in phosphate adsorption.

2. Experimental

2.1. Materials

Cotton linter was supplied by Jinhanjiang refined cotton Co., Ltd., (Jingmen, China). NaH_2PO_4, Pentaethylene hexamine (PEHA) was bought from Aladdin Chemical Co., Ltd. (Shanghai, China). HCl, NaOH, N, N-Dimethylformamide (DMF) and 1-Bromohexane were purchased from Macklin reagent Co., Ltd. (Shanghai, China).

2.2. Preparation of Quaternized Cotton Linter Fiber (QCLF)

The QCLF was prepared by electron beam (EB) pre-irradiation grafting technology, and the procedure is illustrated in Figure 1.

The mixture composed of 30 mL GMA, 3 mL Tween 20 and 67 mL deionized water was nitrogen flowed to get rid of oxygen. Dry cotton linter fibers (2 g) were sealed in PE bags and vacuum pumped, then the cotton linter fiber were irradiated at the dose from 10 to 50 kGy (dose rate: 10 kGy/pass) with energy of 1 MeV by an EB accelerator manufactured by Wasik Associates INC, MA, USA. After irradiation, the sample was immersed into the GMA emulsion solution for grafting reaction at 50 °C for 3 h. Then the grafted cotton linter fiber were washed and dried.

The degree of grafting (DOG) was determined by Equation (1):

$$DOG = (\frac{W_2 - W_1}{W_1}) \times 100\% \tag{1}$$

where W_1 and W_2 were the weights of cotton linters before and after grafting, respectively.

CLF-g-GMA samples (2.0 g) with DOG 255% were immersed into 20 mL 50% PEHA with DMF solution for ring-opening reaction with the reaction condition of 80 °C and 24 h. Then, the samples were immersed into 50% 1-bromohexane with DMF solution for quaternization with the reaction condition of 70 °C and 24 h. Finally the quarterized cotton linter were washed and dried, thus QCLF was obtained.

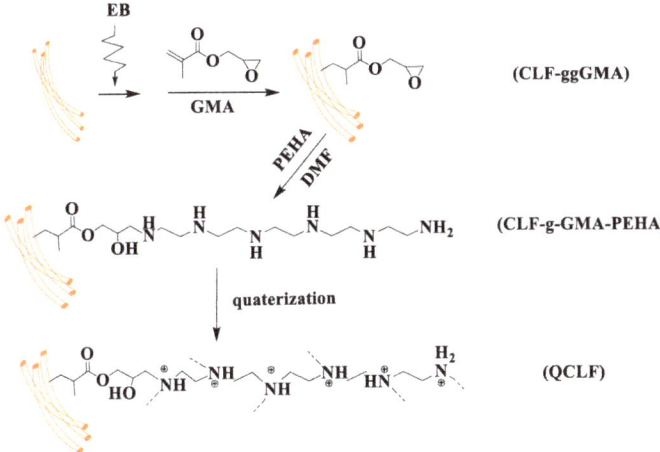

Figure 1. Synthesis route of quaternized cotton linter fiber (QCLF) adsorbent.

2.3. Characterization

The FTIR analysis was performed in the transmittance mode on Nicolet 6700 spectrophotometer (Thermo, Waltham, MA, USA). Surface morphologies were observed by Tescan Vega3, Brno, Czech at the accelerate voltage 10 kV. Thermogravimetric Analysis (TGA) was performed by TG290F3 of Netzsch (Selb, Germany) with the temperature from room temperature to 800 °C.

2.4. Batch Adsorption Experiments

QCLF (0.05 g) were added into 50 mL phosphate solution shaken at 25 °C. The pH was adjusted with 0.1 mol/L HCl or 0.1 mol/L NaOH. In the adsorption kinetics study, the concentration of phosphate was 20 mg/L and the adsorption time was 2, 5, 10, 20, 30, 45, 60, 90 and 120 min. In the adsorption isotherm experiment, the concentration of phosphate was ranging from 100 to 500 mg/L with the variation intervals of 50 mg/L.

The adsorbed amount (Q_t) of total phosphate onto QCLF was determined using Equation (2):

$$Q_t = \frac{(C_0 - C_t) \times V}{m} \qquad (2)$$

where C_0 and C_t were the phosphate concentration before and after adsorption at certain time, V was the volume of the phosphate solution and m was the mass of QCLF.

The phosphate concentration (PO_4^-, HPO_4^- and $H_2PO_4^-$) were determined using Ion chromatograph (MagIC Net 883, Metrohm, Switzerland).

2.5. Column Experiments

The column experiments were conducted with a 3 mL organic glass at a bed height of 50 mm and the inner diameter 9 mm which was fitted with QCLF (1.45 g). The inlet concentrations of phosphate passed through the column were 50 and 100 mg/L and the space velocities (SVs) were 20 to 100 h^{-1}, which corresponded to flow rate 1 to 5 mL/min. After saturated adsorption, 1 mol/L HCl was used to regenerate the QCLF at flow rate 0.25 mL/min.

3. Results and Discussion

3.1. Synthesis of the QCLF

GMA was grafted onto cotton linter by radiation grafting technology. Figure 2 shows the effect of radiation dose on the DOG in 30 wt% GMA solution. The DOG was firstly increased with dose up to maximum value of 255% at 30 kGy, and then maintained at a high platform. This phenomena can be explained by the decay mechanism of the trapped radicals. The grafting of GMA onto cotton linter by EB pe-irradiation graft polymerization was mainly initiated by the free radical mechanism. A high adsorbed dose can initiate more amounts of free radicals and induce high DOG. However, a higher absorbed dose may also result in the decomposition of cellulose substrate. So in this study, the grafted fibers with the DOG of 255% at 30 kGy were used for further experiments.

The grafted cotton linter fiber (CLF-g-GMA) with DOG 255% was immersed into PEHA solution for epoxy ring-opening reaction. The conversion rate calculated by the mass increase was 45%. The quaternization rate in the quaternization reaction was 48% calculated by mass increase.

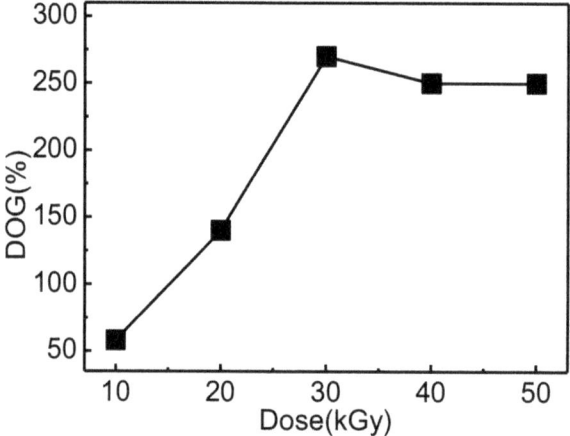

Figure 2. Effect of radiation dose on the degree of grafting (DOG).

3.2. Characterization

3.2.1. FT-IR Analysis

Figure 3 shows the FT-IR spectra of the original cotton linter (a), CLF-g-GMA (b), CLF-g-GMA-PEHA (c), and QCLF (d). The bands of cellulose observed in curve (a) included O–H, C–H, H–O–H, C–O, C–O–C bonding at 3345, 2920, 1635, 1060 and 898 cm^{-1}, respectively [22]. The grafting of GMA was confirmed by the adsorption bands at 1726 cm^{-1} and 908 cm^{-1} which was assigned to the stretching of the carbonyl and epoxy group of GMA. After the ring-opening reaction, the bands at 1564 cm^{-1} and 1465 cm^{-1} can be assigned to N–H and C–H bands of PEHA [23]. After the quaternization reaction, the peaks at 1656 cm^{-1} and 3415 cm^{-1} attributed to the appearance of quaternary nitrogen and then the intensity of N–H peaks sharply decreased, meaning that tertiary amines were converted to quaternary ammonium [24,25]. The bands at 2960 cm^{-1} were assigned to C–H antisymmetric and symmetric stretching of –CH_2–, indicating successful introduction of alkyl chain from 1-bromohexane.

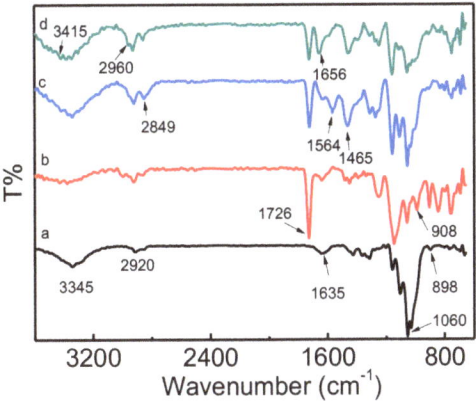

Figure 3. FT-IR spectra of original cotton linter (a), grafted cotton linter fiber (CLF-g-GMA) (b), CLF-g-GMA-Pentaethylene hexamine (PEHA) (c) and QCLF (d).

3.2.2. SEM Photographs

The surface morphologies of cotton linter (a), CLF-g-GMA (b), CLF-g-GMA-PEHA (c) and QCLF (d) are shown in Figure 4. The diameters of these fiber sample increased after grafting and further modification, which might be due to the large DOG of GMA on the cotton linter.

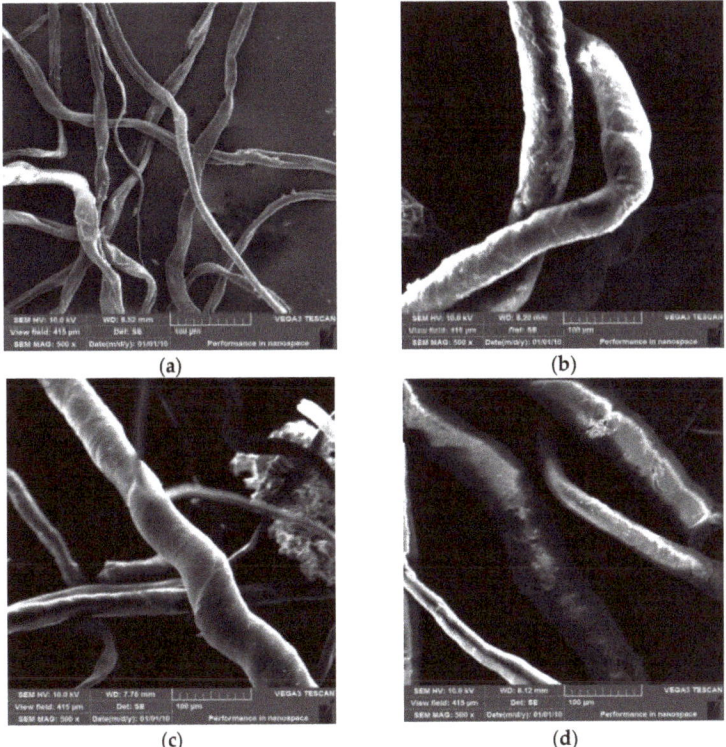

Figure 4. SEM photographs of original cotton linter (**a**), CLF-g-GMA (**b**), CLF-g-GMA-PEHA (**c**) and QCLF (**d**).

3.2.3. TG Analysis

Figure 5 shows the thermal stability of cotton linter, CLF-g-GMA, CLF-g-GMA-PEHA and QCLF. The weight loss below 100 °C was due to the loss of water in the samples. The weight loss of the origin cotton linter happened in the range from 340 to 420 °C, which showed a one-step weight loss. The weight loss of CLF-g-GMA occurred at 220 °C and terminated at 420 °C, which was due to the complex thermal decomposition of the cotton linter and grafted epoxy groups and ester in GMA. The decrease of the degradation temperature for cellulose after grafting of GMA also happened in reference [26]. But for CLF-g-GMA-PEHA and QCLF samples, the thermal degradation curves have two districts. The main decomposition temperature occurred between 220 and 450 °C, the second loss weight took place at 450 °C, which might be due to the residual mass of solid carbon. It suggested that the adsorbents have good thermal resistance for application in phosphate removal.

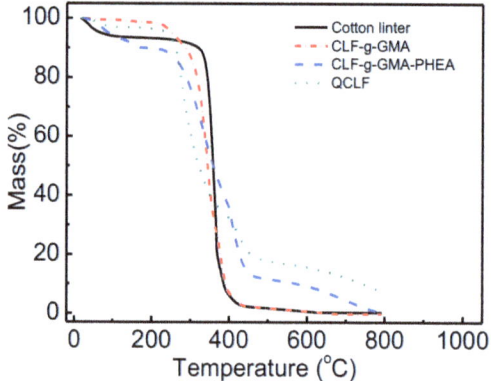

Figure 5. TG analysis for cotton-linter, CLF-g-GMA, CLF-g-GMA-PEHA and QCLF.

3.3. Phosphate Adsorption in Batch Experiments

3.3.1. pH Study

The adsorption of phosphate onto QCLF at different initial pH values are shown in Figure 6. The adsorption capacity was little affected at pH ranging from 4 to 8. The adsorption capacity of QCLF for phosphate was very low at pH < 4 because the dominant phosphate species at lower pH were transformed to neutron H_3PO_4, which had lower affinity to the adsorption sites of the QCLF [27,28]. At pH between 4 and 8, the dominant species was transformed into $H_2PO_4^-$ and HPO_4^{2-}. The adsorption capacity of QCLF for phosphate was dropped significantly at pH > 8 because there would be a competition adsorption between OH^- and phosphate. It was reported that the pH value in the eutrophic lake was in the range 7.5 to 8.5 [2]. So QCLF was very suitable for phosphate removal in eutrophic lakes. The phosphate concentration at pH 7 without pH adjustment was used for further batch and column adsorption tests.

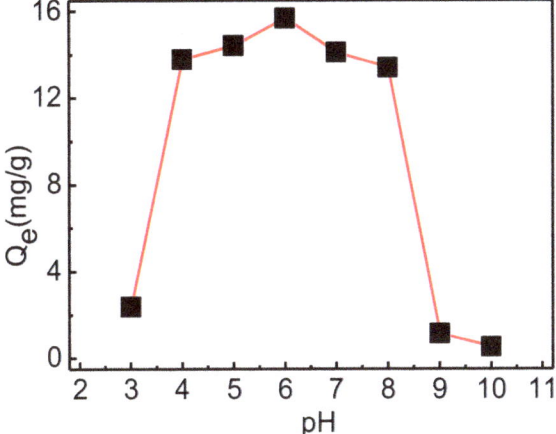

Figure 6. Effect of pH on phosphate adsorption by QCLF adsorbents (initial phosphate concentration 20 mg/L, volume 50 mL, mass 0.05 g, adsorption time 24 h).

3.3.2. Effect of Adsorbent Dosage

Adsorption dosage was a vital parameter influencing the adsorption performance. Figure 7 shows the removal efficiency and adsorption capacity of phosphate at different adsorbent dosages. It was evident that Q_e decreases while the removal % increases significantly when the amount of QCLF increases. This can be explained by the fact that more adsorption sites of adsorbent were worked for electrostatic attractive forces between QCLF and phosphate. The removal efficiency reached 90% at 1 and 92% at 1.4 g/L, which was a smaller change compared to when the dosage was greater than 1 g/L. Therefore, the effectiveness decreased with an increase of adsorbent beyond 1 g/L. So the adsorbent dosage of 1 g/L (0.05 g QCLF in 50 mL solution) was regarded as the optimal dosage for further batch experiment condition.

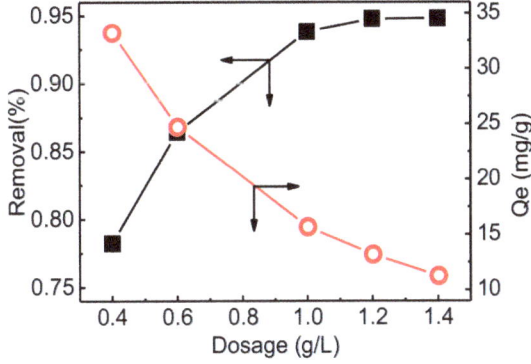

Figure 7. Effect of adsorbent dosage on the removal efficiency and adsorption capacity of phosphate by QCLF adsorbent (initial concentration 20 mg/L, volume 50 mL, pH 7, adsorption time 24 h).

3.3.3. Adsorption Kinetics

The adsorption equilibrium time is of great importance for studying the affinity of the adsorbents to phosphate. The experimental data was fitted by three kinetics models. The pseudo-first-order, pseudo-second-order and intra-particle diffusion kinetic equations were expressed by Equations (3)–(5), respectively [29].

$$\ln(q_e - q_t) = \ln q_e - k_1 t \qquad (3)$$

$$\frac{t}{q_t} = \frac{1}{k_2 q_e^2} + \frac{t}{q_e} \qquad (4)$$

$$q_t = k_{id} t^{1/2} + I \qquad (5)$$

where q_t (mg/g) was the adsorption capacity of phosphate at time t (min) and q_e was the adsorption capacity at equilibrium time. k_1 and k_2 were the rate constant. K_{id} parameter was the reaction rate constant (mg/g·min$^{1/2}$) and I was the intercept.

The adsorption capacity at different adsorption times is shown in Figure 8a. It can be seen that the adsorption capacity increased rapidly and reached equilibrium within 120 min. The pseudo-first-order and pseudo-second-order were used to analyze the experimental data. The parameters q_e, k_1 and k_2 were calculated and are listed in Table 1. Figure 8b shows the plots of pseudo-second-order kinetic model. The higher correlation coefficient R^2 (0.9998) indicated that pseudo-second-order kinetic model can describe the adsorption kinetics well, which indicated the adsorption of phosphate onto QCLF was a monolayer adsorption and chemical adsorption.

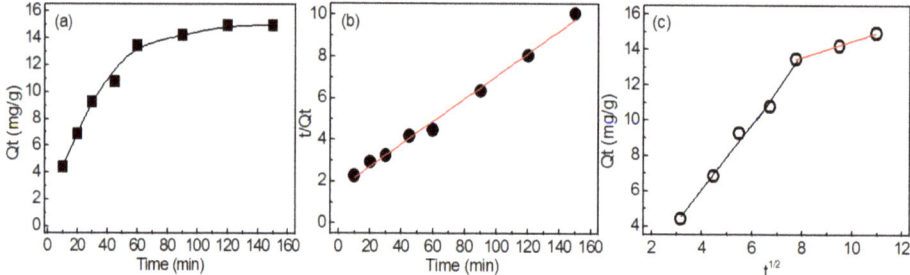

Figure 8. Adsorption kinetics of phosphate removal onto QCLF in 50 mL, 20 mg/L, 0.05 g, pH: 7, (**a**) effect of adsorption time; (**b**) pseudo-second-order kinetic model; (**c**) intra-particle diffusion model.

Intra-particle diffusion model was also used to study the transportation of phosphate to QCLF. Figure 8c shows the plot of intra-particle diffusion model. The plot were divided into two distinct regions. The first region was a fast adsorption and the second region corresponded to a slow equilibrium adsorption. And the fitted plot did not pass through the origin, indicating that intra-particle diffusion was not the only rate-controlling step [29]. The parameters of the intra-particle kinetic model are listed in Table 1.

Table 1. Kinetic parameters obtained from pseudo-first-order, pseudo-second-order kinetic and intra-particle diffusion model.

Model	Parameters	20 mg/L HPO$_4^{2-}$
pseudo-first-order kinetics	k_1 (h^{-1})	0.0305
	q_e (mg/g)	15.314
	R^2	0.9872
pseudo-second-order kinetics	k_2 (g/(mg·min))	0.0047
	q_e (mg/g)	19.409
	R^2	0.9967
Weber–Morris	K_{id1}	1.8415
	I_1	1.3091
	R^2	0.97891
	K_{id2}	0.4543
	I_2	9.9282
	R^2	0.9973

3.3.4. Adsorption Isotherms

The adsorption isotherm of QCLF was conducted at 25 °C with the phosphate concentration of 100–500 mg/L. Figure 9a shows the equilibrium adsorption capacity at different equilibrium concentrations after adsorption. The equilibrium adsorption capacity increased and finally attained the maximum value.

The Langmuir model was applicable to adsorption on homogeneous surface, which was expressed by Equation (6). Equilibrium parameter (R_L) reflected the nature of the dominant adsorption mechanisms in the studied system. The favorability of the adsorption was given by the dimensionless separation faction R_L, which were calculated from Equation (7):

$$\frac{C_e}{q_e} = \frac{C_e}{q_m} + \frac{1}{K_L q_m} \qquad (6)$$

$$R_L = 1/(1 + K_L C_0) \qquad (7)$$

where C_e (mg/L) was the phosphate concentration after adsorption, q_e and q_m (mg/g) were the equilibrium and theoretical maximum adsorption capacity. K_L (L/mg) was the Langmuir constant related to the adsorption affinity [30].

The Freundlich model assumed a heterogeneous surface and its linear form was expressed by Equation (8).

$$\ln q_e = \ln K_F + \frac{1}{n} \ln C_e \qquad (8)$$

where, K_F was the Freundlich constant related to adsorption capacity (mmol/g) and 1/n represent the adsorption intensity of adsorbate on adsorbent [30].

The linear equation for Temkin model was given as Equation (9) [31].

$$q_e = B_t \ln(K_t) + B_t \ln(C_e) \qquad (9)$$

where B_t was related to the heat of sorption and K_t (L/mg) was the equilibrium binding constant.

The linear fitted plot of Langmuir, Freundlich isotherm and Temkin model are shown in Figure 9b–d, whose key parameters are summarized in Table 2. Comparatively, the Langmuir isotherm model with high R^2 value (0.9952) was a better fit for the adsorption of phosphate on QCLF, which denoted the monolayer adsorption of phosphate onto QCLF. The adsorption capacity Q_m were calculated to be 152.44 mg/g, which was higher than other literature reports regarding adsorbents, especially at pH 6–7 (see Table 3). The calculated R_L in this study were between 0 and 1, which reflected the favorable adsorption of phosphate [31,32].

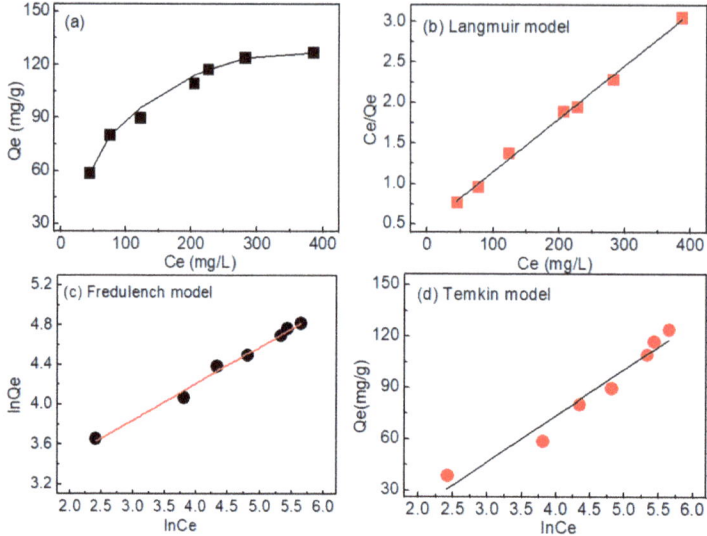

Figure 9. Equilibrium studies of phosphate adsorption: relation of the adsorption capacity with the equilibrium concentration (**a**); Langmuir isotherm model (**b**); Freundlich isotherm model (**c**) and Tekmin isotherm model (**d**).

Table 2. Langmuir, Freundlich, Tekmin isotherm model parameters and correlation coefficients for the adsorption of phosphate.

Adsorbent	Langmuir			Freundlich			Temkin		
	Q_m (mg/g)	K_L	R^2	K_F (mg·L^{-1})	n	R^2	B_T	K_T	R^2
QCL	152.44	0.0139	0.9952	15.52	2.720	0.9899	26.9827	0.2799	0.9476

Table 3. Comparison of adsorption capacity of QCLF with other available different adsorbents.

Adsorbent	The max Adsorption Capacity (mg/g)	pH	Reference
carbonized sludge adsorbent	4.792	7	[32]
diethylamine modified Cellulose	22.88	6.8	[33]
humic acid coated magnetite nanoparticles	28.9	6.6	[34]
quaternized pectin	31.07	7	[30]
wheat straw anion exchanger	52.80	-	[35]
amine-crosslinked Shaddock Peel	59.89	3	[27]
Zirconium (IV) loaded cross-linked chitosan particles	71.68	3	[36]
modified sugarcane bagasse fibers-Fe	152	3	[37]
quaternized cotton linter fiber	152.44	7	This paper

3.4. Column Experiments

Adsorption was an accumulation process of the adsorbate species on the absorption sites of the adsorbent. In the fixed-bed column test, the solution flows continuously through a column of adsorbent. The adsorbent near the column inlet will be saturated first, then the adsorption zone is moved further toward the exit of the bed. When all the adsorbents are saturated, the value of leakage concentration is increased and finally reaches the influent concentration. A plot of leakage concentration of the outlet as a function of proceed time was known as the breakthrough curve. In this study, column tests were carried out in a fixed-bed column by varying different initial concentrations and flow rate. The aim of this experiment was to find the optimal flow rate and influent phosphate concentration to maximize the productivity and efficiency.

3.4.1. Effect of Influent Concentration

The effect of different phosphate concentration on breakthrough performance is shown in Figure 10a,b. The mass of the adsorbent and flow rate were kept constant. The efficiency of phosphate removal was described by the percent of (C_t/C_0). C_t/C_0 in all cases was increased with the time increased, until it reached 1.0. The breakthrough point and the saturated time of phosphate adsorption depended on various influent phosphate concentration. At higher influent concentration, an earlier breakthrough point was observed. Breakthrough point was decreased from 2610 to 1260 min with the influent concentration increased from 50 to 100 mg/L. At higher influent concentration, the function groups were rapidly combined with phosphate which resulted in a decrease in breakthrough time. The driving force increased with the influent concentration increased, then the adsorption points were saturated faster with higher influent concentration.

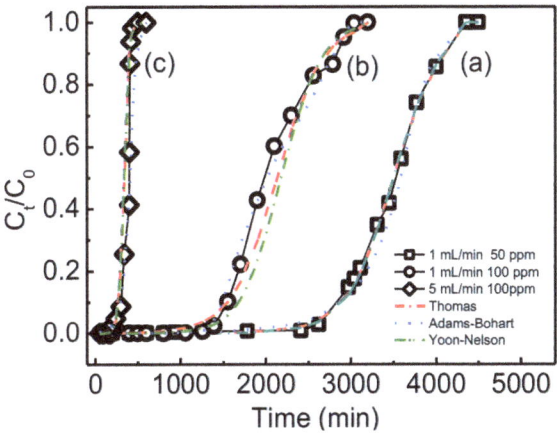

Figure 10. Breakthrough curves with concentration 50 mg/L at flow rate 1 mL/min (a), with concentration 100 mg/L at flow rate 1 mL/min (b), and with concentration 100 mg/L at flow rate 5 mL/min (c).

3.4.2. Effect of Flow Rate

The effect of flow rate at the same influent concentration is compared in Figure 10b,c. It can be seen that the breakthrough points happened at 1260 and 200 min at flow rate of 1 and 5 mL/min, respectively. Their saturation adsorption was reached at 3200 and 500 min at flow rate of 1 and 5 mL/min, respectively, which showed that a high flow rate will saturate the fixed column faster. The difference in the slope of the breakthrough curve and adsorption capacity between the two plots may be explained on the basis of mass transfer fundamentals. At a higher flow rate, phosphate has a short diffusion time and interacts predominately with the functional groups on QCLF which have the highest availability. In regard to a lower flow rate of influent, phosphate had more time to contact with larger numbers of function groups, and this resulted in the higher removal of phosphate in the column.

3.4.3. Estimation of Breakthrough Curve

Bohart–Adams, Thomas and Yoon–Nelson models were used for describing and analyzing the laboratory-scale column, which linear forms were presented in the following Equations (10)–(12), respectively [4,38,39].

$$\ln\left(\frac{C_t}{C_0}\right) = k_{AB}C_0 t - k_{AB}N_0\frac{Z}{F} \tag{10}$$

$$\ln(\frac{C_t}{C_0} - 1) = \frac{K_{TH}q_0 m}{v} - K_{TH}C_0 t \qquad (11)$$

$$\ln(\frac{C_t}{C_0-C_t}) = k_{YN}t - \tau k_{YN} \qquad (12)$$

where C_0 was the influent and C_t was the leakage concentration (mg/L) at time t; k_{AB}, k_{TH} and k_{YN} were the model constants; t was the processing time; q_0 was the adsorption capacity; v was the flow rate; m was the adsorbent mass; N_0 is the saturation concentration; Z was the bed depth of column and F was the superficial velocity defined as the ratio of the volumetric flow rate Q to the cross-sectional area of the bed A. τ was the time for 50% adsorbent saturated. The key parameters of the three models are fitted, calculated and given in Table 4. From the value of the linear fitting correlation coefficient R^2 in Table 4, both the Thomas and Yoon–Nelson models can be shown to have better predicted the adsorption performance for adsorption of phosphate in a fixed-bed column [38].

The Bohart–Adams model was established by assuming that the adsorption equilibrium is not instantaneous and that the rate of adsorption is proportional to both the residual capacity of the adsorbent and the concentration of the adsorbate species. It can be seen from Table 4 that the kinetic constant k_{AB} decreased as the phosphate concentration increased but increased as the flow rate increased. The saturation concentration (N_0) of the column also increased as the phosphate concentration increased but decreased as the flow rate increased. Because higher N_0 and lower k_{AB} means less adsorption resistance, the optimal performance of column experiment were then obtained at higher initial phosphate concentrations and lower flow rate [39].

According to the Thomas model, at the same flow rate of 1 mL/min, the adsorption capacity of phosphate reached 120.61 and 141.58 mg/g at 50 and 100 mg/L, respectively. The results demonstrate that the maximum adsorption capacity was increased with the increasing of the phosphate concentration. This was mainly because the adsorption force increased when the concentration of phosphate increased. On the other hand, the value of Thomas constant K_{TH} was high at lower influent concentration. The adsorption capacity also decreased significantly when the flow rate increased from 1 to 5 mL/min. Furthermore, the adsorption capacity (q_0 = 141.58 mg/g) was slightly smaller than that of theoretical maximum capacity in batch experiment (Q_m = 152.44 mg/g), which showed the column test condition (100 mg/L, 1 mL/min) was suitable to practical use. According to the Yoon–Nelson model, the time τ for penetration to reach 50% decreased significantly with the influent concentration from 50 to 100 mg/L at the same flow rate or with the flow rate from 1 to 5 mL/min at the same concentration of 100 mg/L. All three τ values were very close to the experimental results. So it can be concluded that a higher maximum adsorption capacity can be obtained with a high concentration and low flow rate from both the Thomas and Yoon–Nelson model.

Table 4. Parameters obtained from three dynamic adsorption models.

C_0 (mg/L)	v (mL/min)	Thomas Model			Bohart-Adams Model			Yoon-Nelson Model		
		k_{TH}	q_0	R^2	k_{BA}	N_0	R^2	k_{YN}	τ	R^2
50	1	0.0738	120.61	0.9889	0.0462	61.1088	0.8930	0.0037	3497.81	0.9889
100	1	0.0376	141.58	0.9352	0.0102	89.229545	0.8009	0.0038	2123.74	0.9352
100	5	0.3458	22.419	0.9399	0.1363	14.4303	0.9399	0.0346	336.63	0.9368

3.4.4. Elution Experiment

After the saturated adsorption at concentration 100 mg/L and flow rate 1 mL/min, the phosphate loaded QCLF was regenerated to realize repeated use. Figure 11 shows the elution curves with 1 M HCl at flow rate 0.25 mL/min, the elution process was almost completed within 60 min, which showed that QCLF can be effectively regenerated and for further reuse.

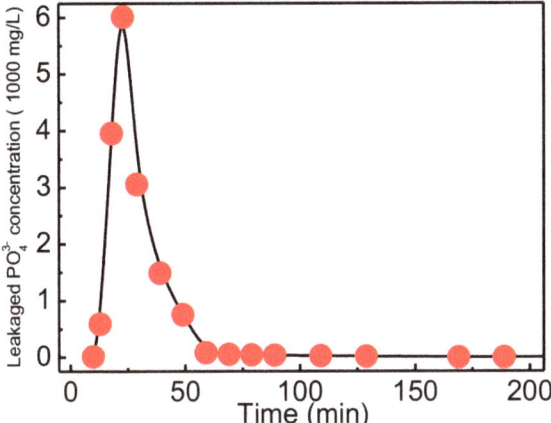

Figure 11. The elution curve at flow rate 0.25 mL/min.

4. Conclusions

A QCLF adsorbent was successfully synthesized by radiation induced grafting technique. The adsorption kinetics of QCLF for phosphate reached equilibrium within 120 min and the kinetics was well obeyed pseudo-second-order mode. The adsorption isotherms were well obeyed Langmuir model with the maximum adsorption capacity 152.44 mg/g. Column experiments showed that the breakthrough curves were dependent on initial phosphate concentration and flow rate. The saturated adsorbents could be efficiently generated by eluting with 1 mol/L HCl. The Thomas and Yoon–Nelson models were both successfully used to predict the breakthrough curves. The adsorption capacity of phosphate at 100 mg/L and flow rate 1 mL/min reached 141.58 mg/g according to the Thomas model. The results showed that the modified cotton linter fibers can be used for phosphate removal. On the other hand, the performance of the fibers after repeated use or magnified use need further investigation in the future.

Author Contributions: Methodology, Z.D.; Validation, X.Y.; Investigation, J.D. and Z.L.; Writing—original draft preparation, J.D. and Z.D.; Writing—review and editing, L.Z.

Funding: This work was supported by the National Natural Science Foundation of China (11875138, 11905070), the Nuclear Technology Special Fund of Hubei University of Science and Technology (2018-19KZ02, 2018-19 × 048) and the Key Project of Technological Innovation of Hubei Province (2017AEA107).

Conflicts of Interest: The authors declare no conflicts of interest.

References

1. Bulgariu, D.; Axinte, O.; Badescu, I.S.; Stroe, C.; Neacsu, V.; Bulgariu, L. Evolution of trophic parameters from amara lake. *Environ. Eng. Manag. J.* **2015**, *14*, 559–565. [CrossRef]
2. Hossain, M.E.; Ritt, C.L.; Almeelbi, T.B.; Bezbaruah, A.N. Biopolymer Beads for Aqueous Phosphate Removal: Possible Applications in Eutrophic Lakes. *J. Environ. Eng.* **2018**, *144*, 04018030. [CrossRef]
3. Loganathan, P.; Vigneswaran, S.; Kandasamy, J.; Bolan, N.S. Removal and Recovery of Phosphate from Water Using Sorption. *Crit. Rev. Environ. Sci. Technol.* **2014**, *44*, 847–907. [CrossRef]
4. Ye, Y.; Jiao, J.; Kang, D.; Jiang, W.; Kang, J.; Ngo, H.H.; Guo, W.; Liu, Y. The adsorption of phosphate using a magnesia–pullulan composite: Kinetics, equilibrium, and column tests. *Environ. Sci. Pollut. Res.* **2019**, *26*, 13299–13310. [CrossRef] [PubMed]
5. Li, R.; Wang, J.J.; Zhou, B.; Awasthi, M.K.; Ali, A.; Zhang, Z.; Lahori, A.H.; Mahar, A. Recovery of phosphate from aqueous solution by magnesium oxide decorated magnetic biochar and its potential as phosphate-based fertilizer substitute. *Bioresour. Technol.* **2016**, *215*, 209–214. [CrossRef]

6. Sud, D.; Mahajan, G.; Kaur, M. Agricultural waste material as potential adsorbent for sequestering heavy metal ions from aqueous solutions–A review. *Bioresour. Technol.* **2008**, *99*, 6017–6027. [CrossRef]
7. Zou, H.; Lv, P.-F.; Wang, X.; Wu, D.; Yu, D.-G. Electrospun poly(2-aminothiazole)/cellulose acetate fiber membrane for removing Hg(II) from water. *J. Appl. Polym. Sci.* **2017**, *134*, 44879. [CrossRef]
8. Monier, M.; Ayad, D.M.; Sarhan, A.A. Adsorption of Cu(II), Hg(II), and Ni(II) ions by modified natural wool chelating fibers. *J. Hazard. Mater.* **2010**, *176*, 348–355. [CrossRef]
9. Du, J.; Dong, Z.; Pi, Y.; Yang, X.; Zhao, L. Fabrication of Cotton Linter-Based Adsorbents by Radiation Grafting Polymerization for Humic Acid Removal from Aqueous Solution. *Polymers* **2019**, *11*, 962. [CrossRef]
10. Hassan, M.S.; Zohdy, M.H. Adsorption Kinetics of Toxic Heavy Metal Ions from Aqueous Solutions onto Grafted Jute Fibers with Acrylic Acid by Gamma Irradiation. *J. Nat. Fibers* **2018**, *15*, 506–516. [CrossRef]
11. Duan, C.; Zhao, N.; Yu, X.; Zhang, X.; Xu, J. Chemically modified kapok fiber for fast adsorption of Pb, Cd, Cu from aqueous solution. *Cellulose* **2013**, *20*, 849–860. [CrossRef]
12. Nguyen, D.D.; Vu, C.M.; Vu, H.T.; Choi, H.J. Micron-Size White Bamboo Fibril-Based Silane Cellulose Aerogel: Fabrication and Oil Absorbent Characteristics. *Materials* **2019**, *12*, 1407. [CrossRef] [PubMed]
13. Dong, C.; Zhang, H.; Pang, Z.; Liu, Y.; Zhang, F. Sulfonated modification of cotton linter and its application as adsorbent for high-efficiency removal of lead(II) in effluent. *Bioresour. Technol.* **2013**, *146*, 512–518. [CrossRef] [PubMed]
14. Li, Y.; Xiao, H.; Pan, Y.; Wang, L. Novel Composite Adsorbent Consisting of Dissolved Cellulose Fiber/Microfibrillated Cellulose for Dye Removal from Aqueous Solution. *ACS Sustain. Chem. Eng.* **2018**, *6*, 6994–7002. [CrossRef]
15. Qu, R.; Sun, C.; Wang, M.; Ji, C.; Xu, Q.; Zhang, Y.; Wang, C.; Chen, H.; Yin, P. Adsorption of Au(III) from aqueous solution using cotton fiber/chitosan composite adsorbents. *Hydrometallurgy* **2009**, *100*, 65–71. [CrossRef]
16. Araga, R.; Sharma, C.S. Amine Functionalized Electrospun Cellulose Nanofibers for Fluoride Adsorption from Drinking Water. *J. Polym. Environ.* **2019**, *27*, 816–826. [CrossRef]
17. Tian, Y.; Wu, M.; Liu, R.; Wang, D.; Lin, X.; Liu, W.; Ma, L.; Li, Y.; Huang, Y. Modified native cellulose fibers—A novel efficient adsorbent for both fluoride and arsenic. *J. Hazard. Mater.* **2011**, *185*, 93–100. [CrossRef]
18. Yu, X.; Tong, S.; Ge, M.; Wu, L.; Zuo, J.; Cao, C.; Song, W. Synthesis and characterization of multi-amino-functionalized cellulose for arsenic adsorption. *Carbohydr. Polym.* **2013**, *92*, 380–387. [CrossRef]
19. Hayashi, N.; Chen, J.; Seko, N. Nitrogen-Containing Fabric Adsorbents Prepared by Radiation Grafting for Removal of Chromium from Wastewater. *Polymers* **2018**, *10*, 744. [CrossRef]
20. Le Moigne, N.; Sonnier, R.; El Hage, R.; Rouif, S. Radiation-induced modifications in natural fibres and their biocomposites: Opportunities for controlled physico-chemical modification pathways? *Ind. Crop. Prod.* **2017**, *109*, 199–213. [CrossRef]
21. Du, J.; Dong, Z.; Yang, X.; Zhao, L. Facile fabrication of sodium styrene sulfonate-grafted ethylene-vinyl alcohol copolymer as adsorbent for ammonium removal from aqueous solution. *Environ. Sci. Pollut. Res.* **2018**, *25*, 27235–27244. [CrossRef] [PubMed]
22. Chattopadhyay, D.; Umrigar, K. Chemical Modification of Waste Cotton Linters for Oil Spill Cleanup Application. *J. Inst. Eng. (India) Ser. E* **2017**, *98*, 103–120. [CrossRef]
23. Anbia, M.; Salehi, S. Removal of acid dyes from aqueous media by adsorption onto amino-functionalized nanoporous silica SBA-3. *Dyes Pigments* **2012**, *94*, 1–9. [CrossRef]
24. Lv, X.; Li, Y.; Yang, M. Humidity sensitive properties of copolymer of quaternary ammonium salt with polyether-salt complex. *Polym. Adv. Technol.* **2009**, *20*, 509–513. [CrossRef]
25. Deng, S.; Zheng, Y.; Xu, F.; Wang, B.; Huang, J.; Yu, G. Highly efficient sorption of perfluorooctane sulfonate and perfluorooctanoate on a quaternized cotton prepared by atom transfer radical polymerization. *Chem. Eng. J.* **2012**, *193*, 154–160. [CrossRef]
26. Chen, I.; Xu, C.; Peng, J.; Han, D.; Liu, S.; Zhai, M.L. Novel Functionalized Cellulose Microspheres for Efficient Separation of Lithium Ion and Its Isotopes: Synthesis and Adsorption Performance. *Molecules* **2019**, *24*, 2762. [CrossRef] [PubMed]
27. Duan, P.; Xu, X.; Shang, Y.; Gao, B.; Li, F. Amine-crosslinked Shaddock Peel embedded with hydrous zirconium oxide nano-particles for selective phosphate removal in competitive condition. *J. Taiwan Inst. Chem. Eng.* **2017**, *80*, 650–662. [CrossRef]

28. Naushad, M.; Sharma, G.; Kumar, A.; Sharma, S.; Ghfar, A.A.; Bhatnagar, A.; Stadler, F.J.; Khan, M.R. Efficient removal of toxic phosphate anions from aqueous environment using pectin based quaternary amino anion exchanger. *Int. J. Boil. Macromol.* **2018**, *106*, 1–10. [CrossRef]
29. Muhammadj, A.; Shah, A.A.; Bilal, S.; Rahman, G. Basic Blue Dye Adsorption fromWater Using Polyaniline/Magnetite (Fe_3O_4) Composites: Kinetic and Thermodynamic Aspects. *Materials* **2019**, *12*, 1764. [CrossRef]
30. Shaban, M.; Abukhadra, M.R.; Khan, A.A.P.; Jibali, B.M. Removal of Congo red, methylene blue and Cr(VI) ions from water using natural serpentine. *J. Taiwan Inst. Chem. Eng.* **2018**, *82*, 102–116. [CrossRef]
31. Zhang, L.; Liu, J.; Guo, X. Investigation on mechanism of phosphate removal on carbonized sludge adsorbent. *J. Environ. Sci.* **2018**, *64*, 335–344. [CrossRef]
32. Silva, F.; Nascimento, L.; Brito, M.; Da Silva, K.; Paschoal, W.; Fujiyama, R.; Silva, D. Biosorption of Methylene Blue Dye Using Natural Biosorbents Made from Weeds. *Materials* **2019**, *12*, 2486. [CrossRef]
33. Fan, C.H.; Zhang, Y.C. Adsorption isotherms, kinetics and thermodynamics of nitrate and phosphate in binary systems on a novel adsorbent derived from corn stalks. *J. Geochem. Explor.* **2018**, *188*, 95–100. [CrossRef]
34. Rashid, M.; Price, N.T.; Pinilla, M.; Ángel, G.; O'Shea, K.E. Effective removal of phosphate from aqueous solution using humic acid coated magnetite nanoparticles. *Water Res.* **2017**, *123*, 353–360. [CrossRef]
35. Xu, X.; Gao, B.-Y.; Yue, Q.-Y.; Zhong, Q.-Q. Preparation of agricultural by-product based anion exchanger and its utilization for nitrate and phosphate removal. *Bioresour. Technol.* **2010**, *101*, 8558–8564. [CrossRef]
36. Liu, Q.; Hu, P.; Wang, J.; Zhang, L.; Huang, R. Phosphate adsorption from aqueous solutions by Zirconium (IV) loaded cross-linked chitosan particles. *J. Taiwan Inst. Chem. Eng.* **2016**, *59*, 311–319. [CrossRef]
37. Carvalho, W.S.; Martins, D.F.; Gomes, F.R.; Leite, I.R.; Da Silva, L.G.; Ruggiero, R.; Richter, E.M. Phosphate adsorption on chemically modified sugarcane bagasse fibres. *Biomass Bioenergy* **2011**, *35*, 3913–3919. [CrossRef]
38. Chen, S.; Yue, Q.; Gao, B.; Li, Q.; Xu, X.; Fu, K. Adsorption of hexavalent chromium from aqueous solution by modified corn stalk: A fixed-bed column study. *Bioresour. Technol.* **2012**, *113*, 114–120. [CrossRef]
39. Bulgariu, D.; Bulgariu, L. Sorption of Pb(II) onto a mixture of algae waste biomass and anion exchanger resin in a packed-bed column. *Bioresour. Technol.* **2013**, *129*, 374–380. [CrossRef]

© 2019 by the authors. Licensee MDPI, Basel, Switzerland. This article is an open access article distributed under the terms and conditions of the Creative Commons Attribution (CC BY) license (http://creativecommons.org/licenses/by/4.0/).

Article

The Influence of Lignin Diversity on the Structural and Thermal Properties of Polymeric Microspheres Derived from Lignin, Styrene, and/or Divinylbenzene

Marta Goliszek [1,*], Beata Podkościelna [1], Olena Sevastyanova [2,3], Barbara Gawdzik [1] and Artur Chabros [1]

1. Department of Polymer Chemistry, Faculty of Chemistry, Maria Curie-Sklodowska University, M. Curie-Sklodowska Sq. 3, 20-031 Lublin, Poland
2. Department of Fibre and Polymer Technology, KTH Royal Institute of Technology, Teknikringen 56-58, SE-10044 Stockholm, Sweden
3. KTH Royal Institute of Technology, Wallenberg Wood Science Center, Teknikringen 56-58, SE-10044 Stockholm, Sweden
* Correspondence: marta.goliszek@poczta.umcs.lublin.pl

Received: 19 July 2019; Accepted: 2 September 2019; Published: 4 September 2019

Abstract: This work investigates the impact of lignin origin and structural characteristics, such as molecular weight and functionality, on the properties of corresponding porous biopolymeric microspheres obtained through suspension-emulsion polymerization of lignin with styrene (St) and/or divinylbenzene (DVB). Two types of kraft lignin, which are softwood (*Picea abies* L.) and hardwood (*Eucalyptus grandis*), fractionated by common industrial solvents, and related methacrylates, were used in the synthesis. The presence of the appropriate functional groups in the lignins and in the corresponding microspheres were investigated by attenuated total reflectance Fourier transform infrared spectroscopy (ATR/FT-IR), while the thermal properties were studied by differential scanning calorimetry (DSC). The texture of the microspheres was characterized using low-temperature nitrogen adsorption. The swelling studies were performed in typical organic solvents and distilled water. The shapes of the microspheres were confirmed with an optical microscope. The introduction of lignin into a St and/or DVB polymeric system made it possible to obtain highly porous functionalized microspheres that increase their sorption potential. Lignin methacrylates created a polymer network with St and DVB, whereas the unmodified lignin acted mainly as an eco-friendly filler in the pores of St-DVB or DVB microspheres. The incorporation of biopolymer into the microspheres could be a promising alternative to a modification of synthetic materials and a better utilization of lignin.

Keywords: lignin; microspheres; composites; polymeric material; fractionation; porosity

1. Introduction

Interest in producing bio-polymer-based materials from renewable resources has increased recently as a step toward sustainable development, by utilizing technologies that are safe for the environment [1–5]. In response to the increasing environmental concern and diminishing resources of petrochemicals, the use of polymers from renewable resources is a promising alternative [6,7].

Lignocellulosic biomass contains one of the most abundant renewable forms of carbon and is, thus, regarded as a logical feedstock to replace traditional fossil resources [8,9]. Biopolymers derived from lignocellulosic biomass, agricultural crops, or wood are currently considered to be the main resource for developing biodegradable and renewable polymeric materials [10,11]. Lignin as a phenolic component of biomass is the most abundant natural substance composed of aromatic moieties and the second most abundant naturally occurring terrestrial polymer after cellulose [12,13]. It is a linker between cellulose bundles, giving rigidity and strength to the cell walls and resistance toward an attack by microorganisms, which provides a route for the transport of water in plant stems. Structurally, lignin is an amorphous macromolecule formed from three phenylpropanoid units: p-coumaryl alcohol, sinapyl alcohol, and coniferyl alcohol connected by ether and carbon-carbon bonds [14], while the functional groups present in lignin (namely hydroxyl, methoxyl, carbonyl, and carboxylic groups) [15] make it possible to be functionalized through various chemical modifications [16–20]. Large quantities of lignin are released by the kraft process during the production of pulp and paper from wood. Lignin has been mainly regarded as a cheap fuel for the process, but recent concerns for more efficient utilization of natural resources have drawn attention to its further valorization [14,21–29]. However, more effort is needed for the engineering of lignin-based value-added materials to make them financially competitive.

To overcome the above-mentioned limitations of using natural polymers, they can be used in composites and blends with synthetic polymers, which is an effective way to obtain a desirable combination of properties that are absent in the individual components [30,31]. In the last few decades, lignin has been added to various polymers and the materials obtained have sometimes been called blends, and, in other cases, composites, but it is not always clear whether lignin forms a blend or acts as a filler in the composite material [32]. For example, Yin et al. [33] prepared a cross-linked biomass-polymer composite by mixing lignin with an epoxy resin and polyamine using a hot press molding process. They showed that the epoxy resin could be cured by lignin, and that a good interfacial combination was formed between the components. Jesionowski et al. [34] obtained advanced multifunctional silica/lignin composite materials. The composite powders obtained were blended with multiwalled carbon nanotubes. The electrochemical activity assessed by cyclic voltammetry revealed the presence of a redox system assigned to a lignin-derived quinone/hydroquinone couple. He et al. [35] prepared a polyaniline-lignin composite and demonstrated its good adsorption of silver ions, which indicates that the lignin unit could play a vital role in the chelation of silver ions, while the polyaniline unit was important for electrical conduction. Ballner et al. [36] obtained lignocellulose nanofiber-reinforced polystyrene from composite microspheres produced by suspension polymerization. They reported superior impact toughness and improved bending strength for the composite in comparison with unreinforced polystyrene.

One of the main issues in the commercial production of value-added lignin products is the heterogeneous nature of lignin, which makes it difficult to standardize the qualities and properties of the products from lignin. Increased attention has recently been given to the fractionation into more homogeneous preparations using organic solvents, ionic liquids, or ultrafiltration [37–45]. For example, Li et al. [38] investigated sequential solvent fractionation of heterogeneous bamboo organosolv lignin. The starting lignin and obtained fractions were compared in terms of functional groups and molecular weight distribution. It was concluded that sequential solvent fractionation can be a useful method to prepare homogeneous lignin with desirable functional groups for further processing. Jääskeläinen et al. [40] have reported a precipitation fractionation method for kraft lignin based on aqueous organic solvents. In this approach, moisture containing lignin can be subjected to fractionation directly without initial drying, which is a great advantage for large-scale applications. It was shown that the molar mass and functionalities of the precipitated lignin fractions depended on the ratio between solvent and water in the precipitation step. Tagami et al. [41] compared the impact of sequential solvent fractionation on the molecular weight, composition, and contents of functional groups of industrial softwood and hardwood kraft lignins. In this work, it was demonstrated that the antioxidant activity, chemical structure, heating values, and thermal and adsorption properties of lignin can be tuned by such processing. The results obtained provide useful information for targeted uses of lignin raw material. Toledano et al. [43] investigated ultrafiltration and selective precipitation as two different methods for lignin fractionation and confirmed that both fractionation processes influence the properties of the obtained lignin. Lauberts et al. [45] investigated fractionation of technical lignin using ionic liquids for further application as antioxidants. It was shown that fractionated lignins with improved polydispersity resulted in higher antioxidant activity as compared to non-fractionated lignins.

Sevastyanova et al. [46] produced a wide range of lignin fractions by ultrafiltration of black liquor using ceramic membranes with a different Mw cut-off. Investigation of the thermal properties of such fractions revealed the possibility to tailor certain parameters by choice of membrane. Lignin fractions with well-defined characteristics could be a predictive tool for the development of high-performance lignin-based systems [47–49].

Polymer porous microspheres obtained by a suspension-polymerization method have great potential in the sorption processes [17]. The addition of lignin into the polymeric system results in more eco-friendly materials [31,50]. In our previous work [50] on the synthesis of microspheres, we used a commercial kraft lignin and low molecular weight kraft lignin obtained by ultrafiltration of industrial black liquor using a ceramic membrane with Mw cut-off of 5 kD. The specific surface area of these materials decreased with increasing lignin content. The main aim of this work is to investigate the impact of lignin fractionation on the properties of the biopolymeric microspheres obtained.

2. Materials and Methods

2.1. Chemicals and Materials

Softwood (*Picea abies* L.) and hardwood (*Eucalyptus grandis*) lignins were obtained by the kraft process and purified according to the LignoBoost technology protocol [51]. The moisture content was 6.4% and 5.1%, respectively, and ash content was 0.6% and 1.2%, respectively. Methacryloyl chloride, styrene, bis(2-ethylhexyl)sulfosuccinate sodium salt, methylene chloride, trimethylamine, magnesium sulfate, and benzyl alcohol were purchased from Sigma-Aldrich. α,α'-bis-isobutyronitrile (AIBN) and divinylbenzene (DVB) were obtained from Merck (Darmstadt, Germany) (62.2% of 1,4-divinylbenzene, 0.2% of 1,2-divinylbenzene, and ethylvinylbenzene were washed with 3% aqueous sodium hydroxide solution before use). Ethyl acetate (reagent grade) and ethanol (absolute) for the lignin fractionation were purchased from Sigma-Aldrich (Stockholm, Sweden), while methanol (analytical grade) was purchased from VWR (Stockholm, Sweden). Acetone, methanol, tetrahydrofuran (THF), chloroform, toluene, decan-1-ol, and acetonitrile for swelling studies were purchased from Avantor Performance Materials (Gliwice, Poland).

2.2. Lignin Fractionation, Modification, and Characterization

Lignins were fractionated with ethyl acetate, ethanol, or methanol by a two-step solvent fractionation, as described in Reference [52]. Additionally, 20 g of lignin (95% dryness) was added to 200 mL of corresponding solvent and stirred for 2 h at room temperature. After that, the insoluble fraction was separated with filter paper, dried overnight in a vacuum oven VACUCELL, MMM Medcenter EINRICHTUNGEN GmbH (Planegg, Germany) at 40 °C, and subjected to the treatment with the next solvent. Hardwood lignin was fractionated by ethyl acetate followed by ethanol, while softwood lignin extraction with ethyl acetate was followed by methanol. The lignin fraction dissolved in the corresponding solvent was recovered by concentrating the solution on a rotary evaporator IKA TV10 basic, VWR (Stockholm, Sweden) substituting the solvent with water by subsequent freeze-drying of the aqueous suspension.

The molecular weight characteristics of lignin samples were carried out using a size exclusion chromatography (SEC) using a Waters instrument consisting of a 515 HPLC pump, 2707 autosampler, and 2998 photodiode array detector (Waters Sverige AB, Sollentuna, Sweden).

The contents of functional groups were calculated by the ^{31}P NMR method using Bruker Avance 400 MHz spectrometer, as described in References [41,52].

Lignin fractions were modified with methacryloyl chloride, as described in Reference [53]. The lignin sample and methylene chloride were placed with triethylamine in an ice bath and stirred. Then, methacryloyl chloride was added dropwise and a reaction proceeded for 1 h at 5 °C and for an additional hour at room temperature. The obtained material was filtered off, washed three times with water to remove trimethylamine hydrochloride, extracted with methylene chloride, and purified using a chromatographic column.

The successful chemical modification was confirmed using the Fourier transform spectroscopy with attenuated total reflection (ATR/FT-IR). ATR/FT-IR spectra were recorded using a Bruker TENSOR 27 spectrometer containing a diamond crystal (Ettlingen, Germany). The spectra were recorded in the range of 600–4000 cm^{-1} with 32 scans per spectrum at a resolution of 4 cm^{-1}.

2.3. Synthesis of Microspheres

Styrene (St) and/or divinylbenzene (DVB) were copolymerized in an aqueous medium with different solvent fractions of lignin: Spruce-ethyl acetate (SL-ea), Spruce-methanol (SL-m), Eucalyptus-ethyl acetate (EL-ea), and Eucalyptus-ethanol (EL-e) using a suspension-emulsion polymerization method, which has previously been described in detail [50]. Unmodified lignin (L) and lignin modified with methacryloyl chloride (L-Met) were used in the synthesis of microspheres. Experimental parameters are presented in Table 1. A constant amount of benzyl alcohol as pore-forming diluent (14 mL) and DVB as a crosslinking agent (5 g) were used in every synthesis.

Table 1. Experimental parameters of the syntheses.

Polymer	Monomers (g)		
	L [1]	L-Met [2]	St [3]
SL-ea-Met+St+DVB	0	2	4
SL-ea +St+DVB	2	0	4
1SL-ea-Met+DVB	0	1	0
2SL-ea-Met+DVB	0	2	0
3SL-ea-Met+DVB	0	3	0
2SL-ea+DVB	2	0	0
SL-m-Met+St+DVB	0	2	4
SL-m +St+DVB	2	0	4
1SL-m-Met+DVB	0	1	0
2SL-m-Met+DVB	0	2	0
3SL-m-Met+DVB	0	3	0
EL-ea-Met+St+DVB	0	2	4
EL-ea +St+DVB	2	0	4
1EL-ea-Met+DVB	0	1	0
2EL-ea-Met+DVB	0	2	0
3EL-ea-Met+DVB	0	3	0
2EL-ea+DVB	2	0	0
EL-e-Met+St+DVB	0	2	4
EL-e +St+DVB	2	0	4
1EL-e-Met+DVB	0	1	4
2EL-e-Met+DVB	0	2	0
3EL-e-Met+DVB	0	3	0
2EL-e+DVB	2	0	0

[1] unmodified lignin, [2] modified lignin, and [3] styrene.

2.4. Characterization Methods of Microspheres

The ATR/FT-IR spectra were obtained as described in Section 2.2.

The calorimetric measurements were carried out in a Netzsch DSC 204 calorimeter (Selb, Germany) operated in a dynamic mode. The dynamic scans were performed at a heating rate of 10 °C min^{-1}, the first scan being from 20 °C to a maximum of 110 °C to remove any adsorbed moisture, and the second from 25 °C to 550 °C in a nitrogen atmosphere (30 cm^3 min^{-1}). The mass of the sample was about 5–10 mg. An empty aluminum crucible was used as a reference.

Characterization of the porous structure was performed using Micrometrics Inc., ASAP 2405 adsorption analyzer (Norcross, GA, USA). Before the analysis, all the materials were degassed at

120 °C. The specific surface area was calculated according to the Brunauer-Emmett-Teller (BET) method, by assuming that the area of a single nitrogen molecule is 16.2 Å^2. The pore volumes and pore size distributions were determined by the Barrett-Joyner-Halenda (BJH) method.

The swelling coefficients (B) were determined from the equilibrium swelling in chosen organic solvents and distilled water, which are calculated using the equation below.

$$B = \frac{V_s - V_d}{V_d} \cdot 100\% \tag{1}$$

where V_s is the volume after swelling and V_d is the volume of the dry sample.

The appearances and morphologies of the microspheres were studied using a Malvern, MORPHOLOGI G3 optical microscope (Malvern, UK).

3. Results and Discussion

By using a fractionation protocol as described in Reference [52], the kraft lignin fractions of different molecular weight but with very similar content of functional groups were produced for hardwood and softwood lignins (SL-ea vs. EL-ea and SL-m vs. EL-e) (Table 2). Generally, for each type of lignin, the fractions with higher molecular weight, SL-m and EL-e, had a higher content of aliphatic -OH groups. Hardwood lignin fractions, EL-ea and EL-e, had half as many condensed guaiacyl Ph-OH groups as corresponding softwood lignin fractions, SL-ea and SL-m, which is in accordance with previously reported results [29]. Gordobil et al. [29] concluded that this structural feature most likely affects the chemical reactivity of lignin as a higher content of vinyl groups per lignin phenylpropanoid unit was introduced on methacrylation for lignin samples that have a lower degree of condensation.

Table 2. Molecular weight characteristics and content of functional groups in lignin solvent fractions.

Lignin Fraction	Mn [1] (g/mol)	Mw [2] (g/mol)	Ð [3]	Aliphatic-OH (mmol/g)	Carboxyl-OH (mmol/g)	Phenolic-OH (mmol/g)		
						Condensed G [4]	Non-Condensed (G+S [5])	Total
Spruce-ethyl acetate (SL-ea)	720	1160	1.6	0.7	0.7	1.8	3.2	5.0
Spruce-methanol (SL-m)	1400	2900	2.1	1.8	0.4	2.0	2.5	4.5
Eucalyptus-ethyl acetate (EL-ea)	630	940	1.5	0.6	0.3	0.8	4.1	4.9
Eucalyptus-ethanol (EL-e)	870	1420	1.6	1.4	0.4	0.9	3.5	4.4

[1] Number average molecular weight, [2] weight average molecular weight, [3] and polydispersity index. [4] Guaiacyl, [5] syringyl.

Four solvent fractions of lignin were successfully modified with methacryloyl chloride, as shown in their ATR/FT IR spectra (Figure 1a–d). After modification, a reduction in the signal intensity of hydroxyl groups at 3391–3393 cm^{-1} [54] was observed, and new strong signals from stretching vibrations of carbonyl groups at 1726–1737 cm^{-1} were clearly visible. The new signal at 1636 cm^{-1} corresponding to C=C bonds [29] indicates the presence of methacrylate units in the lignin molecule, particularly in the SL-ea-Met lignin. The signals at 1120–1125 cm^{-1} and 1032–1041 cm^{-1} are attributed to C-O and C-O-C stretching vibrations in acrylates. The bands at 946 cm^{-1} can be attributed to a terminal C=CH$_2$ bending vibration from the methacrylate groups in the lignin samples [55].

ATR/FT-IR spectra of the materials are presented in Figure 2a–d. The spectroscopic evaluation proved that the synthesis of lignin-containing microspheres had been successful, manifested by the presence of characteristic bands of appropriate functional groups. The increasing lignin content reflected in the increasing intensity of these bands agreed entirely with the expectations. In the spectra of the materials containing styrene, a signal was visible at 3010 cm^{-1}, which can be attributed to stretching and deformation vibrations of C-H and confirm that styrene has been incorporated into the structure of the microspheres. Bands at 2920 and 2850 cm^{-1} can be assigned to the asymmetric and symmetric stretching vibrations of C-H in -CH$_2$- groups [56] present in lignin, as well as in DVB and/or St. For the materials with St, signals between 3000–2780 cm^{-1} assigned to C(sp^3)-H bonds are more intense due to the formation of a polymer backbone [57]. Signals at 1740 cm^{-1} attributed to stretching vibrations of carbonyl groups were observed for the materials containing modified lignin.

The absorption peaks at 1596 and 1510 cm^{-1} can be assigned to C = C of aromatic skeletal vibrations from lignin, DVB, and/or St. The slight increase in intensity of these signals with increased lignin content indicate the incorporation of aromatic moieties from lignin into the synthesized materials. An additional increase in intensity of these signals indicate the presence of St in the polymer system. The signals ranging from 1200 to 1000 cm^{-1} are due to C-O-C stretching vibrations in acrylates. Their intensity increased with increasing content of modified lignin in the materials. The bands visible at around 830 cm^{-1} on all the presented spectra are attributed to aromatic C-H out-of-plane deformation vibrations, which are present in all the components used.

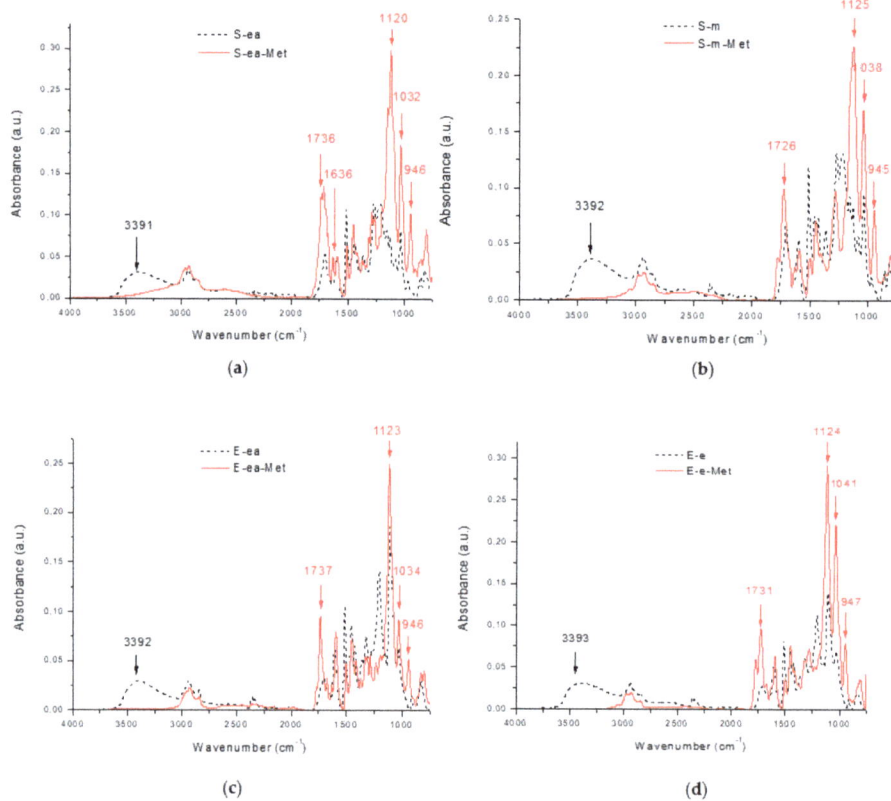

Figure 1. ATR/FTIR spectra of lignins: (**a**) SL-ea, (**b**) SL-m, (**c**) EL-ea, and (**d**) EL-e, before and after modification with methacryloyl chloride.

Thermal properties of the lignin-containing materials were studied by means of differential scanning calorimetry (DSC). The DSC curves are presented in Figure 3 and the maximum decomposition values (T_d) and enthalpy of decomposition (ΔH_d) are given in Table 3. In general, the synthesized materials are characterized by good thermal resistance. The DSC analysis showed that the thermal behavior of the materials obtained is similar, with a well-shaped calorimetric profile containing a single endothermic peak. A small exothermic effect at around 170 °C associated with the post-crosslinking process was also observed, particularly for the DVB-based materials containing SL-ea and EL-ea lignin, even though it was invisible for those with EL-e lignin, which can suggest that microspheres with EL-e lignin had the highest degree of crosslinking. The maximum decomposition temperatures for pure DVB and St-DVB microspheres are 447.2 °C and 426.9°, correspondingly, as shown in our previous paper [50]. The use of methacrylated lignin resulted in a slight deterioration of the thermal properties of the materials and, with increasing lignin content, the maximum decomposition temperature decreased. These values differed slightly (±5–6 °C) among microspheres with different lignin types and different fractions. Among the materials with the highest content of lignin (3 g), the highest maximum decomposition temperature was observed for the microspheres containing EL-ea-Met lignin. The use of lignin in St-DVB microspheres resulted in the improvement of thermal stability, especially when samples with molecular weight above 1000 g/mol were used [50]. The enthalpy of decomposition (ΔH_d) depends on the amount of lignin in the materials and decreases with increasing lignin content.

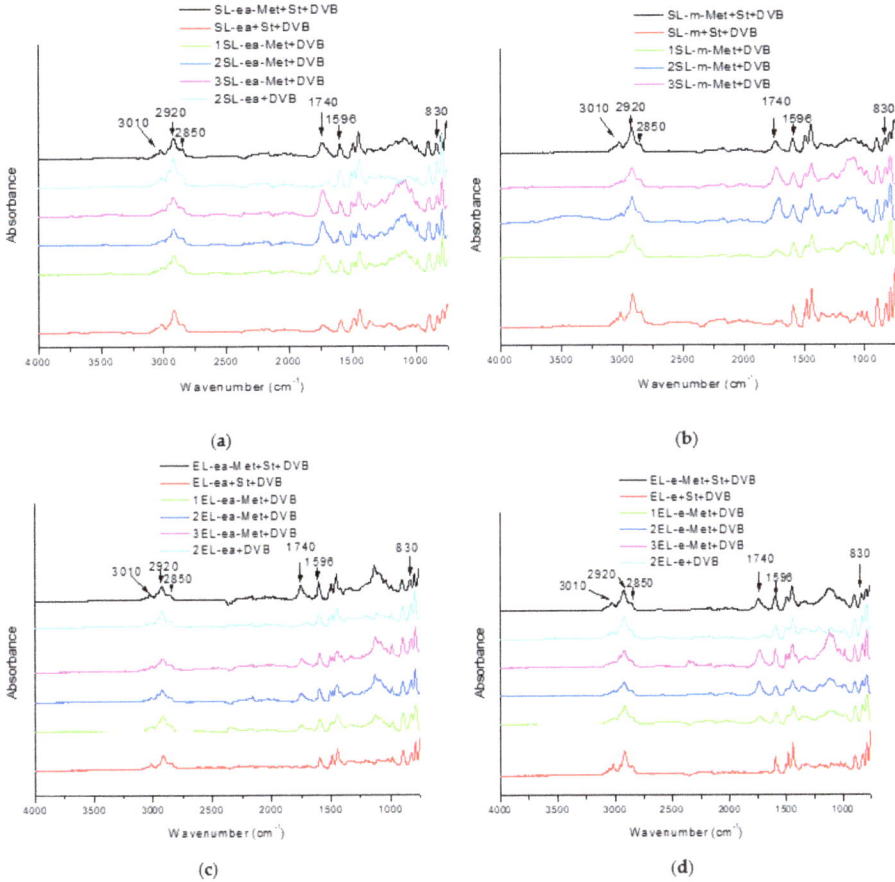

Figure 2. ATR/FTIR spectra of polymers with (a) SL-ea, (b) SL-m, (c) EL-ea, and (d) EL-e.

The surface area and porosity results are presented in Table 4. The addition of a lignin component slightly decreased the porosity of the DVB-containing or St-DVB-containing microspheres [50]. On the other hand, lignin additive enables the incorporation of various functionalities, which is essential for sorption processes. Among the materials containing DVB and modified lignin, the higher molecular weight of initial lignin fraction resulted in more developed porosity of corresponding microspheres within a series (SL-ea vs. SL-m and EL-ea vs. EL-e). The highest S_{BET} values were obtained for the materials with 2 g of modified lignin irrespective of lignin origin. Polymer microspheres with EL-e-Met lignin had the highest specific surface area and the largest pore volume among all studied materials, which can be related to a formation of a highly cross-linked network. EL-e lignin fraction had a less condensed macromolecular structure with a high content of phenolic and aliphatic hydroxyl groups in phenylpropanoid units, as shown in Table 1. The higher percentage of these groups as compared to softwood lignin fraction could react with methacryloyl chloride [29], and this could give a more homogeneous distribution of methacrylic groups in the lignin macromolecules and a more crosslinked network in the final microspheres. The inclusion of styrene, however, led to a decrease in porosity. Figure 4 shows the pore size distribution of the materials, which reveals more information about the porosity. All the materials are mesoporous. The pore size distributions are bimodal with a narrow peak with a maximum at 3 nm and a broader peak with a maximum at 12–50 nm. The presence of modified lignin in the structure of microspheres results in a shift of the second peak toward a lower value suggesting the creation of narrower, deeper, and more uniform pores in the material. The larger the amount of modified lignin used, the narrower and deeper the pores are in the materials. The fractionated lignin results in a greater porosity of the synthesized materials than in the microspheres where commercially available and low molecular weight lignin was applied [50].

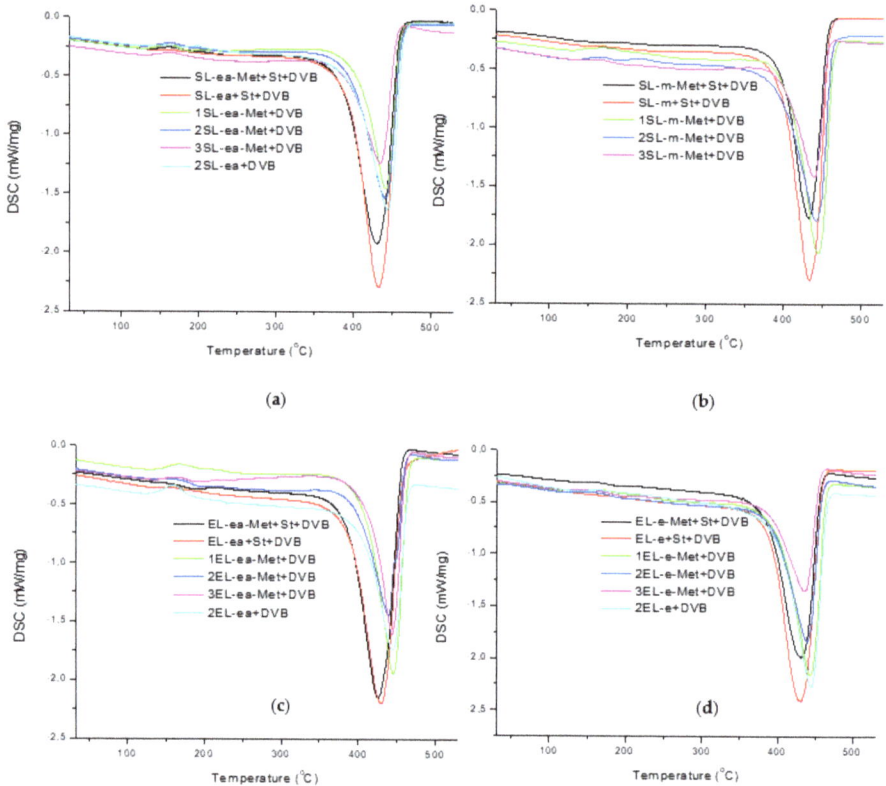

Figure 3. DSC curves of polymers containing (a) SL-ea, (b) SL-m, (c) EL-ea, (d) EL-e.

The results of swelling studies of the polymer materials are presented in Table 5. The lowest swellability coefficients were obtained for the polymers with EL-e lignin, which can be related to its crosslinking ability. Ethyl acetate fractions of lignin (SL-ea and EL-ea) had a higher swellability coefficient than EL-e and SL-m. Except for the materials with EL-e lignin, the greatest swelling tendency was observed in the materials with styrene and unmodified lignin. An increase in lignin content usually led to an increased swelling. In general, crosslinked materials have a low tendency to swell, but the presence of functional groups from lignin caused networks to interact with the solvents. None of the microspheres swelled in distilled water. The highest swelling coefficients were obtained in acetone and chloroform.

Table 3. DSC data.

Polymer	Td [1] (°C)	ΔHd [2] (J/g)
SL-ea-Met + St + DVB	430.2	539.8
SL-ea + St + DVB	432.7	595.1
1SL-ea-Met + DVB	441.6	321.0
2SL-ea-Met + DVB	440.6	378.0
3SL-ea-Met + DVB	434.3	296.8
2SL-ea + DVB	444.1	428.3
SL-m-Met + St + DVB	432.2	568.9
SL-m + St + DVB	434.3	529.9
1SL-m-Met + DVB	444.4	409.3
2SL-m-Met + DVB	442.0	420.5
3SL-m-Met + DVB	439.5	251
EL-ea-Met + St + DVB	425.3	558.7
EL-ea + St + DVB	428.6	556.3
1EL-ea-Met + DVB	445.1	407.7
2EL-ea-Met + DVB	438.3	318.9
3EL-ea-Met + DVB	443.7	315.2
2EL-ea + DVB	444.2	326.7
EL-e-Met + St + DVB	431.5	505.1
EL-e + St + DVB	430.5	579.6
1EL-e-Met + DVB	443.6	446.9
2EL-e-Met + DVB	439.4	377.2
3EL-e-Met + DVB	435.2	281.6
2EL-e + DVB	444.7	479.0

[1] Temperature of maximum decomposition, and [2] enthalpy of decomposition.

Table 4. Pore structure parameters of the polymers.

Polymer	S_{BET} [1] (m^2/g)	V_{TOT} [2] (cm^3/g)	D_A [3] (nm)
SL-ea-Met + St + DVB	51	0.104	8.2
SL-ea + St + DVB	139	0.597	17.1
1SL-ea-Met + DVB	103	0.126	4.9
2SL-ea-Met + DVB	396	0.700	7.1
3SL-ea-Met + DVB	314	0.542	6.9
2SL-ea + DVB	416	0.996	9.6
SL-m-Met + St + DVB	291	0.983	13.5
SL-m + St + DVB	23	0.077	13.2
1SL-m-Met + DVB	474	1.405	11.9
2SL-m-Met + DVB	442	0.969	8.8
3SL-m-Met + DVB	462	0.803	6.9
EL-ea-Met + St + DVB	229	0.517	9.0
EL-ea + St + DVB	30	0.080	10.6
1EL-ea-Met + DVB	384	0.937	9.7
2EL-ea-Met + DVB	434	0.953	8.8
3EL-ea-Met + DVB	62	0.083	5.4
2EL-ea + DVB	212	0.802	15.1
EL-e-Met + St + DVB	410	0.938	9.1
EL-e + St + DVB	195	0.785	16.1
1EL-e-Met + DVB	394	0.996	10.1
2EL-e-Met + DVB	506	1.342	10.6
3EL-e-Met + DVB	483	0.981	8.1
2EL-e + DVB	299	1.135	15.2

[1] specific surface area, [2] total pore volume, [3] and average pore diameter.

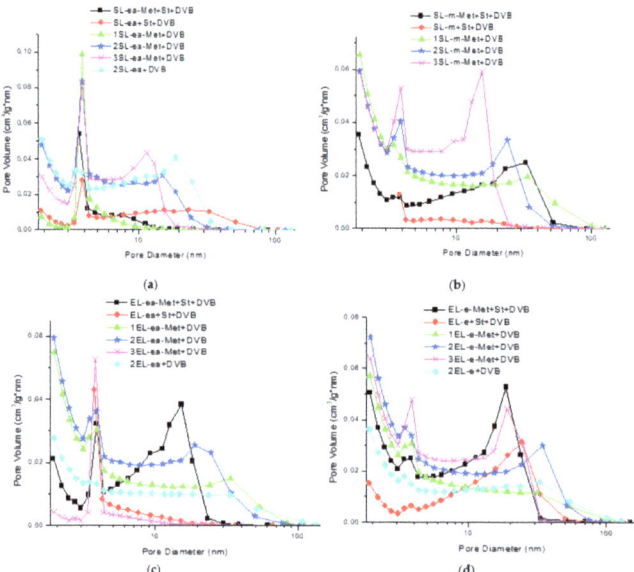

Figure 4. Pore size distribution curves of polymers containing (**a**) SL-ea, (**b**) SL-m, (**c**) EL-ea, and (**d**) EL-e.

Table 5. Swelling studies.

Polymer	Swellability Coefficient, B (%)						
	Acetone	THF [1]	Chloroform	ACN [2]	Methanol	Toluene	Aqua dest.
SL-ea-Met + St + DVB	122	122	122	122	100	100	0
SL-ea + St + DVB	100	30	91	67	58	100	0
1SL-ea-Met + DVB	113	78	75	63	63	63	0
2SL-ea-Met + DVB	67	46	85	82	64	75	0
3SL-ea-Met + DVB	58	83	45	55	80	45	0
2SL-ea + DVB	73	47	67	67	60	83	0
SL-m-Met + St + DVB	0	0	0	10	0	0	0
SL-m + St + DVB	100	60	120	100	80	209	0
1SL-m-Met + DVB	8	8	8	15	0	0	0
2SL-m-Met + DVB	0	0	0	0	0	11	0
3SL-m-Met + DVB	22	10	11	11	11	10	0
EL-ea-Met + St + DVB	23	25	55	36	27	55	0
EL-ea + St + DVB	109	120	136	91	118	127	0
1EL-ea-Met + DVB	7	13	7	7	7	22	0
2EL-ea-Met + DVB	20	27	6	8	15	7	0
3EL-ea-Met + DVB	40	40	80	70	40	70	0
2EL-ea + DVB	0	6	0	0	0	6	0
EL-e-Met + St + DVB	22	10	11	0	0	0	0
EL-e + St + DVB	0	9	0	0	0	0	0
1EL-e-Met + DVB	6	0	0	10	0	0	0
2EL-e-Met + DVB	22	0	0	11	0	0	0
3EL-e-Met + DVB	25	0	13	13	13	14	0
2EL-e + DVB	0	0	0	0	0	0	0

[1] Tetrahydrofuran, and [2] acetronitrile.

The spherical shape of the obtained materials was confirmed by photomicrographs, as shown in Figure 5. All the samples contain microspheres of different sizes. Materials with unmodified lignin had much smaller diameters (9–21 µm) than those with modified lignin (35–63 µm). Materials with modified lignin and DVB had the largest diameters in the range of 45–63 µm, and the microspheres with modified lignin from hardwood (EL-ea-Met and EL-e-Met) had larger diameters (49–63 µm) than those with modified lignin from softwood (SL-ea-Met and SL-m-Met). The diameters ranged from 45 to 52 µm. All the synthesized microspheres had a tendency to agglomerate. The most homogeneous size distribution was obtained for the materials using EL-e-Met lignin and DVB.

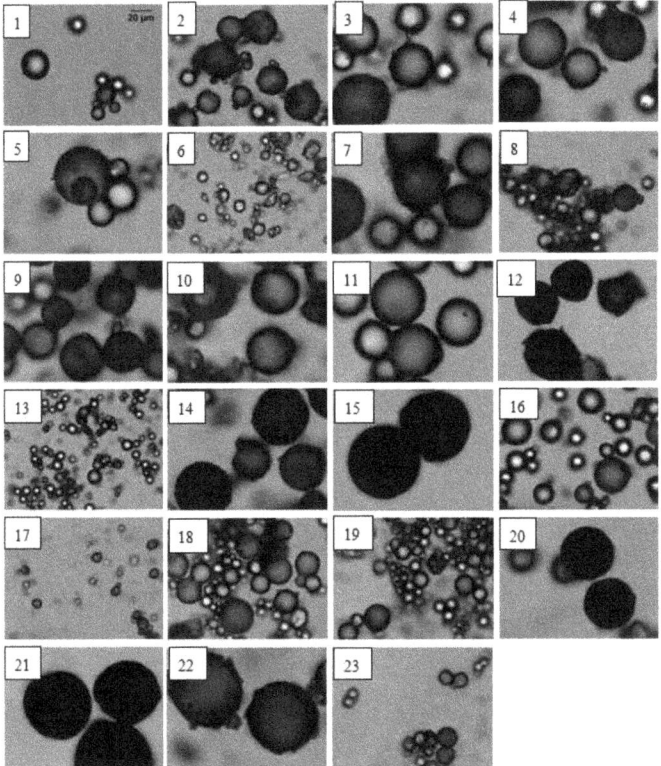

Figure 5. Photomicrographs of the microspheres. 1-SL-ea-Met + St + DVB, 2-SL-ea + St + DVB, 3-1SL-ea-Met + DVB, 4-2SL-ea-Met + DVB, 5-3SL-ea-Met + DVB, 6-SL-ea + DVB, 7-SL-m-Met + St + DVB, 8-SL-m + St + DVB, 9-1SL-m-Met + DVB, 10-2SL-m Met + DVB, 11-3SL-m-Met + DVB, 12-EL-ea-Met + St + DVB, 13-EL-ea + St + DVB, 14-1EL-ea-Met + DVB, 15-2EL-ea-Met + DVB, 16-3EL-ea-Met + DVB, 17-EL-ea + DVB, 18-EL-e-Met + St + DVB, 19-EL-e + St + DVB, 20-1EL-e-Met + DVB, 21-2EL-e-Met + DVB, 22-3EL-e-Met + DVB, 23-EL-e + DVB. The scale bar (see sample 1) is the same for all images.

4. Conclusions

Two types of kraft lignin, including softwood (SL) and hardwood (EL) fractionated in common industrial solvents, were successfully modified with methacryloyl chloride. The modified and unmodified lignins were then used for the synthesis of polymeric microspheres with St and/or DVB. The structural characteristics of the lignin fractions, such as degree of condensation, number and type of various hydroxyl groups, and molecular weight, influenced the modification process and the properties of the materials. Thermal properties of materials were studied by DSC and the materials were found to have a high thermal resistance. The incorporation of methacrylated lignin into the microspheres resulted in the greatest specific surface area and porosity. The largest specific surface area and the largest total pore volumes were found in the materials with SL-m and EL-e fractions, which had a higher molecular weight than the corresponding ethyl acetate fractions SL-ea and EL-ea. SL-m and EL-e fractions had a higher content of aliphatic hydroxyl groups than the ethyl acetate fractions.

These groups can also react with methacryloyl chloride, and, thus, can give a more homogeneous distribution of methacrylic groups in the lignin macromolecules and, therefore, better characteristics of the final products. The microspheres swelled most in acetone and chloroform. They did not swell at all in distilled water. Materials with EL-e had the lowest swellability coefficients, which may be related to their high degree of crosslinking. The most homogeneous size distribution of microspheres was obtained with EL-e-Met lignin and DVB.

Author Contributions: Conceptualization, M.G. and B.P. Formal analysis, M.G., B.P., and O.S. Investigation, M.G., B.P., O.S., and A.C. Methodology, M.G. and B.P. Project administration, M.G. and B.P. Resources, M.G., B.P., O.S., and B.G. Supervision, B.P., O.S., and B.G. Visualization, M.G., B.P., O.S., and A.C. Writing—Original draft, M.G. Writing—Review & editing, B.P., O.S., and B.G.

Funding: This research received no external funding.

Acknowledgments: The authors would like to thank Ayumu Tagami (Research Laboratory, Nippon Paper Industries Co., Ltd., 5-21-1 Oji, Kita-ku, Tokyo 114-0002, Japan) for the preparation and characterization of lignin fractions used in the synthesis of microspheres and the COST Action CA17128 Establishment of a Pan-European Network on the Sustainable Valorization of Lignin for making it possible to exchange an experience with the other scientists.

Conflicts of Interest: The authors declare no conflict of interest.

References

1. Manjarrez Nevárez, L.A.; Ballinas Casarrubias, L.; Celzard, A.; Fierro, V.; Torres Muñoz, V.; Camacho Davila, A.; Torres Lubian, J.R.; González Sánchez, G. Biopolymer-based nanocomposites: Effect of lignin acetylation in cellulose triacetate films. *Sci. Technol. Adv. Mater.* **2011**, *12*, 045006. [CrossRef] [PubMed]
2. Yu, L.; Dean, K.; Li, L. Polymer blends and composites from renewable resources. *Prog. Polym. Sci.* **2006**, *31*, 576–602. [CrossRef]
3. Saito, T.; Brown, R.H.; Hunt, M.A.; Pickel, D.L.; Pickel, J.M.; Messman, J.M.; Baker, F.S.; Keller, M.; Naskar, A.K. Turning renewable resources into value-added polymer: Development of lignin-based thermoplastic. *Green Chem.* **2012**, *14*, 3295–3303. [CrossRef]
4. Meier, M.A.R.; Meier, M. Renewable Resources for Polymer Chemistry: A Sustainable Alternative? *Macromol. Rapid Commun.* **2011**, *32*, 1297–1298. [CrossRef] [PubMed]
5. Imre, B.; Pukánszky, B. Compatibilization in bio-based and biodegradable polymer blends. *Eur. Polym. J.* **2013**, *49*, 1215–1233. [CrossRef]
6. Tănase, E.E.; Râpă, M.; Popa, O. Biopolymers based on renewable resources—A review. In Proceedings of the International Conference Agriculture for Life, Life for Agriculture, Bucharest, Romania, 5–7 June 2014; Scientific Bulletin, Series F, Biotechnologies: Bucharest, Romania; Volume XVIII.
7. Kaplan, D.L. Introduction to Biopolymers from Renewable Resources. In *Biopolymers from Renewable Resources. Macromolecular Systems—Materials Approach*; Kaplan, D.L., Ed.; Springer: Berlin/Heidelberg, Germany, 1998; pp. 1–29.
8. Zhang, Z. Lignin Modification and Degradation for Advanced Composites and Chemicals. Ph.D. Thesis, Georgia Institute of Technology, Atlanta, GA, USA, December 2017.
9. Saini, J.K.; Saini, R.; Tewari, L. Lignocellulosic agriculture wastes as biomass feedstocks for second-generation bioethanol production: Concepts and recent developments. *3 Biotech* **2015**, *5*, 337–353. [CrossRef]
10. Väisänen, T.; Haapala, A.; Lappalainen, R.; Tomppo, L. Utilization of agricultural and forest industry waste and residues in natural fiber-polymer composites: A review. *Waste Manag.* **2016**, *54*, 62–73. [CrossRef] [PubMed]
11. Treinyte, J.; Bridziuviene, D.; Fataraite-Urboniene, E.; Rainosalo, E.; Rajan, R.; Cesoniene, L.; Grazuleviciene, V. Forestry wastes filled polymer composites for agricultural use. *J. Clean Prod.* **2018**, *205*, 388–406. [CrossRef]
12. Whetten, R.; Sederoff, R. Lignin Biosynthesis. *Plant Cell* **1995**, *7*, 1001–1013. [CrossRef]
13. Sen, S.; Patil, S.; Argyropoulos, D.S. Thermal Properties of Lignin in Copolymers, Blends, and Composites: A Review. *Green Chem.* **2015**, *17*, 4862–4887. [CrossRef]
14. Hatakeyama, H.; Hatakeyama, T. Lignin Structure, Properties, and Applications. In *Biopolymers Advances in Polymer Science*; Abe, A., Dusek, K., Kobayashi, S., Eds.; Springer: Berlin/Heidelberg, Germany, 2009; Volume 232, pp. 1–63.

15. Crestini, C.; Melone, F.; Sette, M.; Saladino, R. Milled wood lignin: A linear oligomer. *Biomacromolecules* **2011**, *12*, 3928–3935. [CrossRef] [PubMed]
16. Sette, M.; Wechselberger, R.; Crestini, C. Elucidation of Lignin Structure by Quantitative 2D NMR. *Chem. Eur. J.* **2011**, *17*, 9529–9535. [CrossRef] [PubMed]
17. Goliszek, M.; Sobiesiak, M.; Fila, K.; Podkościelna, B. Evaluation of sorption capabilities of biopolymeric microspheres by the solid-phase extraction. *Adsorption* **2019**, *25*, 289–300. [CrossRef]
18. Laurichesse, S.; Avérous, L. Chemical modification of lignins: Towards biobased polymers. *Prog. Polym. Sci.* **2014**, *39*, 1266–1290. [CrossRef]
19. Thielemans, W.; Wool, R.P. Lignin Esters for Use in Unsaturated Thermosets: Lignin Modification and Solubility Modeling. *Biomacromolecules* **2005**, *6*, 1895–1905. [CrossRef] [PubMed]
20. Li, X.; Weng, J.K.; Chapple, C. Improvement of biomass through lignin modification. *Plant J.* **2008**, *54*, 569–581. [CrossRef] [PubMed]
21. Bernier, E.; Lavigne, C.; Robidoux, P. Life cycle assessment of kraft lignin for polymer applications. *Int. J. Life Cycle Assess.* **2013**, *18*, 520–528. [CrossRef]
22. Atifi, S.; Miao, C.; Hamad, W.Y. Surface modification of lignin for applications in polypropylene blends. *J. Appl. Polym. Sci.* **2017**, *134*, 45103. [CrossRef]
23. Alekhina, M.; Ershova, O.; Ebert, A.; Heikkinen, S.; Sixta, H. Softwood kraft lignin for value-added applications: Fractionation and structural characterization. *Ind. Crops Prod.* **2015**, *66*, 220–228. [CrossRef]
24. Ghaffar, S.H.; Fan, M. Lignin in straw and its applications as an adhesive. *Int. J. Adhes. Adhes.* **2014**, *48*, 92–101. [CrossRef]
25. Thakur, V.K.; Thakur, M.K.; Raghavan, P.; Kessler, M.R. Progress in Green Polymer Composites from Lignin for Multifunctional Applications: A Review. *ACS Sustain. Chem. Eng.* **2014**, *2*, 1072–1092. [CrossRef]
26. Gopalakrishnan, K.; Ceylan, H.; Kim, S. Renewable biomass-derived lignin in transportation infrastructure strengthening applications. *Int. J. Sustain. Eng.* **2013**, *6*, 316–325. [CrossRef]
27. Berlin, A.; Balakshin, M. Industrial Lignins: Analysis, Properties, and Applications. In *Bioenergy Research: Advances and Applications*; Vijai, G., Maria Tuohy, G., Kubicek, C.P., Saddler, J., Xu, F., Eds.; Elsevier: New York, NY, USA, 2014; pp. 315–336.
28. Duval, A.; Lawoko, M. A review on lignin-based polymeric, micro- and nano-structured materials. *React. Funct. Polym.* **2014**, *85*, 78–96. [CrossRef]
29. Gordobil, O.; Moriana, R.; Zhang, L.; Labidi, J.; Sevastyanova, O. Assesment of technical lignins for uses in biofuels and biomaterials: Structure-related properties, proximate analysis and chemical modification. *Ind. Crops Prod.* **2016**, *83*, 155–165. [CrossRef]
30. Naseem, A.; Tabasum, S.; Zia, K.M.; Zuber, M.; Ali, M.; Noreen, A. Lignin-derivatives based polymers, blends and composites: A review. *Int. J. Biol. Macromol.* **2016**, *93*, 296–313. [CrossRef]
31. Podkościelna, B.; Sobiesiak, M.; Gawdzik, B.; Zhao, Y.; Sevastyanova, O. Preparation of lignin-containing porous microspheres through the copolymerization of lignin acrylate derivatives with St and DVB. *Holzforschung* **2015**, *69*, 769–776. [CrossRef]
32. Kun, D.; Pukánszky, B. Polymer/lignin blends: Interactions, properties, applications. *Eur. Polym. J.* **2017**, *93*, 618–641. [CrossRef]
33. Yin, Q.; Yang, W.; Sun, C.; Di, M. Preparation and properties of lignin-epoxy resin composite. *BioResources* **2012**, *7*, 5737–5748. [CrossRef]
34. Jesionowski, T.; Klapiszewski, Ł.; Milczarek, G. Kraft lignin and silica as precursors of advanced composite materials and electroactive blends. *J. Mater. Sci.* **2014**, *49*, 1376–1385. [CrossRef]
35. He, Z.W.; Lü, Q.F.; Zhang, J.Y. Facile preparation of hierarchical polyaniline-lignin composite with a reactive silver-ion adsorbability. *ACS Appl. Mater. Interfaces* **2012**, *4*, 369–374. [CrossRef]
36. Ballner, D.; Herzele, S.; Keckes, J.; Edler, M.; Griesser, T.; Saake, B.; Liebner, F.; Potthast, A.; Paulik, C.; Gindl-Altmutter, W. Lignocellulose Nanofiber-Reinforced Polystyrene Produced from Composite Microspheres Obtained in Suspension Polymerization Shows Superior Mechanical Performance. *ACS Appl. Mater. Interfaces* **2016**, *8*, 13520–13525. [CrossRef]
37. Wang, K.; Xu, F.; Sun, R. Molecular characteristics of Kraft-AQ pulping lignin fractionated by sequential organic solvent extraction. *Int. J. Mol. Sci.* **2010**, *11*, 2988–3001. [CrossRef]
38. Li, M.F.; Sun, S.N.; Xu, F.; Sun, R.C. Sequential solvent fractionation of heterogeneous bamboo organosolv lignin for value-added application. *Sep. Purif. Technol.* **2012**, *101*, 18–25. [CrossRef]

39. Park, S.Y.; Kim, J.Y.; Youn, H.J.; Choi, J.W. Fractionation of lignin macromolecules by sequential organic solvents systems and their characterization for further valuable applications. *Int. J. Biol. Macromol.* **2018**, *106*, 793–802. [CrossRef]
40. Jääskeläinen, A.S.; Liitiä, T.; Mikkelson, A.; Tamminen, T. Aqueous organic solvent fractionation as means to improve lignin homogeneity and purity. *Ind. Crops Prod.* **2017**, *103*, 51–58. [CrossRef]
41. Tagami, A.; Gioia, C.; Lauberts, M.; Budnyak, T.; Moriana, R.; Lindström, M.E.; Sevastyanova, O. Solvent fractionation of softwood and hardwood kraft lignins for more efficient uses: Compositional, structural, thermal, antioxidant and adsorption properties. *Ind. Crops Prod.* **2019**, *129*, 123–134. [CrossRef]
42. Duval, A.; Vilaplana, F.; Crestini, C.; Lawoko, M. Solvent screening for the fractionation of industrial kraft lignin. *Holzforschung* **2016**, *70*, 11–20. [CrossRef]
43. Toledano, A.; Serrano, l.; Garcia, A.; Mondragon, I.; Labidi, J. Comparative study of lignin fractionation by ultrafiltration and selective precipitation. *Chem. Eng. J.* **2010**, *157*, 93–99. [CrossRef]
44. Jönsson, A.S.; Wallberg, O. Cost estimates of kraft lignin recovery by ultrafiltration. *Desalination* **2009**, *237*, 254–267. [CrossRef]
45. Lauberts, M.; Sevastyanova, O.; Ponomarenko, J.; Dizhbite, T.; Dobele, G.; Volperts, A.; Lauberte, L.; Telysheva, G. Fractionation of technical lignin with ionic liquids as a method for improving purity and antioxidant activity. *Ind. Crops Prod.* **2017**, *95*, 512–520. [CrossRef]
46. Sevastyanova, O.; Helander, M.; Chowdhury, S.; Lange, H.; Wedin, H.; Zhang, L.; Ek, M.; Kadla, J.F.; Crestini, C.; Lindström, M.E. Tailoring the Molecular and Thermo-Mechanical Properties of Kraft Lignin by Ultrafiltration. *J. Appl. Polym. Sci.* **2014**, *131*, 9505–9515. [CrossRef]
47. Zhao, Y.; Tagami, A.; Dobele, G.; Lindström, M.E.; Sevastyanova, O. The Impact of Lignin Structural Diversity on Performance of Cellulose Nanofiber (CNF)-Starch Composite Films. *Polymers* **2019**, *11*, 538. [CrossRef]
48. Aminzadeh, S.; Lauberts, M.; Dobele, G.; Ponomarenko, J.; Mattsson, T.; Lindström, M.E.; Sevastyanova, O. Membrane filtration of kraft lignin: Structural properties and anti-oxidant activity of the low-molecular-weight fraction. *Ind. Crop. Prod.* **2018**, *112*, 200–209. [CrossRef]
49. Passoni, V.; Scarica, C.; Levi, M.; Turri, S.; Griffini, G. Fractionation of Industrial Softwood Kraft Lignin: Solvent Selection as a Tool for Tailored Material Properties. *ACS Sustain. Chem. Eng.* **2016**, *4*, 2232–2242. [CrossRef]
50. Goliszek, M.; Podkościelna, B.; Fila, K.; Riazanova, A.; Aminzadeh, S.; Sevastyanova, O.; Gun'ko, V. Synthesis and structure characterization of polymeric nanoporous microspheres with lignin. *Cellulose* **2018**, *25*, 5843–5862. [CrossRef]
51. Tomani, P. The LignoBoost process. *Cell. Chem. Technol.* **2010**, *44*, 53–58.
52. Tagami, A. Towards Molecular Weight-Dependent uses of Kraft Lignin. Ph.D. Thesis, KTH Royal Institute of Technology, Stockholm, Sweden, 2018. Available online: http://kth.diva-portal.org/smash/get/diva2:1240150/FULLTEXT01.pdf (accessed on 14 September 2018).
53. Podkościelna, B.; Goliszek, M.; Sevastyanova, O. New approach in the application of lignin for the synthesis of hybrid materials. *Pure Appl. Chem.* **2017**, *89*, 161–171. [CrossRef]
54. Faix, O. Classification of lignins from different botanical origins by FT-IR spectroscopy. *Holzforschung* **1991**, *45*, 21–28. [CrossRef]
55. Jaswal, S.; Gaur, B. Green methacrylated lignin model compounds as reactive monomers with low VOC emission for thermosetting resins. *Green Process. Synth.* **2015**, *4*, 191–202. [CrossRef]
56. Silverstein, R.M.; Webster, F.X.; Kiemle, D.J. *Spectrometric Identification of Organic Compounds*, 7th ed.; John Wiley & Sons, Inc.: Hoboken, NJ, USA, 2005.
57. Hermán, V.; Takacs, H.; Duclairoir, F.; Renault, O.; Tortai, J.H.; Viala, B. Core double–shell cobalt/graphene/polystyrene magnetic nanocomposites synthesized by in situ sonochemical polymerization. *RSC Adv.* **2015**, *5*, 51371–51381. [CrossRef]

© 2019 by the authors. Licensee MDPI, Basel, Switzerland. This article is an open access article distributed under the terms and conditions of the Creative Commons Attribution (CC BY) license (http://creativecommons.org/licenses/by/4.0/).

MDPI
St. Alban-Anlage 66
4052 Basel
Switzerland
Tel. +41 61 683 77 34
Fax +41 61 302 89 18
www.mdpi.com

Materials Editorial Office
E-mail: materials@mdpi.com
www.mdpi.com/journal/materials

www.ingramcontent.com/pod-product-compliance
Lightning Source LLC
LaVergne TN
LVHW070404100526
838202LV00014B/1384